Here are your

2007 SCIENCE YEAR
Cross-Reference Tabs

For insertion in your WORLD BOOK

Each year, SCIENCE YEAR, THE WORLD BOOK ANNUAL SCIENCE SUPPLEMENT, adds a valuable dimension to your WORLD BOOK set. The Cross-Reference Tab System is designed especially to help you link SCIENCE YEAR's major articles to the related WORLD BOOK articles that they update.

How to use these Tabs:

First, remove this page from SCIENCE YEAR.

Begin with the first Tab, **Carson, Rachel**. Take the C-Ch volume of your WORLD BOOK set and find the **Carson, Rachel,** article. Moisten the **Carson, Rachel,** Tab and affix it to that page.

Glue all the other Tabs to the corresponding WORLD BOOK articles.

2007 SCIENCE YEAR

The World Book Annual Science Supplement

A review of science
and technology
during the
2006 school year

World Book, Inc.

a Scott Fetzer company
Chicago

www.worldbook.com

World Book, Inc.
233 N. Michigan Ave.
Chicago, IL 60601

ISBN: 0-7166-0560-0
ISSN: 0080-7621
Library of Congress Control Number: 65-21776
Printed in the United States of America.

STAFF

▪ EDITORIAL

Editor in Chief
Paul A. Kobasa

**Associate Director,
Supplementary Publications**
Scott Thomas

**Managing Editor,
Supplementary Publications**
Barbara A. Mayes

Senior Editor
Kristina A. Vaicikonis

Staff Editors
Heather McShane
S. Thomas Richardson
Marty Zwikel

Editorial Assistant
Ethel Matthews

Cartographic Services
H. George Stoll, Head
Wayne K. Pichler, Manager,
Digital Cartography
John M. Rejba,
Senior Cartographer

Indexing Services
David Pofelski, Head
Aamir Burki, Staff Indexer

Permissions Editor
Janet Peterson

▪ GRAPHIC DESIGN

Associate Director
Sandra M. Dyrlund

Associate Manager, Design
Brenda B. Tropinski

Senior Designers
Don Di Sante
Isaiah W. Sheppard, Jr.

Contributing Designer
Lucy Lesiak

**Associate Manager,
Photography**
Tom Evans

Photographs Editor
Kathryn Creech

**Contributing Photographs
Editors**
Carol Parden
Amor Montes de Oca

**Production and
Administrative Support**
John Whitney

▪ LIBRARY SERVICES

Stephanie Kitchen,
Information Services Coordinator

▪ PRODUCTION

**Director, Manufacturing
and Pre-Press**
Carma Fazio

Manufacturing Manager
Barbara Podczerwinski

Production Manager
Anne Fritzinger

Print Promotional Manager
Marco Morales

Proofreading
Anne Dillon

Text Processing
Curley Hunter
Gwendolyn Johnson

▪ MARKETING

Director, Direct Marketing
Mark R. Willy

Marketing Analyst
Zofia Kulik

CONTRIBUTORS AND CONSULTANTS

Baine, Celeste, B.S., M.Ed.
Director, Engineering Education
Service Center. [*Engineering*]

Berenstein, Angela, B.S.
Nanotechnology Education Programs
Coordinator, National Nanotechnology
Infrastructure Network, University of
California at Santa Barbara.
[*Science Studies, Nanotechnology*]

Bernick, Jeanne, B.S.
Crops and Issues Editor,
Farm Journal Media. [*Agriculture*]

Bolen, Eric G., B.S., M.S., Ph.D.
Professor Emeritus, Department of
Biological Sciences, University of North
Carolina at Wilmington. [*Conservation*]

Breeding, David C., B.S., M.S., M.B.A.
Director, Environmental Health, Safety,
and Security Services. [**Consultant—
Consumer Science,** *The Skinny on
Tattooing*]

Brett, Carlton E., M.S., Ph.D.
Professor, Department of Geology,
University of Cincinnati.
[**Special Report,** *Tiktaalik: A Landmark
Discovery; Fossil Studies*]

Brown, Michael, B.A., M.S., Ph.D.
Professor of Planetary Astronomy,
Division of Geological and Planetary
Sciences, California Institute of
Technology. [**Special Report,** *Exploring
the Suburban Solar System*]

Chiras, Daniel, B.A., Ph.D.
Visiting Professor, Colorado College
at Colorado Springs.
[*Environmental Pollution*]

Cohen, Tom, B.A., M.A.
Director, Media Relations,
Conservation International
[**Consultant—Special Report,**
New Guinea's Hidden Paradise]

Connaughton, Dennis, B.A.
Free-Lance Medical Writer. [**Special
Report,** *Turning Up the Volume on
Hearing Loss*]

Cooper, Irene, B.J., M.L.S.
Children's Books Editor, *Booklist.* [*Books
About Science for Younger Readers*]

DeMarco, Patricia M., B.S., Ph.D.
Executive Director, Rachel Carson
Homestead Association. [*Environmental
Pollution* (**Close-Up**)]

Despres, Renée, Ph.D.
Free-Lance Writer. [**Consumer Science,**
*The Skinny on Tattooing; Medical
Research*]

Dutch, Steven I., B.A., Ph.D.
Professor, Natural and Applied Sciences,
University of Wisconsin at Green Bay.
[**Special Report,** *The Early Earth*]

Graff, Gordon, B.S., M.S., Ph.D.
Free-Lance Science Writer. [*Chemistry*]

Hay, William W., B.S., M.S., Ph.D.
Professor Emeritus, Geological Sciences,
University of Colorado at Boulder.
[*Geology*]

Haymer, David S., M.S., Ph.D. Professor,
Department of Cell and Molecular
Biology, John A. Burns School of
Medicine, University of Hawaii at Manoa.
[*Genetics*]

Hester, Thomas R., B.A., Ph.D. Professor
Emeritus of Anthropology, University of
Texas at Austin. [*Archaeology*]

Houtman, Jacqueline, B.A., M.S., Ph.D.
Free-Lance Science Writer and Editor.
[*Public Health* (**Close-Up**)]

Johnson, Christina S., B.A., M.S. Science
Writer, California Sea
Grant College Program, Scripps
Institution of Oceanography.
[*Oceanography*]

Klein, Catherine J., Ph.D., R.D.
Senior Staff Scientist, Life Sciences
Research Office. [*Nutrition*]

Knight, Robert, B.A., M.M.
Free-Lance Writer. [**Consumer Science,**
The Perks of Coffee]

Konrad, Rachel, B.A.
Silicon Valley Correspondent,
The Associated Press. [*Computers
and Electronics*]

Kowal, Deborah, M.A., P.A.
Adjunct Assistant Professor, Emory
University Rollins School of Public Health.
[*Public Health*]

Kramer, Thomas, A.M., M.D.
Associate Professor of Psychiatry; Director, Student Counseling and Resource Service, University of Chicago. [*Psychology*]

Lunine, Jonathan I., B.S., M.S., Ph.D.
Professor of Planetary Science and Physics, University of Arizona Lunar and Planetary Laboratory. [*Astronomy*]

March, Robert H., A.B., M.S., Ph.D.
Professor Emeritus of Physics and Liberal Studies, University of Wisconsin at Madison. [*Physics*]

Marschall, Laurence A., B.S., Ph.D.
Professor of Physics, Gettysburg College. [*Books About Science*]

Merz, Beverly, B.A.
Free-Lance Science Writer. [**Special Report,** *Chronic Inflammation*]

Milius, Susan, B.A.
Life Sciences Writer, *Science News*. [*Biology*]

Mills, Claudia E., B.A., Ph.D.
Independent Research Scientist, Friday Harbor Laboratories, University of Washington. [**Special Report,** *The Secret Lives of Jellyfish*]

Milo, Richard G., B.A., M.A., Ph.D.
Associate Professor of Anthropology, Chicago State University. [*Anthropology*]

Morring, Frank, Jr., A.B.
Senior Space Technology Editor, *Aviation Week & Space Technology*. [*Space Technology*]

Owsley, Douglas W., B.S., Ph.D.
Curator, National Museum of Natural History, Smithsonian Institution. [*Archaeology* (**Close-Up**)]

Peres, Judy, B.A., M.S.L.
Specialist Reporter, *Chicago Tribune*. [*Drugs*]

Pichler, Cindy B., B.S., M.A., Au.D.
Doctor of Audiology. [**Consultant— Special Report,** *Turning Up the Volume on Hearing Loss*]

Podojil, Joseph R., B.S., Ph.D.
Postdoctoral Fellow, Microbiology-Immunology, Northwestern University. [**Consultant—Special Report,** *Chronic Inflammation*]

Rademaker, Kurt, B.A., M.S.
Research Assistant, Climate Change Institute, University of Maine. [**Special Report,** *First Civilization in the Americas*]

Sandweiss, Daniel H., B.S., Ph.D.
Dean and Associate Provost for Graduate Education, Professor of Anthropology and Quaternary and Climate Studies, University of Maine. [**Special Report,** *First Civilization in the Americas*]

Sforza, Pasquale M., B.Ae.E., M.S., Ph.D.
Professor of Mechanical and Aerospace Engineering, University of Florida. [*Energy*]

Snow, John T., B.S.E.E., M.S.E.E., Ph.D.
Dean, College of Geosciences, Professor of Meteorology, University of Oklahoma. [*Atmospheric Science; Atmospheric Science* (**Close-Up**)]

Snow, Theodore P., B.A., M.S., Ph.D.
Professor of Astrophysics, University of Colorado at Boulder. [*Astronomy*]

Tamarin, Robert H., B.S., Ph.D.
Dean of Sciences, University of Massachusetts Lowell. [*Ecology*]

Teich, Albert H., B.S., Ph.D.
Director, Science and Policy Programs, American Association for the Advancement of Science. [**Science Studies,** *Nanotechnology; Science and Society*]

Thibeault, Brian, Ph.D.
Project Scientist, University of California at Santa Barbara. [**Consultant—Science Studies,** *Nanotechnology*]

Toomey, Keith, B.A.
Director of Communications, Lighting Research Center, Rensselaer Polytechnic Institute. [**Consumer Science,** *"LED"ing the Way*]

Wilson, Dave, B.A.
Producer, Cable News Network. [**Consumer Science,** *Removable Storage: Memory on the Move*]

CONTENTS

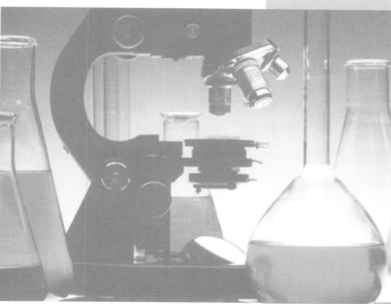

Discoveries of what may be the 10th planet and the fossils of what was an ancestor of the first land animals were among the developments that made the year eventful in science and technology. These two pages present highlights of the stories chosen by the editors of *Science Year* as the most memorable or important of the year, along with information on where in the book to find details about them.

FIRST SUCCESSFUL LABORATORY-GROWN ORGANS

The successful implantation of the first laboratory-grown organs in human patients—bladders grown using the patients' own cells—was reported in April 2006. In the Science News Update section, see **MEDICAL RESEARCH,** page 227.

FROM WATER TO LAND

The fossils of a 375-million-year-old animal, whose discovery was reported in April 2006, show more clearly than any other fossils ever found a nearly even blend of characteristics found in fish and in the first four-legged land animals. In the Special Reports section, see *TIKTAALIK:* **A LANDMARK DISCOVERY,** page 62.

THE 10TH PLANET?

An object nicknamed Xena and known formally as 2003 UB313 is the largest object discovered in the solar system since 1846. Some astronomers believe Xena, which is slightly larger than Pluto, proves that the solar system has more than nine planets. But does it? In the Special Reports section, see **EXPLORING THE SUBURBAN SOLAR SYSTEM,** page 26.

Rita
23 September

Dennis
10 July

Katrina
28 August

Wilma
21 October

Emily
17 July

RECORD-BREAKING SEASON

The 15 hurricanes that swept through the Atlantic Ocean, Caribbean Sea, and Gulf of Mexico in 2005 produced the most active hurricane season on record. The tally included seven hurricanes, including Katrina and three other major storms, that made landfall in the United States. In the Science News Update section, see the Close-Up **TRACKING HURRICANES AND IMPROVING FORECASTS,** page 177.

LANDMARK RULING

In December 2005, a federal judge in Pennsylvania ruled in a closely watched case that intelligent design theory is not science and, thus, has no place in a public-school science classroom. In the Science News Update section, see **SCIENCE AND SOCIETY,** page 250.

DOLPHIN SNOUT-GUARDS

The discovery in June 2005 that bottle-nosed dolphins living off the Australian coast had learned to use sponges to protect their snouts while searching for food may represent the first known instance of culture in marine mammals. In the Science News Update section, see **OCEANOGRAPHY,** page 236.

CHIMPANZEE GENOME DECODED

The first version of the complete genetic code of chimpanzees, published in September 2005, revealed that chimpanzees and human beings share about 98.5 percent of the genetic material found in each group. In the Science News Update section, see **GENETICS,** page 220.

SUPPORT FOR INFLATION THEORY

An analysis of microwave radiation emitted soon after the formation of the universe, reported in March 2006, provided the first direct evidence supporting the theory that the early universe went through a brief period of rapid expansion. In the Science News Update section, see **PHYSICS,** page 240.

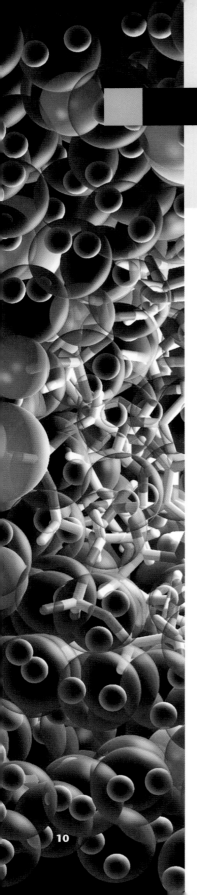

SPECIAL REPORTS

These feature articles take an in-depth look at significant and timely subjects in science and technology.

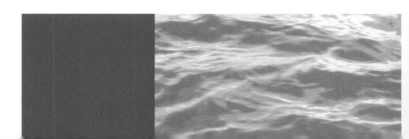

The city of Caral in western Peru may have been the first city in the Americas.

FIRST CIVILIZATION IN THE AMERICAS

Ancient Peru is one of six areas on Earth where a civilization developed without influence from other societies.

By Daniel H. Sandweiss and Kurt Rademaker

The inhospitable river valleys and nearby Pacific coast of western Peru are an unlikely location for a civilization to develop. The region, an extension of the Atacama Desert of northern Chile, is one of the driest places in the world. Nevertheless, nestled in four narrow valleys that run down from the Andes Mountains to the Pacific Ocean lies a long-overlooked archaeological treasure—the ruins of the oldest civilization in the Americas. Known as the North-Central Coast (NCC) civilization, it dates from about 3000 to 1800 B.C., roughly the same time that the first civilizations of the Old World were flourishing in ancient Egypt, Mesopotamia, and the Indus Valley.

Like these civilizations—and those of Mesoamerica (Central America) and ancient China—the NCC civilization arose without influence from outside groups. Like other early civilizations, the people of the NCC built monumental buildings, created art, and seem to have had a system of social classes. Unlike all other early civilizations, however, the people of the NCC did not have a written language. Most significantly, the NCC civilization, sometimes known as Norte Chico (Spanish for *Little North*), appears to have been the only early civilization that developed without *pottery* (baked clay). Because later Andean civilizations developed pottery, the period during which the NCC civilization existed is known as the Late Preceramic Period of Andean history.

The term *civilization* should not be confused with *culture*. A culture is the way of life of a group of people. Culture includes a society's arts, beliefs, customs, institutions, inventions, language, technology, and values. *Civilization* implies an advanced social, political, and economic structure. Architectural ruins, signs of a religious tradition, and other archaeological evidence leave no doubt that the people who inhabited the Huaura, Fortaleza, Pativilca, and Supe river valleys created a civilization.

TERMS AND CONCEPTS

Civilization: The advanced social, political, and economic structure of a people.

Culture: The way of life of a people, including their beliefs, technologies, and values.

Midden: A garbage mound.

Quipu: A cord with knotted strings or threads used for record keeping.

Radiocarbon dating: A method of determining an object's approximate age by measuring its remaining levels of radioactive carbon.

Shicra: A woven reed bag that was filled with rocks for use in construction. Shicras were also used for storage and transport.

Despite a century of excavation and study, archaeologists still have many unanswered questions about the NCC civilization. One of the most intriguing is how this civilization arose from the *foraging* (food-gathering) societies that inhabited the Andes Mountains and nearby coastal desert for thousands of years. Some archaeologists believe that these societies developed the only ancient civilization based more on fishing than on agriculture, though other archaeologists strongly disagree with this theory. Another major question focuses on the unexplained disappearance of the NCC civilization in about 1800 B.C. What happened to this civilization, which consisted of at least 20 settlements, was probably home to thousands of people, and may have been the center of a vast trading network?

The history of NCC excavations

Excavations in the NCC region began in the early 1900's with the work of German archaeologist Max Uhle at Aspero, a settlement located where the Rio Supe meets the Pacific Ocean. In the 1940's, American archaeologist Gordon Willey also worked at Aspero, and American geographer Paul Kosok published photographs and a description of Caral, another NCC site upriver from Aspero. None of these researchers, however, realized that they were looking at evidence of a complex civilization or just how old that civilization was.

In the 1970's, archaeologists excavating several of Aspero's small platform mounds estimated that the site dated from about 3200 to 2000 B.C. They based these dates on the techniques used in making the textiles they discovered, on the absence of pottery, and, some time later, on radiocarbon dates. Since the 1990's, archaeologists have discovered and dated other NCC sites farther from the ocean. One of the most important of these archaeologists is Ruth Shady Solis of San Marcos National University in Lima. Shady is known chiefly for her pioneering excavations of Caral. In the early 2000's, archaeologist Jonathan Haas of the Field Museum in Chicago discovered in adjacent valleys several previously unknown NCC sites dating from the Late Preceramic Period. Because researchers had excavated so few NCC sites, however, they were just beginning to understand the extent of this civilization.

In 2001, Shady asked Haas and anthropologist Winifred Creamer of Northern Illinois University in Dekalb to aid her in conducting the first radiocarbon dating tests of Caral's major structures. Radiocarbon dating is a process used to determine the age of an *organic* (carbon-containing) specimen by measuring the amount of a radioactive form of carbon in it. Scientists have determined that, after an organism dies, its level of radioactive carbon decreases over time at a constant rate. However, radiocarbon data can be misleading if a sample is composed

The authors:

Daniel H. Sandweiss is Dean and Associate Provost for Graduate Studies and a Professor of Anthropology and Quaternary & Climate Studies at the University of Maine at Orono.

Kurt Rademaker is a graduate student in the Climate Change Institute at the University of Maine.

THE RIVER VALLEYS OF THE NORTH-CENTRAL COAST CIVILIZATION

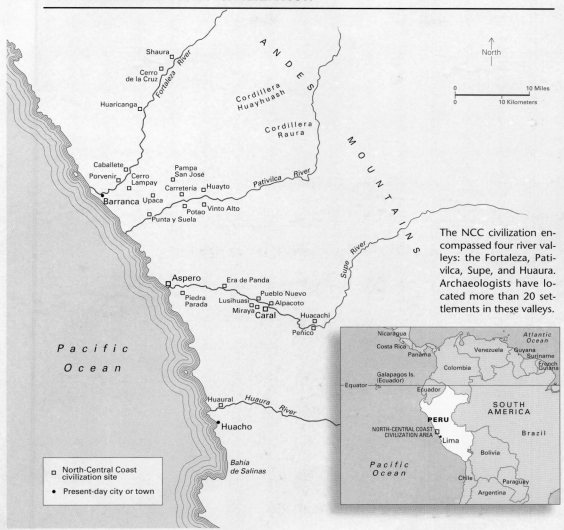

The NCC civilization encompassed four river valleys: the Fortaleza, Pativilca, Supe, and Huaura. Archaeologists have located more than 20 settlements in these valleys.

of objects from a variety of time periods. Remains of a building made of wood from a number of time periods, for example, can skew radiocarbon data. Fortunately for modern archaeologists, the people of the NCC built their structures using woven reed bags called *shicras* that were filled with stones and river rocks. All the reeds in a given bag come from the same time period, so the process was able to provide specific age estimates for the bags and, consequently, the buildings themselves. After radiocarbon dating shicra samples, as well as samples of trash and materials contemporary with the buildings of the site, Shady, Haas, and

Six large platform mounds rise from Caral (above). The stone pyramids were constructed in tiers (right) and were often topped by a complex of rooms. Archaeologists believe these buildings were used for large, public religious ceremonies.

Creamer determined that the major structures at Caral dated from about 2090 to 1640 B.C. However, the archaeologists also concluded that people had probably lived there for several hundred years before the mounds were constructed because it would have taken some time for a society to develop the ability to design and carry out such massive projects.

In fact, Caral is the only NCC site that has been extensively excavated and studied, so archaeologists use it as a model for interpreting other NCC settlements. Caral differs from other settlements chiefly in size—it is the largest well-known Late Preceramic site anywhere in the Andes, covering more than 110 hectares (270 acres) of land. For this reason, some researchers believe that the settlement may have been the political or religious center of the NCC civilization.

Architecture of the NCC civilization

The buildings at Caral exhibit a fascinating variety of ceremonial, religious, and domestic architecture. Dominating the site are six large platform mounds. The largest mound, Piramide Mayor, has a base measuring 160 meters (525 feet) by 150 meters (490 feet), an area equal to four football fields. It stands 18 meters (60 feet) tall, as high as a four-story building. The other mounds are somewhat smaller, with some less than half the size of Piramide Mayor.

All the mounds were built using the same technique. First, workers built retaining walls of cut stone. Then they placed stone-filled shicras within the retaining walls. Finally, the retaining walls were carefully covered with layers of colored plaster.

Atop each mound lies a complex of rooms. There, archaeologists have found the remains of plants and other objects often associated with religious rites in Andean traditions. These artifacts suggest that the rooms were used for ceremonial purposes.

Near each of the six large platform mounds lies another complex of large rooms with plaster-covered stone walls. The trash and debris found in these rooms and nearby *middens* (garbage mounds) included the bones of sea lions and sea birds, which were relatively uncommon foods among the NCC people. These findings led scientists to conclude that the rooms were residences for high-status individuals. Each of these complexes covers an area of from 450 to 800 square meters (from 4,840 to 8,610 square feet), less than half the size of a football field.

On the southern edge of Caral's central zone, archaeologists have found what appears to be a residential complex with a different kind of room. Although the complex itself is much larger than the complexes near the platform mounds, the individual rooms are smaller. In addition, their walls were made of mud, cane, and wood poles. The smaller size, less sturdy construction, and absence of special foods suggest that these rooms were residences for people with a lower social standing.

Near two of Caral's platform mounds, archaeologists have found sunken circular plazas ranging from 20 to 40 meters (65 to 130 feet) in diameter, and from 1 to 3 meters (3 to 10 feet) deep. Because of the types of artifacts found in and around the platform mounds and circular plazas, some archaeologists believe that the structures had a religious significance. Among these artifacts were musical instruments and the remnants of plants, including coca, a stimulant. Researchers also found pipelike inhalers, which they believe the NCC people used to ingest *psychotropic* (mind-altering) substances for religious purposes.

At least 15 other Late Preceramic sites in the Supe Valley feature platform mounds built near sunken circular plazas. However, Caral's mounds

Stone steps climb the side of one of Caral's pyramid mounds.

and plazas are among the largest in the NCC, suggesting that this architectural combination was deeply important to the site's builders and that the site may have been a center for the region.

Archaeologists also have uncovered a number of other architectural features at Caral, including hearths and terraces. The scarcity of artifacts found in or near these features has made it difficult for archaeologists to determine what the structures were used for. Some researchers speculated that they may have served some religious or administrative function.

The people of the NCC

Archaeologists know little about the people of NCC. Excavators have located only a few human remains, and these have yet to be analyzed. Because the NCC people had no pottery, thieves have looted few NCC cemeteries. As a result, few human remains have been brought to the surface, where they would come to the attention of archaeologists. In addition, the people of the NCC left no written accounts of their history and way of life that can be read by modern archaeologists. Finally, excavators have found few artifacts.

Nevertheless, archaeologists have been able to theorize about some aspects of NCC life. For example, building the enormous platform mounds must have required a large amount of labor. Their existence suggests that Caral had some sort of leadership that planned and directed the construction. Archaeologists are unsure whether these leaders were political or religious figures, however. The mounds also indicate that religion played a central role in NCC society.

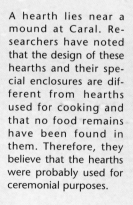

A hearth lies near a mound at Caral. Researchers have noted that the design of these hearths and their special enclosures are different from hearths used for cooking and that no food remains have been found in them. Therefore, they believe that the hearths were probably used for ceremonial purposes.

By comparing the residential architecture in various areas of a site, archaeologists gather clues about the living standards of its inhabitants. Differences in construction methods or in the degree or style of decoration, for example, generally indicate the presence of different social classes. At Caral, researchers believe, the residents of the large, plastered, stone-walled rooms near the platform mounds probably enjoyed a higher social standing than the people who lived farther away in rooms made of wood poles, cane, and mud. The inhabitants of the finer rooms may have been priests, political leaders, or other elites of

Structures at Caral include residential buildings believed to have been inhabited by workers (left) as well as monumental buildings and sunken circular platforms that were probably used for religious ceremonies (below). Because the NCC people did not leave any written records, archaeologists rely on artifacts to help them determine the function of these structures.

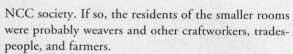

An archaeologist at Caral (right) points out one of the woven reed bags called *shicras* that served as the foundation of NCC structures. NCC workers filled shicras with rocks (below) and placed them between retaining walls of cut stone.

NCC society. If so, the residents of the smaller rooms were probably weavers and other craftworkers, tradespeople, and farmers.

Like the people of other early civilizations, the people of the NCC created art, which consisted almost exclusively of carved gourds, unbaked clay figurines, and woven textiles. Excavators have found no trace of pottery, a common feature of all first civilizations elsewhere. Apparently, the people of the NCC relied solely on gourds and shicras for transport and storage. Nor have excavators found any evidence of metallurgy, which usually develops along with pottery, as both require the use of kilns that can reach high temperatures. The people of the NCC relied on stone as well as wood and such other plant-based tools as gourds and shicras for daily activities, craft production, and construction.

The NCC civilization appears to have been the only early civilization that did not develop a system of writing based on representations of words or sounds. Shady's 2005 discovery of a type of *quipu* at Caral, however, has raised intriguing questions about this issue. A quipu is a cord with knotted strings of various lengths, colors, weaves, and designs used for record keeping. Quipus were highly developed by the Inca in the 1400's. The quipu found at Caral indicates that the NCC people may have had a system for counting and for conveying information. Some archaeologists have proposed that Andean quipus also were used to record poems, legends, and historical accounts, though they have yet to determine how such information was encoded using the knots and other elements of the quipu. Before Shady's discovery, the oldest known quipus dated from about A.D. 600 to 1000 during the Wari Empire. The Caral quipu is as much as 3,000 years older than the Wari examples, suggesting that the people of the NCC may have developed the quipu system.

In 2003, Haas and Creamer announced the discovery of a fascinating object in a looted burial site in the Patavilca Valley. They found a gourd etched with the image of a splay-footed, fanged figure holding a staff and what appears to be the head of a snake. The image is strikingly similar to that of the "staff god" who dominated Andean religious practices from about 1000 B.C. until about A.D. 1000. Haas and Creamer radiocarbon dated the gourd to 2250 B.C., making it the oldest image of this god ever found in the Americas. Previously, the earliest known image of the "staff god" dated from about 500 B.C. Haas and Creamer contended that the figure indicates that the NCC culture may have directly influenced the religious practices of later Andean cultures. However, some archaeologists speculated that the significant difference in age between the two images suggests that in about 500 B.C. an artist carved the image into an ancient gourd.

Some archaeologists also believe that the NCC civilization may have influenced the development of a religious tradition called Kotosh that flourished throughout the central Andes in the Late Preceramic Period. The open-air sunken circular plazas and vented ceremonial hearths found at Caral and other NCC sites resemble those found at sites linked to the Kotosh tradition. Kotosh practices, however, took place in small, private rooms. New discoveries in the NCC region may shed light on the connection between these two religious traditions.

Food in the NCC region

In a normal year, the river valleys and arid coastal lands of western Peru get no rain at all. Human life—let alone a civilization—would be almost impossible in this extreme environment if not for two factors. The Pacific Ocean off Peru's shores is one of the richest fisheries in the world. The ocean offered ancient settlers—and continues to give modern Peruvians—abundant fish, mollusks, and crustaceans. In addition, rivers flowing west from the Andes to the Pacific cut through the coastal desert. During the Preceramic Period, as now, the rivers supplied water for plants and animals living along their banks as well as for the region's farmers.

Excavations of middens throughout the NCC region indicate that seafood was the mainstay of the NCC diet. American archaeologist Michael Moseley, who excavated on the coast in the 1970's, found numerous remains of clam and mussel shells along with the bones of small fish, including sardines and anchovies. Moseley proposed that the NCC people caught fish and other marine creatures using nets

NCC people use cotton nets to catch seafood, which scientists believe was the main source of food for these ancient people. Some archaeologists believe the NCC civilization began with fishing communities on the coast and moved inland.

made of woven cotton and floats made of gourds, the remains of which he also found.

Archaeologists also have discovered the remains of such food crops as squash, beans, and guava at NCC sites. The extreme dryness of the NCC region posed a major challenge to ancient farmers. Farming at Caral was especially difficult because many of the fields are far from the river. To obtain a water supply for their crops, Caral's farmers would have needed a network of irrigation canals connected to the river. In 2001, Shady, Haas, and Creamer determined, on the basis of the terrain, that a canal that still flows near the site probably lies in the same location as one used by the NCC people. In January 2006, archaeologist Tom Dillehay of Vanderbilt University in Nashville, Tennessee, reported finding the remnants of irrigation canals in the highlands of Peru about 480 kilometers (300 miles) northwest of Caral. These canals, which date from about 3400 B.C., are more than 1,000 years older than major buildings in Caral. This finding suggests that, because of the communication and sharing of ideas known to have existed between early Andean societies, the people of Caral may have learned to build and maintain canals from an earlier highland society.

Theories of NCC development

Researchers believe that the close relationship between Caral's agriculture and fishing industries was the key to the city's success. Farmers provided the cotton and gourds for fishing nets, and the fishers provided food for the farmers. But which came first? That question has triggered a major debate among archaeologists studying the NCC civilization.

Some archaeologists, including Haas and Creamer, believe that the NCC civilization, like all other early civilizations, began with irrigation farming in the river valleys and then expanded to the Pacific coast. Other archaeologists, led by Michael Moseley, argue that the NCC civilization developed first in coastal fishing villages and then spread inland. If Moseley is correct, the NCC people created the only early civilization based primarily on fishing.

The debate began in 1975 with the publication of Moseley's book *The Maritime Foundations of Andean Civilization.* In the book, Moseley proposed that seafood—not crops grown using irrigation agriculture—served as the foundation of the NCC civilization. Moseley knew that the coastal people also possessed a knowledge of agriculture and grew a variety of edible crops. Because most of the plant remains found in NCC settlements came from cotton and gourds, he concluded that the majority of their crops were "industrial crops" that were used for tools rather than food. Moseley's theory flew in the face of conventional archaeological theory, which held that irrigation agriculture was the foundation of all early civilizations. He proposed that the

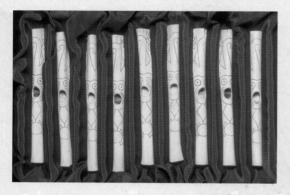

A *quipu* (top), a cord with knotted threads or strings for record keeping, and flutes made of pelican bones demonstrate the artistry and complexity of the NCC civilization.

fishing settlements traded with the local farming villages. Moseley also contended that it was only after the discovery of pottery in the so-called Initial Period (from about 1800 to 800 B.C.) that the coastal people began to practice irrigation agriculture. Moseley believed that the development of irrigation agriculture allowed coastal groups such as the NCC people to move further inland. (Traditionally, archaeologists believed that agriculture and pottery developed hand in hand, with the pottery used to store the abundance of crops harvested from irrigated fields.)

For many years, Moseley's theory dominated archaeologists' thinking about the NCC people. Since the late 1990's, Shady has argued that Caral was the birthplace of NCC civilization, though Moseley and others suggest that it was founded for the purpose of supplying coastal sites such as Aspero with cotton nets and gourd floats. In fact, radiocarbon dates from Aspero (acquired from charcoal remains) and Caral indicate that people lived at Aspero before they did at Caral.

In the early 2000's, Shady found evidence to support her theory that the foundations of NCC civilization included long-distance trade as well as irrigation agriculture and fishing. Shady and her team uncovered 32 flutes made of pelican and condor bones dating from the Late Preceramic Period as well as 37 *cornets* (musical instruments that resemble a shortened trumpet) made of deer and llama bones. The bones came from animals native to Peru's mountainous highlands, not its coastal area. She also discovered plant remains from the Amazon and Ecuador. Shady contended that Caral was the center of an enormous trade network that spread across Peru and into Ecuador to the north. Fishermen and weavers could have traded their wares for foods they could not grow locally, she argued.

In 2004, after performing further radiocarbon testing, Haas and Creamer proposed that agriculture, not fishing, stimulated the growth of the NCC civilization. They proposed that NCC settlements appeared first in valleys and that Aspero and other coastal sites were of minor importance. As evidence for their theory, they cited the relatively smaller size of the coastal settlements and their lack of major architecture. Haas and Creamer also disagreed with Shady that Caral was the center of the NCC civilization, noting that other, unexcavated inland sites are at least as large. For archaeologists, the debate is far from over.

The end of the NCC civilization is as mysterious as its beginning. The civilization seems to have ended around 1800 B.C. Haas suggested that the NCC people may have become part of another culture or may have been overrun by warlike neighbors. Archaeologists have found no evidence of warfare or large-scale destruction, however. Some archaeologists argued that changes in the climate of the region may have affected the availability of fish, forcing the people of the NCC to move elsewhere.

Ongoing research

As more sites from the NCC civilization are excavated, archeologists will gain a clearer picture of this fascinating society. However, archaeologists worry that many NCC sites will be damaged or lost before they can be investigated. Some landowners have indicated a desire to build on NCC sites. Groups of squatters have settled on other sites, damaging or otherwise disturbing them. In addition, though archaeologists have found relatively few artifacts at NCC sites, looters continue to damage sites as they look for ancient objects. Peru has laws protecting archaeological sites, but the government allocates little money for enforcement. With each new discovery filling holes in archaeologists' understanding of the NCC culture, protecting these sites is vitally important to the understanding of this unusual Andean civilization.

■ FOR ADDITIONAL INFORMATION

Articles and books

Haas, J., W. Creamer, and A. Ruiz, "Dating the Late Archaic occupation of the Norte Chico region in Peru." *Nature*, December 23, 2004, pp. 1020-1023.

Moseley, M. E., *The Maritime Foundations of Andean Civilization.* Cummings, 1975.

Richardson, J. B., *People of the Andes.* Smithsonian Institution, 1994.

Sandweiss, D. H., K. A. Maasch, R. L. Burger, J. B. Richardson III, H. B. Rollins, and A. Clement, "Variation in Holocene El Niño frequencies: Climate records and cultural consequences in ancient Peru," *Geology,* July 2001, pp. 603-606.

Shady Solis, R., J. Haas, and W. Creamer, "Dating Caral, a Preceramic Site in the Supe Valley on the Central Coast of Peru," *Science,* April 27, 2001, pp. 723-726.

Web sites

Caral project Web site—www.caralperu.gob.pe/principal_ing.htm

Norte Chico project Web site—www.fieldmuseum.org/research_collections/anthropology/anthro_sites/PANC

QUIPU Web site (Andean archaeology)—infodome.sdsu.edu/research/guides/quipu

Exploring the Suburban Solar System

By Michael E. Brown

Scientists searching the outer reaches of the solar system have discovered thousands of new objects, including what may be a new planet.

One stormy night near the turn of the millennium, a friend and I were sitting beneath the massive Hale Telescope at Palomar Observatory near San Diego. The telescope was standing idle inside its closed dome because of bad weather, and so we passed the time discussing recent discoveries about the edge of the solar system. I stated my conviction that beyond Pluto—the most distant of the known planets since its discovery in 1930—more planets were waiting to be found. In fact, I was so certain that I was willing to bet on it. We set a deadline of Dec. 31, 2004, for the discovery of a 10th planet. Then we did something that might seem surprising. We chose our own definition for the word *planet*. In 1999—and even as of mid-2006—astronomers had not yet agreed on an official definition for this familiar object. We did not think we could calculate immediately how big any newly discovered object was. As a result, we decided our criterion should be the amount of sunlight the object reflected, a rough estimate of size.

The idea that there might be planets beyond Pluto had been percolating in astronomers' minds since 1992, when David Jewitt of the University of Hawaii at Honolulu and Jane Luu, then at the University of California at Berkeley, spotted a small, icy object circling the sun well beyond Pluto's orbit. The object they discovered was the first identified member of a vast population of such objects orbiting the sun in the outer reaches of the solar system. This area is called the Kuiper *(pronounced KY pur)* belt, for Dutch-born American astronomer Gerard P. Kuiper, who in 1951 theorized that the belt may be the source of *short-period* comets (comets that take fewer than 200 years to orbit the sun). Astronomers believe that objects in the Kuiper belt are "building blocks" left over from the formation of the planets 4.6 billion years ago. By the end of 2000, astronomers looking at small areas of the sky had discovered nearly 500 Kuiper belt objects (KBO's), with the largest being about one-third the size of Pluto. Pluto's diameter is about 2,300 kilometers (1,430 miles). Because we were looking at much larger segments, of the sky, it seemed just a matter of time—and hard work—until we would find more objects at least that size.

My confidence about the discovery of new planets was bolstered by the project that I was working on at the time. Chad Trujillo at the Gemini Observatory on Mauna Kea, Hawaii; David Rabinowitz at Yale University in New Haven, Connecticut; and I had just begun the broadest search ever made of the Kuiper belt to find new large bodies in the solar system. Our work involved taking pictures of the skies using the 1.2-meter (48-inch) Samuel Oschin Telescope at Palomar Observatory. Night after night, we searched a different section of the sky that—from Earth—covered an area about the size of a hand extended at arm's length. We were looking for faint objects moving at a certain speed across the background of stars so distant that they appear to be *fixed* (unmoving).

THE SOLAR SYSTEM AS WE KNOW IT*

| Kilometers from the sun | 150 million | 779 million | 1.4 billion | 2.9 billion | 4.5 billion |
| (Miles) | (93 million) | (484 million) | (891 million) | (1.8 billion) | (2.8 billion) |

Sun
Mercury
Earth
Venus
Mars
ASTEROID BELT
Jupiter
Saturn
Uranus
Neptune
KUIPER BELT BEGINS

*Distances are not proportional.

The solar system includes the sun; nine "major planets"; small, rocky bodies known as *asteroids* or "minor planets" that circle the sun between the orbits of Mars and Jupiter; a vast region of comets that make up the Kuiper belt (with comets that take fewer than 200 years to orbit the sun) and the Oort cloud (with comets that take more than 200 years to orbit the sun).

The author:
Michael E. Brown is a professor of planetary astronomy at the California Institute of Technology in Pasadena and a member of the team that discovered 2003 UB313, a possible 10th planet.

Every object orbiting the sun moves across the sky. Objects close to the sun—such as the terrestrial planets Mercury, Venus, and Mars and the small, rocky asteroids orbiting between Mars and Jupiter—move across the sky quickly because of their relatively short orbits. Between the orbits of Jupiter and Neptune, in the thinly populated realm of the solar system's giant planets, objects move at more moderate speeds. The objects we were seeking were the most distant—and, therefore, the slowest moving—of all. We took three pictures of each area three hours apart and then compared them (using sophisticated computer programs) with images of millions of fixed stars and galaxies. Occasionally, we found what we were looking for—a faint point of light that looked like any other star in the first picture but had changed position slightly in the second picture and moved again in the third. About once a week, our photographs captured a dim, slow-moving object that we could add to astronomers' collection of known KBO's.

Most of the KBO's we found were small—perhaps one-fourth the size of Pluto. However, on a few occasions, we found an object brighter than anything else in the Kuiper belt. Each time we found such an object, we thought that we had finally discovered something larger than Pluto. But after more detailed examinations using other telescopes, we realized that none of these objects was larger than two-thirds the size of Pluto. In late December 2004, we discovered our largest object yet, but even that body was only about three-fourths of Pluto's size. On Jan. 1, 2005, I sent a note to my friend conceding the bet. My prediction that new planets would be found within five years had not come true.

Our search did not end, however. Four days after losing the bet, I was examining an unremarkable part of the sky in the constellation Cetus, the sea monster. Suddenly, the computer picked out a bright object that looked like a star in the first picture but had shifted its position

slightly in the second picture. In the third picture, it had shifted again. This object was one of the brightest that we had ever seen in motion. Moreover, it was moving more slowly than any other KBO we had ever seen, meaning that it was also the most distant. The fact that it was so bright despite being so far away meant that the object was big—bigger than any of the KBO's we had seen to date. A quick calculation on my computer verified what I had initially guessed. This object was not only bigger than Pluto but also the largest object discovered in the solar system since 1846. I had lost the bet four days earlier, but the discovery confirmed my belief that our solar system had more than nine planets. I temporarily nicknamed the object Xena, after the Greek warrior princess in the 1990's television series. But had we really found a new planet? Some astronomers thought not.

Defining planet

Although the word *planet* has existed for thousands of years, its meaning has changed to reflect our understanding of the cosmos. The word comes from an ancient Greek word meaning *wanderer* or *something that moves in the sky.* The ancient Greeks counted seven planets: the five you can see with the unaided eye—Mercury, Venus, Mars, Jupiter, and Saturn—plus the moon and the sun. In 1543, the Polish astronomer Nicolaus Copernicus proposed that the sun, rather than Earth, lies at the center of the solar system, and that Earth and the other planets revolve around the sun. Once scientists had accepted these ideas, the definition of a planet changed to include Earth but exclude the sun as well as the moon, which orbits Earth. In 1610, the Italian astronomer and physicist Galileo observed objects orbiting Jupiter the way the moon orbits Earth. His

TERMS AND CONCEPTS

Asteroid belt: A region of the solar system that lies between the orbits of Mars and Jupiter and is filled with small, rocky bodies called *asteroids* or *minor planets.*

Kuiper belt: A region of the solar system that begins beyond the orbit of Neptune and contains thousands of small, rocky bodies; the source of *short-period comets,* those that take fewer than 200 years to orbit the sun.

Oort cloud: A region at the outermost edge of the solar system; the source of *long-period comets,* those that take longer than 200 years to orbit the sun.

Orbital resonance: A relationship between two orbiting bodies that exert a gravitational pull on each other.

Plutino: The name applied to any of more than 300 objects in the Kuiper belt that have orbits similar to Pluto's.

2003 UB313: An object larger than Pluto that was discovered in 2005 and that may be the 10th planet.

discovery solidified the concept that planets travel around the sun while moons travel around planets. In 1781, the solar family of planets expanded to seven when the British astronomer William Herschel found Uranus, a large body similar to Jupiter and Saturn.

In 1801, the number of planets rose to eight with the discovery of Ceres, a large object that orbits between Mars and Jupiter. Most people have never heard of "planet Ceres," though, because soon afterward, astronomers began to find other objects like it in that region of space. Rather than having discovered new planets, astronomers had discovered the *asteroid belt,* an area filled with many small, rocky bodies like Ceres. By the 1850's, Ceres and similar objects had been demoted from "planet" to "asteroid." (Asteroids are also called *minor planets.)* With this move, the scientific meaning of the word *planet* had changed once again. The word no longer simply meant any object that orbits the sun; a small object that is one of many similar objects no longer qualified.

A teaching chart of the solar system from the 1860's reveals how our knowledge of the solar system—as well as our understanding of what a planet is—has changed through time. The chart shows 8 "principal planets" and 42 "minor planets" or asteroids.

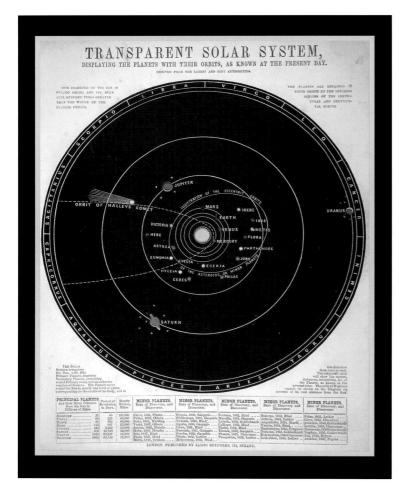

A different kind of planet

Xena (scientifically known as 2003 UB313) is not the first celestial body to puzzle astronomers. When Pluto was discovered, its unusual characteristics led some astronomers to question whether it should be considered a planet. Pluto is, by far, the smallest of the nine planets, measuring only about half the diameter of Mercury, the second smallest. Pluto moves around the sun in an *elliptical* (oval) orbit, rather than the nearly circular orbit followed by the other planets. While the other planets orbit the sun in the *ecliptic plane* (an imaginary surface through Earth's orbit around the

New Horizons, the first spacecraft ever sent to Pluto, flies by that planet in an artist's drawing. Pluto's moon Charon is visible in the upper right, with the sun shining beyond. New Horizons, launched on Jan. 19, 2006, was scheduled to reach Pluto in about 2015. It was to map the surface of Pluto and Charon, document their geology, and analyze Pluto's atmosphere.

sun), Pluto is tilted by almost 20 degrees away from the rest. However, astronomers had no other way to classify it, and so Pluto was—and still often is—referred to as the ninth planet.

The discovery of the Kuiper belt in 1992 allowed astronomers to make better sense of Pluto. It was not a lonely oddball at the edge of the solar system but the then-largest known KBO. Just as discoveries 150 years earlier had forced astronomers to reconsider the status of Ceres and other asteroids, the discovery of the Kuiper belt forced astronomers to reconsider the status of Pluto. If Ceres is the largest member of a certain group of objects—and, therefore, not a planet—why should Pluto be treated any differently?

Some astronomers want Pluto to keep its planetary status because it—like all the other planets—is round. Astronomers are not merely attached to this particular shape. Objects in space are round only if they have enough *mass* (amount of matter) that their gravity overcomes the strength of the material they are made of. That is, if a boulder were dropped into space, it would maintain whatever irregular shape it originally had. However, if a large number of boulders were dropped into a particular region of space, their combined mass—and the gravitational force they produce—would crush the individual boulders into the shape of a single sphere. A rocky object must be approximately one-third the size of Pluto to have enough gravitational pull to become a sphere. Any space object that is massive enough to be round is massive enough to have severely modified whatever materials it was made of. According to this argument, Pluto qualifies as a planet.

One problem with defining a planet as a sphere is that many objects in the solar system are round, including the moons orbiting Earth, Jupiter, Saturn, and other large planets. Ceres, the asteroid formerly known as a planet, is round. The Kuiper belt has at least a dozen objects that are probably round. Should the word *planet* include all these objects?

Popular culture has certainly embraced Pluto as a planet. In 2000, when astrophysicist Neil deGrasse Tyson prepared an exhibit of planets at New York City's Hayden Planetarium (of which Tyson is the director), he excluded Pluto. The decision made the front page of *The New York Times,* and hundreds of letters and e-mails protesting the move poured into the museum. On the other hand, the public would probably not embrace the idea of instantly creating dozens of new planets. How are astronomers to proceed if members of the general public, who have a certain understanding of the word *planet*, are unwilling to accept the scientific definitions astronomers have proposed?

The discovery of 2003 UB313 brought this simmering debate to a boil. Should we call it a planet? No, if we use the definition that also excludes Pluto. Yes, if we use the spherical definition that also includes dozens of other objects. How can we reconcile these definitions and decide what to call the new object?

Some scientists argue that we should abandon any attempt to agree on a scientific definition of the word *planet.* They contend that people have used the word in its current context for more than 1,000 years, while modern science has existed for only a few hundred years. What if astronomers were to simply agree that a planet is any object that is at least the size of Pluto? If so, 2003 UB313 is a planet; and anything else that we find that is at least as large as Pluto would also be a planet. Admittedly, this definition is arbitrary and nonscientific, but it fits what most people think of when they use the word. Moreover, this definition is no less scientific than those of such words as *continent,* which is as strongly influenced by popular culture as it is by the discipline of geography. Australia, for example, is considered a continent, while the islands Iceland and Greenland are not. Europe and Asia comprise one land mass but are considered two continents.

Throughout 2006, the organization responsible for naming planets and other celestial objects—the International Astronomical Union (IAU)—struggled over the decision of whether to consider Xena/2003 UB313 a planet. Until the group decides, an official name for the object cannot be chosen. The IAU planned to announce an official definition of a planet in September 2006.

HOW DO ASTRONOMICAL OBJECTS GET THEIR NAME?

The brightest stars in the sky—those visible to the unaided eye—have ancient names. After the invention of the telescope in the early 1600's, the first person to discover or *catalog* (list) an object had the privilege of naming it. The Dutch-born astronomer J. L. E. Dreyer began the practice of giving galaxies a name that begins with "NGC" followed by a number. NGC stands for "New General Catalog," a list of galaxies Dreyer published in 1888.

With its founding in 1919, the International Astronomical Union (IAU) became the authority for naming everything in the sky, including stars, galaxies, asteroids, and planets, as well as the rocks, hills, and craters on planets and moons. New objects found outside the solar system receive numbers rather than names. Most new objects found inside the solar system are named, and rules govern which names are allowed for which objects. (Most asteroids are not named, because there are so many of them.) The discoverer of a celestial object is generally allowed to propose a name, and an IAU committee approves it. Craters on Mercury, for example, are named for "deceased artists, musicians, painters, and authors who have made outstanding or fundamental contributions to their field." Physical features on Venus are all named after women, including goddesses and famous women.

The early Romans named Mercury, Venus, Mars, Jupiter, and Saturn after Roman deities. When Uranus and Neptune were discovered, these planets received the names of Roman gods as well. However, the IAU has never made an official rule for naming new planets. Pluto's discoverer named that planet for a Greek god, with the IAU's approval.

For now, the newly discovered object that may be the 10th planet is known as 2003 UB313, a temporary catalog number assigned by the IAU's Minor Planet Center. 2003 UB313 is a code name indicating that it was discovered in the second half of October 2003. If the IAU decides that the object is a large member of the Kuiper belt, the discoverers will be allowed to propose a name that is related to "a creation deity in any mythology." Previous Kuiper belt object names have included Quaoar (pronounced *KWAH oh wahr*), the creation force of the Native American Tongva tribe, and Ixion (pronounced *ihk SEE uhn*), the father of the race of *centaurs* (creatures that are half-man, half-horse) in Greek mythology. If the IAU determines that 2003 UB313 is a planet, the newly found object will likely be assigned the name of a major god in Greek or Roman mythology.

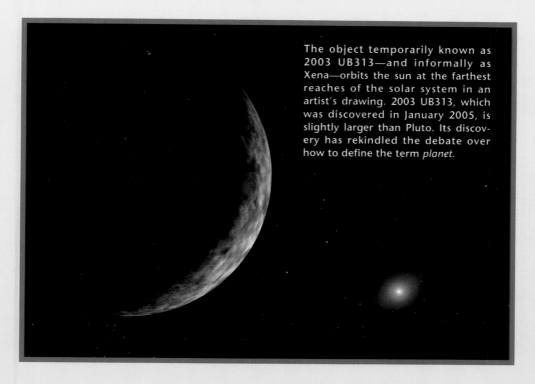

The object temporarily known as 2003 UB313—and informally as Xena—orbits the sun at the farthest reaches of the solar system in an artist's drawing. 2003 UB313, which was discovered in January 2005, is slightly larger than Pluto. Its discovery has rekindled the debate over how to define the term *planet*.

Discovering the secrets of the Kuiper belt

Are there any other planets in either the Kuiper belt or the extreme edge of the solar system beyond it? To know the answer requires an understanding of what we have learned about the Kuiper belt. Between 1930, when Pluto was discovered, and 1992, when Jewitt and Luu found the first KBO, many technological innovations transformed astronomy. Larger telescopes allowed astronomers to see fainter objects; more powerful computers automatically combed through images representing vast areas of the sky; and, most important, digital cameras replaced inefficient photographic plates. By the early 2000's, those cameras were large enough to capture wide expanses of the sky, allowing

WHAT IS A PLANET?

The meaning of the word *planet* has changed over time as scientists' knowledge of the universe has grown. In 2006, the discovery of new objects in the outer solar system was propelling efforts to establish specific criteria for determining whether a particular celestial body is a planet. The International Astronomical Union, a worldwide association of astronomers, planned to announce the official definition of a planet in September. The criteria might include one or more of the following:

■ Size: Should be at least 2,000 kilometers (1,240 miles) in diameter (Pluto's diameter is about 2,300 kilometers [1,430 miles])
■ Shape: Should be round, as the result of geological and mechanical forces
■ Orbit: Should orbit a star rather than another object in space and should have its own individual orbit
■ Mass: Should have a mass that is greater than the combined mass of all other bodies in nearby orbits

The relative sizes of the largest known Kuiper belt objects and their moons are depicted in an artist's drawing. All the objects are smaller than Pluto, with the exception of 2003 UB313—or Xena—whose diameter is slightly larger. Some astronomers argue that if Pluto is considered a planet, Xena should be as well.

astronomers to find greater numbers of objects in the Kuiper belt. By 2006, astronomers had found at least 1,000 objects in orbit beyond Neptune. The objects range in size from 2003 UB313's diameter of about 2,400 kilometers (1,490 miles) to a few faint objects perhaps 50 kilometers (30 miles) across.

Much effort has gone into figuring out where KBO's are and what shape their orbits take around the sun. Such basic information is an important key to understanding the formation and history of the belt. An oceanographer might use clumps of driftwood washed up along a beach to chart ocean winds and currents. In a similar way, astronomers look for patterns of KBO locations and orbits to chart their interactions with planets and their movement during the 4.6 billion years of the solar system's existence.

So far, astronomers have found three such patterns in the Kuiper belt. The first pattern involves a population of objects with orbits similar to Pluto's. In a relationship called *orbital resonance*, Pluto orbits the sun twice for every three times that Neptune travels around the sun. Orbital resonance occurs when two orbiting bodies exert a gravitational pull on each other. In Pluto's case, orbital resonance prevents the planet from crashing into Neptune even though at one point, Pluto crosses the larger planet's orbit. Orbital resonance also has produced Pluto's tilted, elliptical orbit around the sun, which takes 248 years to complete. If Pluto were the only object with these characteristics, it might be an odd coincidence. However, astronomers now know that some 300 KBO's have tilted, elliptical, 248-year orbits in a 3:2 resonance with Neptune. Jewitt dubbed these objects *Plutinos.* A force other than coincidence must be at work.

The power of Neptune

The exact nature of this force was first suggested in 1984 by the Uruguayan astronomer Julio Fernandez and the Taiwanese astronomer Wing-Huen Ip, then of the Max Planck Institute in Germany. In developing theories about the early formation of the solar system, Fernandez and Ip proposed that the planets from Jupiter to Pluto must have migrated to their present locations from orbits closer to the sun. They theorized that every time an object in the then-hypothetical Kuiper belt approached Neptune, Neptune's gravitational field would fling the object either inward or outward. (In the same way, space probes traveling to the outer solar system take advantage of Jupiter's gravitational pull, which slings them farther out into space, saving time and fuel.) Every time Neptune flung an object outward, Neptune itself compensated for the change by moving slightly inward. Every time Neptune flung an object inward, Neptune moved slightly outward. Over time, more objects were flung inward than outward because objects flung outward are still affected by the sun's gravitational pull. The pull of the sun generally drew them back into the inner solar system, where they interacted with Neptune again. However, objects pulled inward also encountered the other giant planets and were never seen again. They may have crashed into the giant planets and been destroyed, or the giants may have flung them back out of the solar system. According to Fernandez and Ip's theory, Neptune moved away from the sun at a snail's pace—about 0.02 kilometer (0.01 mile) per hour.

Fernandez and Ip's idea was not widely accepted until a number of Plutinos had been found. In 1993, the Indian-born astronomer Renu Malhotra at the University of Arizona in Tucson suggested that the Plutinos could have arrived at their present orbits only if Neptune's orbit once had been much closer to the sun. Over about 100 million years, the planet moved from an orbit of about 3 billion kilometers (1.9 billion miles) to an orbit of about 4.5 billion kilometers (2.8 billion miles) away from the sun. As Neptune's orbit expanded, more and more objects found themselves in resonance with Neptune. Once an object was in resonance, it was trapped by Neptune's gravitational attraction and was pushed outward. Neptune's gravitational attraction caused the object's orbit to gradually stretch and tilt. Pluto was only one of many objects swept up into such an orbit. When Neptune reached its current location, all the objects in resonance with it—including Pluto—were left with 248-year orbits.

The discovery of Plutinos proved that Neptune had migrated outward. The orbital patterns of a second group of KBO's called *scattered KBO's* (or *scattered disk objects)* support Fernandez and Ip's explanation of how Neptune had moved. Scattered KBO's have highly elliptical orbits that, at their closest point to the sun, almost intersect Neptune's orbit. These characteristics are precisely the orbits astronomers expected to find in objects that have been flung outward by Neptune, based on their computer models of how the early universe formed. The existence of scattered KBO's following highly elliptical orbits demonstrates that Neptune continues to move outward, though more slowly.

Scientists have discovered a third group of objects in the Kuiper belt, though they do not yet understand the role these objects have played in the history of the solar system. These objects are called *classical KBO's,* because their orbits most closely resemble those that computer models show would have existed in the earliest stages of the solar system's history. The orbits of classical KBO's are more circular than those of other KBO's, and classical KBO's generally travel in the same plane of the ecliptic as the rest of the planets. Unlike Plutinos and scattered KBO's, which have been pushed and flung around by Neptune, most classical KBO's appear to have made no significant movements in the past 4.6 billion years.

Both scattered and classical KBO's seem to come to an abrupt end at a distance of about 7 billion kilometers (4.5 billion miles) beyond Neptune. As astronomers find more objects within the Kuiper belt, they hope to better map out the patterns of orbits and perhaps finally learn what forces brought them to the locations objects occupy today.

The search for planet-sized objects in the Kuiper belt that my colleagues and I are conducting most likely has uncovered all the large classical KBO's and, perhaps, all the large Plutinos. New planets, however, may exist among the scattered KBO's. When scattered KBO's are close to the sun, they appear bright and so are readily visible to astronomers on Earth. However, on the outer parts of their long, elliptical orbits, KBO's are dim and hard to spot. The newly found scattered KBO, 2003 UB313, is currently as far from the sun and Earth—and thus as faint—as it ever gets. By 2010, our survey of the entire Kuiper belt will likely be completed. At that time, if any other new planets exist, my colleagues and I will have to assume that they are in the Oort cloud, the most distant region of the solar system, not in the Kuiper belt.

Thousands of icy bodies swirl through space in an artist's drawing of the Kuiper belt, a region that begins beyond the orbit of Neptune. Astronomers believe that Kuiper belt objects are "building blocks" left over from the formation of the planets about 4.6 billion years ago.

The outermost edge of the solar system

Although a number of astronomers theorized about the existence of a vast sphere of comets at the edge of the solar system, the Dutch astronomer Jan Oort developed the theory that became most widely accepted. In 1950, Oort proposed that *long-period comets*—those that take

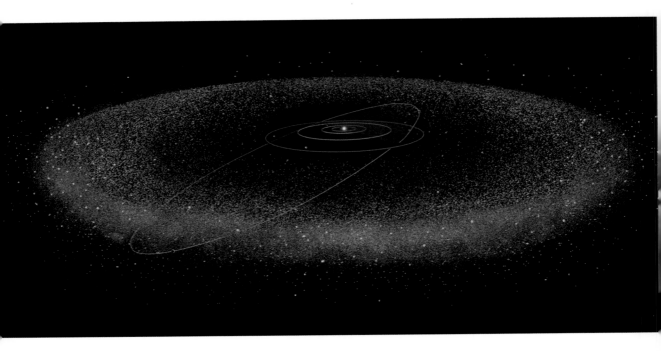

The Oort cloud, depicted in a painting, is a cluster of as many as 1 trillion comets, smaller objects, and perhaps even planets swirling through the outermost reaches of the solar system. At the center of the cloud lies the sun, surrounded by the nearly circular orbits of eight of the planets and Pluto's elliptical orbit. The Oort cloud begins roughly 800 billion kilometers (500 billion miles) from the sun and may extend for another 8 trillion kilometers (5 trillion miles).

at least 200 years to revolve around the sun—originally had been part of the inner solar system. At some point, Jupiter and the other giant planets had flung them out to their current location. Oort realized, however, that even though Jupiter was massive enough to fling objects to the edge of the solar system, another force must have stopped the comets from continuing on into interstellar space. That force, astronomers now believe, comes from nearby stars that exert just the right gravitational tug on fleeing comets to force them to remain within the solar system.

All the comets in the Oort cloud are far too distant and faint to be detected with existing telescopes. Therefore, knowledge of Oort cloud objects comes to us indirectly. Occasionally, a passing star's gravitational tug will send an Oort cloud object back into the inner solar system, where it can be seen as a comet on a highly elliptical orbit.

In early 2004, Trujillo, Rabinowitz, and I discovered one such object during our search of the Kuiper belt. The object—now named Sedna, after the Inuit goddess of the sea—follows a distant elliptical orbit that, unlike those of scattered KBO's, never comes close to Neptune or to any other planet. In fact, Sedna's orbit is unlike anything astronomers have ever seen. Sedna's orbit looks more like that of an object from the Oort cloud (that is, like that of a long-period comet) than that of a KBO, except that it is about a hundred times closer to the sun than it should be.

The simplest explanation for this puzzle is that parts of the Oort cloud may be closer than we thought because the stars that helped create the Oort cloud are hundreds of times closer to the solar system. Astronomers have made fairly accurate measurements for the distances to these stars, and so we know they are not really closer. However, they may have been much closer in the past. Many stars form within vast clusters of stars. Over time, the stars gradually drift away from one another. If our sun formed within a cluster, its companion stars would have been much closer. As a result, the Oort cloud would have been closer to the center of our solar system. After the cluster stars drifted away, this early Oort cloud would have been left, frozen in place as a kind of fossil record of the earliest history of our section of the Milky Way Galaxy.

Although this idea of how the Oort cloud may have formed is scientifically compelling, it cannot be proved simply by the existence of Sedna. Like Pluto before we knew about the Kuiper belt, Sedna is a singular object that for now can be described as an oddball on the outer fringe of the solar system. In the future, however, as astronomers discover more and more objects in the region beyond the Kuiper belt, we should be able to understand Sedna's place in the solar system and how it formed.

Based on the number of comets seen so far, astronomers believe that the inner Oort cloud contains an even larger number of icy bodies than the millions thought to inhabit the Kuiper belt. Because there are so many more comets, the largest ones in the Oort cloud will probably turn out to be several times as large as the largest objects in the Kuiper belt.

Today, we are able to see inside distant galaxies. However, much as we can see the moon on a clear night but not the tiny mosquito buzzing somewhere above our head, our current technology is probably not powerful enough to find new worlds in the Oort cloud. To discover new worlds, we will need to depend on future technological advances—and a new generation of astronomers willing to bet on the possibility.

■ FOR ADDITIONAL INFORMATION

Books and periodicals

Cowen, Ron. "Outer Limits: Solar System at the Fringe." *Science News,* Jan. 14, 2006, pp. 26-28.

Davies, John. *Beyond Pluto: Exploring the Outer Limits of the Solar System.* Cambridge University Press, 2001.

Fussman, Cal. "The Man Who Finds Planets." *Discover,* May 2006, pp. 38-45.

Web sites

Michael E. Brown's home page—http://www.gps.caltech.edu/~mbrown/planetlila/index.html

Lunar and Planetary Institute: About Our Solar System—http://www.lpi.usra.edu/education/explore/solar_system/

National Aeronautics and Space Administration: Solar System Exploration—http://solarsystem.nasa.gov/index.cfm

THE SECRET LIVES OF JELLYFISH

BY CLAUDIA E. MILLS

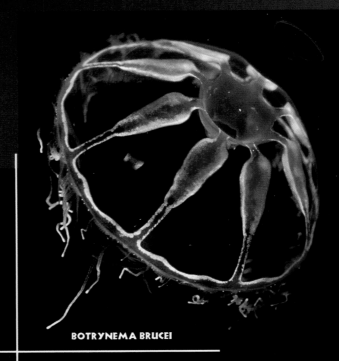

COLOBONEMA SERICEUM

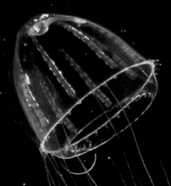

BOTRYNEMA BRUCEI

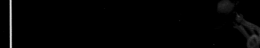

AEGINURA GRIMALDII

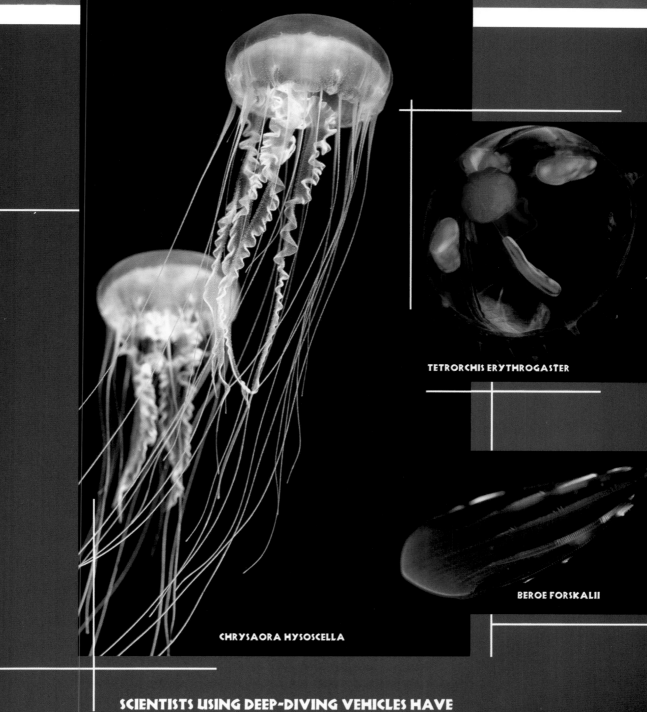

TETRORCHIS ERYTHROGASTER

BEROE FORSKALII

CHRYSAORA HYSOSCELLA

SCIENTISTS USING DEEP-DIVING VEHICLES HAVE
DISCOVERED AMAZING DETAILS ABOUT JELLYFISH
BEHAVIOR AND THE IMPORTANCE OF THESE
GRACEFUL CREATURES IN MARINE ENVIRONMENTS.

TERMS AND CONCEPTS

Bioluminescent: Able to produce light.

Biomass: Total weight of living matter.

Bloom: Sudden increase in a jellyfish population.

Cilia: Tiny hairlike structures.

Cnidarians: One of the main groups of animals; includes jellyfish, also known to biologists as medusae, some of which are bioluminescent.

Colloblast: A microscopic sticky structure on the tentacles of most ctenophores that is used to catch prey.

Ctenophores: One of the main groups of animals; also called comb jellies, most of which are bioluminescent.

Estuary: The broad mouth of a river where its fresh water meets the salt water of the sea.

Invertebrate: An animal without a backbone.

Mesogloea: The jellylike material that lies between the two layers of cells that make up a jellyfish's body.

Nematocyst: A stinging cell on the tentacle of a cnidarian that is used for capturing prey and defending itself.

ROV: Remotely operated vehicle.

Submersible: A piloted or remotely operated vehicle often used for research in the deep sea.

The author:
Claudia E. Mills is an independent research scientist at Friday Harbor Laboratories at the University of Washington in Friday Harbor.

Imagine trying to figure out the details of a jellyfish's life by studying the formless blob that it quickly becomes out of water. How would you determine what this creature looked like when it was alive? What clues would tell you how it caught its food? How could you fathom the way it moves through water? To learn how a particular type of animal lives, you must watch it—as it eats and is eaten, hides from predators, mates, and tends its young. What do you do, however, when the animal lives far below the ocean's surface?

The study of jellyfish evolved with the technology that gave scientists access to the deeper waters of the ocean. In the 1800's and early 1900's, groups of scientists set sail to explore the world's oceans. Using towed nets, these scientists collected at least half of the known *robust* (durable) species of jellyfish. Sometimes their nets brought up specimens from as far down as the *abyssal plain* (flat, deep stretches of the sea floor). Scientists also learned that, in certain places, they could effortlessly collect jellyfish species that normally live in the deep. Near Nice, off the coast of France, for example, seasonal upwelling currents deliver these deep-sea creatures to the surface. With the development of scuba equipment in the 1940's, biologists could closely observe jellyfish up to 30 meters (98 feet) below the surface.

The use of *submersibles* in science, beginning in the 1950's, revolutionized the study of jellyfish. These piloted or remotely operated vehicles (ROV's) provided scientists with exciting opportunities to study deep-dwelling jellyfish, especially the more fragile species, in their own environments. These vessels often record the details of jellyfish life with high-quality video cameras and other photographic equipment. Submersibles also can carry containers and other devices for collecting jellyfish alive and unharmed. For example, the reversible suction sampler—a sort of underwater vacuum cleaner—carefully sucks in a jellyfish as well as some of its surrounding water for transport to a ship.

Jellyfish, also called *jellies,* are transparent or semitransparent *invertebrates* (animals without backbones), most of which swim freely in the ocean. They get their name from the *gelatinous* (jellylike) material, called *mesogloea,* that lies between the two layers of cells that make up their body. This material, which serves as a skeleton, may be thick or thin and varies in density from floppy to stiff. Jellyfish live in all the oceans, from the tropics to the Arctic and Antarctic, and at all depths.

Scientists recognize two main groups of jellyfish—*cnidarians* (pronounced *ny DAIR ee uhnz*) and *ctenophores* (*TEEN noh fohrz* or *TEHN uh fohrz*). Cnidarians include the most familiar type of jellyfish, the bell-shaped creatures known to biologists as *medusae.* Medusae swim by expanding and contracting their body, like an umbrella opening and shutting. This jet action forces water out from beneath the body, and the medusa moves forward, bell first.

Although most medusae cannot see and do not have teeth or brains, they are efficient predators. Covering their tentacles are stinging cells called *nematocysts.* When touched, the nematocysts explode, driving tiny poisoned barbs like harpoons into the unlucky sea creature that came in contact with them. The poison can paralyze or kill the victim. The jelly then uses its tentacles to pass the prey to its mouth. Although all medusae are toxic to their prey, only a small number can seriously harm people.

Ctenophores, another major group of jellies, are also known as *comb jellies.* The body of a ctenophore may be shaped like a ball, a belt, or a thimble. A few deep-sea ctenophores can grow to 91 centimeters (3 feet), but most are closer to the size of a thimble. The comb jelly gets its name from the eight bands of comb-like organs on the sides of its body, which consist of tightly packed rows of *cilia* (tiny hairlike structures). Ctenophores move through the water by beating their combs in a coordinated way. Most ctenophores catch prey using *colloblasts,* sticky structures covering their tentacles. For some ctenophores, transferring food to the mouth is an acrobatic exercise. They spin while contracting the muscles of their tentacles so that they become wrapped in their tentacles. The tentacles eventually sweep across the mouth, transferring the food.

The deep-sea environment of the jellyfish is relentlessly carnivorous, scientists have found. As a submersible maneuvers through the deeper areas of the ocean, its lights reveal jellyfish swimming or drifting, each with its tentacles expanded into a miniature "drift net" for capturing prey. The largest of these nets can extend up to several meters into the dark water. The shape of each net depends on the species and its hunting style. The jellies that create these nets are so abundant that some researchers have estimated that jellyfish may account for up to 40 percent of the *biomass* (total weight of living matter) of the open ocean.

Submersible research has also revealed that many jellyfish and most ctenophores are *bioluminescent* (able to produce light). Scientists have observed an array of lights in blues, greens, and reds, which seem to vary by habitat and, to some extent, by depth. The light may serve to scare off predators, though scientists are not always certain of its function because they rarely see jellyfish bioluminescence in nature.

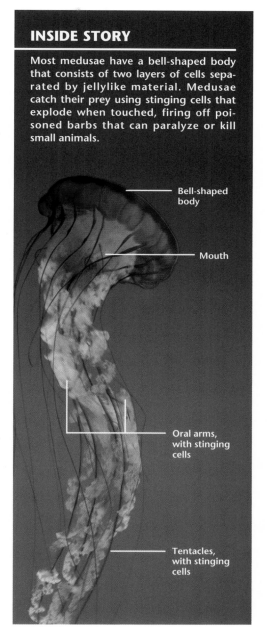

INSIDE STORY

Most medusae have a bell-shaped body that consists of two layers of cells separated by jellylike material. Medusae catch their prey using stinging cells that explode when touched, firing off poisoned barbs that can paralyze or kill small animals.

Bell-shaped body

Mouth

Oral arms, with stinging cells

Tentacles, with stinging cells

Advances in molecular biology have led to important findings about the evolution of jellies and the genetic relationships between various jelly groups. For example, scientists studying jelly DNA (deoxyribonucleic acid; the material that determines heredity) have learned that medusae and ctenophores are not as closely related as their appearance and behaviors might suggest. They actually belong to separate *phyla* (primary divisions of the animal kingdom), which may be even more different than previously thought. Over time, the animals evolved in such ways that they now seem quite similar. In addition, some scientists, including biologist Kevin Peterson of Dartmouth College in Hanover, New Hampshire, have theorized that jellyfish were not just simple forerunners of more complex and varied animals. They argued that the jellyfish's anatomical and genetic

MAGNIFICENT PESTS

A giant jellyfish called *Nemopilema nomurai* dwarfs a diver off the coast of Echizen, Japan. Before the year 2000, *blooms* (sudden population explosions) of *N. nomurai* occurred about every 20 years. Since 2000, this jellyfish species, which can weigh as much as 150 kilograms (330 pounds), has bloomed nearly every August, filling and ripping fishing nets.

complexity indicate that its evolutionary history is as elaborate as that of other animals.

Scientists also have made exciting discoveries about jellyfish anatomy and behavior. They understand that though jellies do not have a brain to receive and interpret sensory information, their nerve cells can still communicate in an organized way. For example, some species react to light, moving down through the water as daylight grows and up as the light fades. The jellyfish may be moving along with small plankton, which they eat.

Some of scientists' most important recent findings have highlighted the importance of medusae and ctenophores in marine *ecosystems* (the living things in a particular place and their interrelated physical and chemical environment). For example, an increase in jellyfish *blooms* (sudden population explosions) may signal environmental problems. In some places, blooms seem to be occurring because pollution—in the form of sewage or runoff from farms and cities—has improved feeding conditions for certain jellyfish species.

In 2004, plate-sized *Sanderia malayensis* bloomed for the first time in the *estuary* of China's Yangtze River, whose flow has been radically changed by the construction of the Three Gorges Dam to the southwest. (An estuary is the broad mouth of a river where its fresh water meets the salt water of the sea.)

LIVE OBSERVATIONS

Evolutionary developmental biologist Casey Dunn examines a type of cnidarian called a siphonophore aboard the research ship *Western Flyer*. Dunn probes the siphonophore, looking for clues about its development. A camera attached to the top of the microscope projects the image seen through the lens onto a computer screen for viewing by colleagues.

The medusae have seriously disrupted summer trawl fishing in the estuary, one of China's richest fishing grounds. In some areas, jellyfish have blanketed the water's surface.

Other jellyfish blooms have resulted from invasions by *exotic* (foreign) species that escaped their natural predators. Among these is an American ctenophore that has nearly destroyed commercial fishing in the Black and Caspian seas. In the eastern Mediterranean, the invasion of a stinging jellyfish from the Red Sea has severely affected swimming beaches.

Some scientists have also linked jellyfish blooms to overfishing. As large predator fish disappear from an ecosystem, some scientists suspect that local jellyfish benefit from a more abundant food supply, and their populations explode. Understanding the impact of these blooms on marine ecosystems ranks as one of scientists' most important tasks.

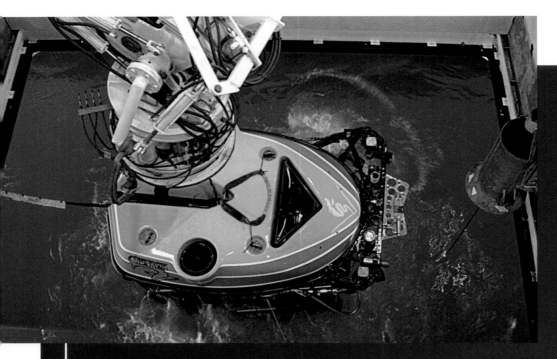

BELOW THE SURFACE

The submersible *Tiburon* (above), a remotely operated vehicle (ROV) about the size of a car, descends through an opening in the hull of the *Western Flyer*. The ROV, which is tethered to a heavy cable, can dive as deep as 4,000 meters (13,000 feet) below the surface. Scientists aboard the *Western Flyer* control the *Tiburon's* movements as well as its cameras, collection devices, lights, and other instruments. Submersibles like the *Tiburon* have revolutionized the study of deep-sea jellyfish.

A siphonophore called *Praya dubia* glides through the water, trailing a "curtain" of tentacles covered with stinging cells. The image, taken by a submersible's camera, captures only about 75 percent of this siphonophore, one of the longest animals on Earth. The animal, which may extend for 40 meters (130 feet), is pulled by a pair of swimming bells [see one in inset, right]. Siphonophores consist of many *polyps* (hollow, cylinder-shaped animals) that have specialized functions. Although many siphonophores produce spots of blue or green light, a camera's flash illuminated the animal in the photograph.

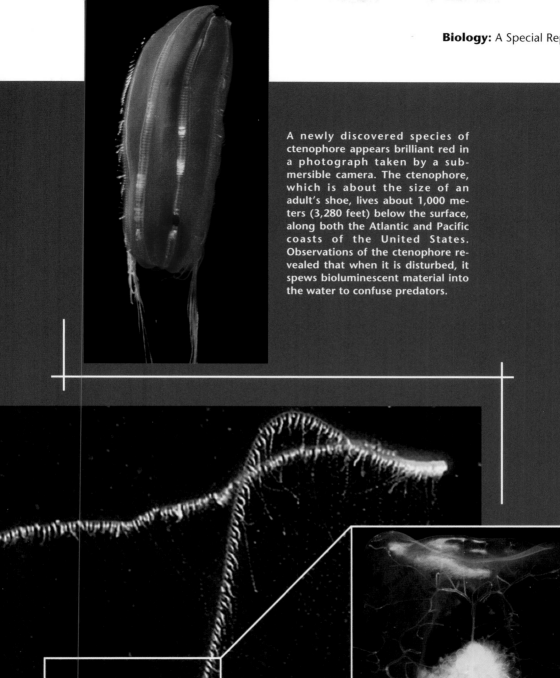

A newly discovered species of ctenophore appears brilliant red in a photograph taken by a submersible camera. The ctenophore, which is about the size of an adult's shoe, lives about 1,000 meters (3,280 feet) below the surface, along both the Atlantic and Pacific coasts of the United States. Observations of the ctenophore revealed that when it is disturbed, it spews bioluminescent material into the water to confuse predators.

ATTACHED BEAUTIES

Giant versions of a type of cnidarian called stauromedusae cluster on hardened lava from a deep-sea vent in the Pacific Ocean. *Lucernaria janetae* measures up to 10 centimeters (4 inches) in length, twice the size of other stauromedusae. Unlike most jellyfish, stauromedusae live attached to rocks or plants instead of swimming freely. Other species of stauromedusae include *Haliclystus salpinx* (below, left), which lives on brown kelp; *Manania handi* (below, middle), which lives on sea grass; and *Manania distincta* (below, right), which lives on seaweed.

BOX JELLYFISH

Box jellyfish, a type of cnidarian called cubomedusae, are named for the the cubical shape of their body. These transparent jellyfish, which can be up to 1 meter (3 feet) in length, live mainly in shallow water along shorelines. The venom of some box jellyfish, particularly the sea wasp and Irukandji jellyfish, can be fatal to human beings.

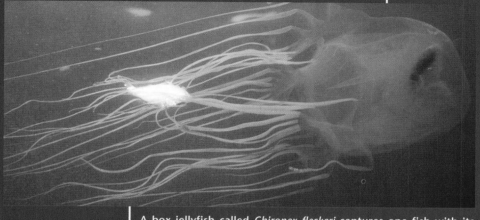

A box jellyfish called *Chironex fleckeri* captures one fish with its deadly tentacles while digesting another fish (dark area) in its bell.

Box jellyfish are the only type of jellyfish with complex eyes. They can have up to 24 eyes grouped into four clusters that are attached to flexible stalks at each corner of their boxy body. In each cluster, one or more of the eyes appear to be simply spots of light-sensitive pigment. However, two of the eyes are more complex, with a cornea and a lens, which focus light, and a retina, which receives images. Scientists think that because the lens focuses images slightly behind—instead of on—the retina, box jellies have blurry vision. As a result, they probably use their eyes for navigating rather than hunting. Scientists do not know how the jellyfish processes images as it does not have a brain.

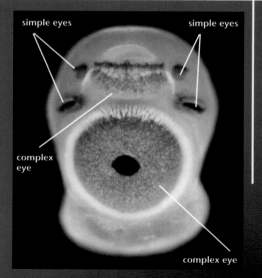

simple eyes

simple eyes

complex eye

complex eye

JELLYFISH FINDS

The discovery of previously unknown species of deep-sea jellyfish and insights into jellyfish behavior are among the remarkable findings made by scientists using submersibles. Cameras aboard these vehicles allow scientists to watch and photograph the fragile animals in their own environment.

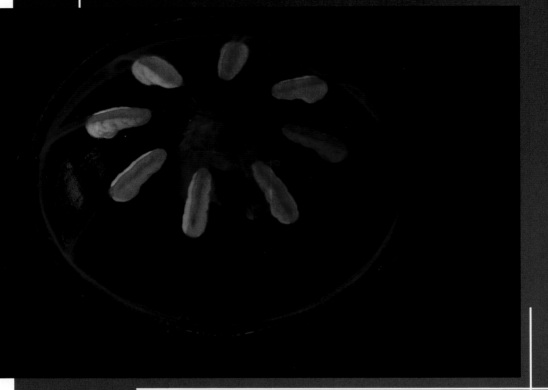

Eight brilliantly colored *gonads* (sexual organs) attached to circulatory canals dominate the bell of a jellyfish called *Crossota millsae.* This species, which measures about 28 millimeters (1.1 inches) in diameter, lives in waters from 1,000 to 3,700 meters (3,280 to 12,140 feet) deep off the coasts of California, Hawaii, and Japan. In 2003, this species was named in honor of Claudia E. Mills, the author of this article.

A newly discovered red cydippid ctenophore appears in fascinating detail in a photograph taken aboard a research ship. Scientists collected the fragile animal using an advanced capture device attached to a submersible. Such devices can transport jellyfish from the deep ocean without damaging their bodies and tentacles. The ctenophore's tightly packed cilia appear in rainbow colors because of light scattered by the camera's flash.

A jellyfish nicknamed "Big Red"—unlike most other jellyfish—has fleshy oral arms but no tentacles. Scientists discovered this medusa, whose bell can measure up to 1 meter (3.2 feet) wide, in 1993 and described it scientifically in 2003. The jellyfish has been found in deep waters off California, Hawaii, and Japan. The number of oral arms varies by individual animal from four to seven. The jellyfish's scientific name is *Tiburonia gran rojo*—*Tiburonia* after the *Tiburon* submersible, and *gran rojo*, Spanish for *big red*.

RED LURES

Spots of glowing red material nestled among transparent tentaclelike structures represent the first known bioluminescent "lures" used by a marine invertebrate to attract prey. The siphonophore inhabits waters so deep that prey are scarce. The lures, which are only a few millimeters long, resemble tiny deep-sea plankton animals that are eaten by small deep-sea fish that are instead attracted to and eaten by this type of siphonophore. The discovery of the bioluminescent lures was reported in 2005 by scientists led by biologist Steven Haddock of the Monterey Bay Aquarium Research Institute in California.

A BLUE-GREEN RING

A cnidarian called *Aequorea victoria* is illuminated underwater by a camera's flash. *A. victoria* produces bioluminescent proteins that, in nature, form a dotted blue-green ring on the rim of its bell. *Cloned* (genetically duplicated) versions of these proteins are widely used by scientists as a "tag" to track the activity of cells, genes, microorganisms, and pollutants. Previously abundant in Washington's Puget Sound, *A. victoria*'s numbers have drastically diminished in the 1990's and early 2000's for unknown reasons.

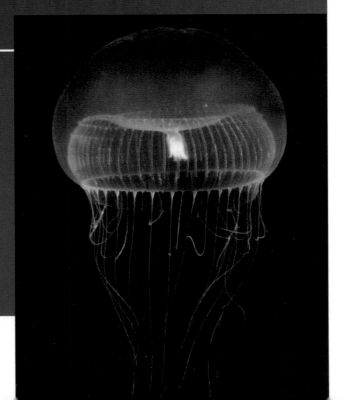

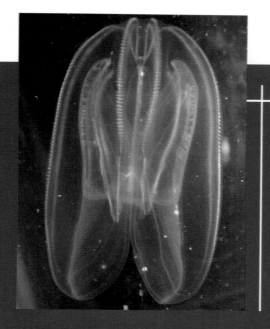

UNWELCOME VISITOR

A ctenophore called *Mnemiopsis leidyi* has caused major damage to fishing industries in the Black and Caspian seas. The ctenophore, which is native to the western Atlantic Ocean, invaded the Black Sea in the 1980's, probably carried by an international cargo ship. There, *M. leidyi* populations bloomed. The ctenophore consumed vast amounts of the small marine organisms normally eaten by native fish, causing dramatic declines in fish populations. In the late 1990's, *M. leidyi* passed through shipping canals from the Black Sea to the Caspian Sea, where it has caused similar ecosystem damage.

UNEXPLAINED BLOOMS

Chrysaora melanaster is one of a number of jellyfish species that have bloomed around the world for reasons that are unclear to scientists. Populations exploded in the Bering Sea during the 1990's, perhaps because of a climate change in the Pacific Ocean. A number of scientists believe some jellyfish blooms result from pollution or other environmental problems.

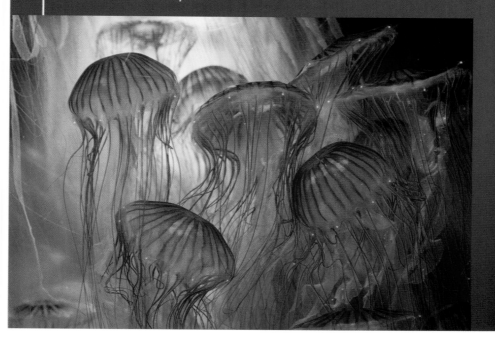

NEW GUINEA'S HIDDEN PARADISE

BY S. THOMAS RICHARDSON

RESEARCHERS SEEKING A
MYSTERIOUS BIRD DISCOVER
AN ECOLOGICAL PARADISE ON
THE ISLAND OF NEW GUINEA.

A Berlepsch's six-wired bird-of-paradise gets its name from the wirelike feathers that extend from its head. Scientists had searched for the bird since it was first described in 1897. In December 2005, an international research team finally discovered the bird on the island of New Guinea.

After almost a century of unsuccessful attempts to locate the home of the rare Berlepsch's six-wired bird-of-paradise, an international research team finally succeeded in December 2005 while searching an almost impenetrable jungle on the island of New Guinea. They also found much, much more.

Otto Kleinschmidt, a German *ornithologist* (scientist who studies birds), first described the bird in 1897 upon discovering a skin in a private collection. Little was known of the bird, other than that it probably came from New Guinea, which lies in the Pacific Ocean north of Australia. After years of frustrating searches for the bird in various locations on New Guinea, scientists turned to the 800,000 hectares (2 million acres) of largely unexplored jungle in the Foja Mountains. The area, classified as a wildlife sanctuary, is located on the western side of the island. Bruce Beehler, vice president of the nonprofit organization Conservation International, based in Washington, D.C., led the month-long expedition of 11 other scientists.

Part of the crew was dropped into the jungle by helicopter, which allowed them to reach an area unexplored by even the area's native population. Almost immediately, they saw a previously unknown species of honeyeater, the first new bird species discovered on New Guinea since 1939. On the second day of the expedition, the astonished team members became the first scientists to see a male six-wired bird-of-paradise, watching as one performed an elaborate mating dance for a nearby female.

Beehler and his team also discovered more than 20 new species of frogs, a rare golden-mantled tree kangaroo, five new species of palms, a previously unknown species of rhododendron, and four new species of butterflies. "It's as close to the Garden of Eden as you're going to find on Earth," Beehler said of the Foja jungle.

The author:
S. Thomas Richardson is an Associate Editor for *Science Year*.

Critically reviewed by Conservation International.

Bruce Beehler of Conservation International holds a female Berlepsch's six-wired bird-of-paradise in the Foja Mountains of New Guinea. Beehler led an international team of researchers on a search for the bird in December 2005.

A smooth-skinned frog represents one of more than 20 previously unknown species of frogs collected in the Foja Mountains of New Guinea by Stephen Richards, a *herpetologist* (scientist who studies reptiles and amphibians) from the South Australia Museum in Adelaide.

One frog discovered in the Foja Mountains is only 14 millimeters (0.6 inch) long. Scientists found the frog only after it emitted a soft call from the floor of the forest.

This 11.4-meter (37.4-foot) tall palm (above) is one of five previously unknown species of palms discovered in the Foja Mountains. The tree produces a rough-skinned fruit (right).

A golden-fronted bowerbird in full display holds up a berry in front of its bower during a mating ritual, in the first known photograph of the bird. A female (not shown) sits on the other side of the bower. The fruit is used for decoration and display, not food.

A previously unknown species of smoky honeyeater is particularly intriguing to scientists because of its bright orange facepatch. It represents the first new bird species found on New Guinea since 1939.

A mountain owlet-nightjar, a member of a widespread but rarely seen species, perches on a branch in the Foja Mountains in December 2005.

An endangered long-beaked echidna scuttles across the jungle floor. The animal was so unaccustomed to people—and thus so unafraid of them—that scientists were able to carry it back to their camp for study. The echidna is one of only two kinds of mammal that lays eggs. (The other is the platypus.)

TIKTAALIK:
A LANDMARK DISCOVERY

By Carlton E. Brett

Scientists have discovered fossils of an ancient creature with features of fish as well as those found in the first land-based animals.

Finding a fossil that represents a "missing link" in the evolution of life on Earth is as rare as it is exciting. In April 2006, a team of paleontologists announced the discovery of bones from one of the most spectacular examples of a transitional creature ever found. The 375-million-year-old fossils, excavated on a remote Arctic island in the Canadian territory of Nunavut, show more clearly than any other fossils ever discovered a nearly even blend of characteristics found in fish and in the first four-legged land animals. The paleontologists named the new species *Tiktaalik (tihk TAH lihk),* an Inuit word meaning *long, flattened fish.*

Many scientists believe the discovery of *Tiktaalik* may represent an event in the history of paleontology as significant as the discovery of *Archaeopteryx* in a stone quarry in the German state of Bavaria in the 1860's. *Archaeopteryx,* one of the most famous of all missing links, shows the combined characteristics of a small predatory reptile—what scientists now consider a dinosaur—and a bird. It provided the first solid evidence that birds descended from reptiles.

The remarkably well-preserved fossils of *Tiktaalik* reveal significant details about the transition of aquatic, gill-respiring fish to land-dwelling, air-breathing amphibians. Specifically, the fossils show how lobe-finned fish (the group that includes lungfish and coelacanths) evolved into *tetrapods* (four-legged, land-dwelling, air-breathing animals). Indeed, the scientists nicknamed the new species a "fishapod." The most primitive tetrapods alive today include such amphibians as frogs, toads, and salamanders. Amphibians typically spend a portion of their lives in moist terrestrial environments breathing air but also have early aquatic stages and generally return to water to breed. Scientists believe that *Tiktaalik,* in contrast, probably spent most of its life in the water but used its short, leglike fins to crawl onto land to find food, escape from predators, or lay eggs.

Discovering *Tiktaalik*

Paleontologists Edward B. Daeschler of the Academy of Natural Sciences in Philadelphia; Neil H. Shubin of the University of Chicago;

and Farish A. Jenkins, Jr., of Harvard University in Cambridge, Massachusetts, found the new fossils on Ellesmere Island, which lies only about 960 kilometers (600 miles) from the North Pole. According to Daeschler, the scientists chose this remote location because a geology textbook noted that the island had excellent *outcrops* (exposed rock) dating from the Devonian Period that had never been explored for ancient *vertebrates* (animals with backbones). Paleontologists have long known that animals first crawled onto land from the sea during the Devonian, which lasted from about 410 million to 360 million years ago. During the Late Devonian, what is now Ellesmere Island lay near the equator in a humid tropical climate. For their fossil search, the scientists looked for rocks that had formed from sediments carried by ancient streams or rivers or by floods, which were common and widespread in the tropics of the Devonian Period. In particular, they focused on *sandstones* and *mudstones* (rocks consisting of compacted sand and mud) that contained the remains of fossilized plants that flourished along Devonian streams and rivers.

Current conditions on Ellesmere Island are so harsh that the rocks are bare of ice and snow and accessible for study for only about two months of the year—during what passes for summer. Even then, the scientists had to contend with cold and snow. They also carried shotguns as protection from possible polar bear attacks. The scientists' hunch about Ellesmere Island was a good one, though. Few paleontologists are fortunate enough to prospect in a little-studied spot and turn up fossils that exceed their wildest hopes. Shubin accidentally discovered the best specimen of *Tiktaalik* when he sat down on a rock to eat his lunch. To his surprise, he saw the front end of a skull protruding from the edge of a nearby sandstone outcrop. The scientists removed an overlying layer of rock to expose the almost complete skeleton of a crocodilelike animal nearly 3 meters (9 feet) long with a large, flattened head and upward-facing eye sockets. The paleontologists turned to the elders from a nearby Inuit village when naming the newly discovered fossil.

The sediments in which *Tiktaalik* was found were laid down by streams that issued from mountains formed by a collision of land masses that are now part of North America and Europe. In fact, the rocks containing the *Tiktaalik* fossils apparently formed from *splay deposits*—that is, deposits laid down by flooding rivers that breached their natural embankments and dumped gravels, sands, and muds, as well as fish onto nearby flood plains. The *Tiktaalik* fossils probably were buried rapidly, as their skeletons were preserved in an *articulated* condition—that is, with all bones in place and connected to one another.

Studying *Tiktaalik*

The scientists used the fossilized spores of ancient land plants found in the sediments to date the *Tiktaalik* fossils as about 375 million years old,

The author:
Carlton E. Brett is a professor of Geology at the University of Cincinnati in Ohio.

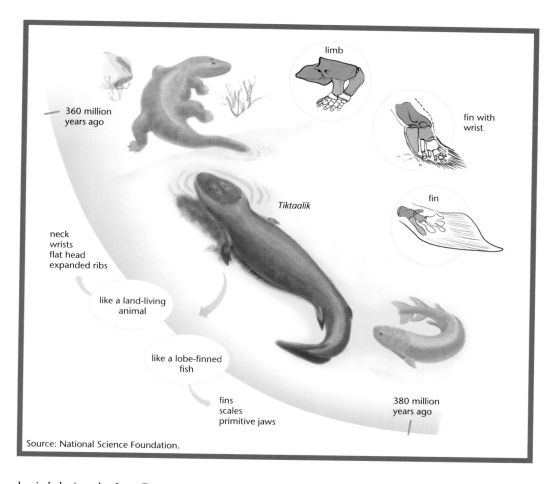

360 million
years ago

limb

fin with
wrist

fin

Tiktaalik

neck
wrists
flat head
expanded ribs

like a land-living
animal

like a lobe-finned
fish

fins
scales
primitive jaws

380 million
years ago

Source: National Science Foundation.

buried during the Late Devonian Period. This date places *Tiktaalik* between the youngest previously known tetrapodlike fishes, which date from about 385 million years ago, and the oldest true land-dwelling tetrapods, which date from about 365 million years ago.

A number of *Tiktaalik's* features show the transition from fish to tetrapods. For example, the animal had fishlike jaws as well as evidence of gills, fins, and scales. It also had features common to amphibians, including a neck and upward-facing eyes. *Tiktaalik's* front fins, however, resembled the limbs of modern land-dwelling animals. Nearly all fish fins consist of a web of skin supported by a skeleton of rods called rays. In contrast, the limbs of all tetrapods, including human beings, consist of a single upper bone joined to two lower bones. Tetrapod limbs also end in wrists or ankles with flexible *digits* (toes and fingers), typically five in number. The wrist and ankle swivel. Hints of all these features appear in the forelimbs of *Tiktaalik.* As of mid-2006, paleontologists had not excavated the creature's hindquarters, and so it is unclear if its rear limbs had similar features.

Tiktaalik's front fins illustrate the transition from aquatic animals to land-based animals. In *Tiktaalik,* what had been the simple fins of fish had started to develop into fingers and a swiveling wrist. In land-dwelling tetrapods, these bones later became fingers, toes, wrists, and ankles.

While fish have a head rigidly attached to the rest of their body, *Tiktaalik* had a skull connected to the rest of its skeleton by a flexible neck. As a result, *Tiktaalik* could raise and lower its head and even swivel it slightly from side to side. This greater mobility was probably an important development in the evolution of land animals. In land animals, the head must be able to move independently of the shoulder bones, which help support the animal as it stands on its limbs.

From gills to ears

Tiktaalik's fossils also provide a fascinating look at the development of the ability to hear airborne sounds, a major shift for animals adjusting to life on land. In fish, nerve endings lying along each side of their body detect vibrations and waves in the water, including those produced by prey or predators. This system does not work in air, however. To compensate, tetrapods developed a different system. Tetrapods possess a short, rodlike bone called the *stapes* that transmits airborne vibrations to the inner ear. The stapes did not develop as a wholly new structure in tetrapods. It evolved from the *hyomandibula,* a bone found in the rear jaw of fish.

Primitive jawless fish seem to have had an extra *gill slit* for breathing that was framed by a pair of bony supports called gill bars. (The gill slit is the visible opening on the outside of a fish's body through which water enters the gills.) According to the prevailing scientific theory, the bones in front of the gill slit developed a full hinge and evolved into jaws. The gill slit evolved into the *spiracle,* a small circular opening between the jaw and the hyomandibula. In January 2006, paleontologists Martin Brazeau and Per Erik Ahlberg of the University of Uppsala in Sweden published an in-depth study of the hyomandibula in *Panderichthys,* a fish from the Middle Devonian that may be the closest known ancestor of *Tiktaalik.* Those scientists found that in *Panderichthys,* the hyomandibula had become shorter and thicker. Shubin and his team discovered that in *Tiktaalik,* the hyomandibula had become shorter still and had started to resemble the stapes. Eventually, in tetrapods, the hyomandibula did become the stapes, and the spiracle became part of what is now the ear canal.

The evolution of the spiracle into the middle ear illustrates a key concept in evolution—*exaptation.* Major evolutionary breakthroughs may result when a structure already present in an organism gains a new function. For example, *Archaeopteryx* could fly thanks to feathers that evolved from scales that had nothing to do with flight. As animals evolved from aquatic creatures to land dwellers, the hyomandibula changed from a support bone in the gills, to a prop for the back of the jaw, to a part of a spiracular breathing apparatus, to a functioning stapes. In *Tiktaalik,* this transition was nearly complete.

The spiracle also may have played a role in *Tiktaalik's* ability to breathe air. The animal probably expanded the floor of its mouth cavity and expanded and contracted its ribcage, as frogs do. Air may have been drawn into simple lungs—themselves modified from pouches of the gut— through the modified spiracle.

Scientists have obtained some clues about *Tiktaalik's* feeding behavior from the position of its eyes. The eyes of most fish look to the side. In contrast, *Tiktaalik's* eyes faced upward, a position that would have enabled it to spot other creatures swimming overhead. *Tiktaalik* may have been a hunter that stood on stream bottoms, waiting in ambush. The animal's elongated, flattened head may have been used for lunging at prey, which it caught with its sharp, backward-directed teeth.

Tiktaalik may have hunted on land as well. The paleontologists believe that *Tiktaalik's* sturdy ribcage, shoulder blades, and limb bones gave the animal the rigidity and support it needed to leave the shallow, rather sluggish streams where it lived to make short trips onto nearby riverbanks. By flexing its shoulders, elbows, and wrists, this fish could have lifted itself onto its short, limblike fins. Once again, structures originally adapted for one purpose took on a new role. In this case, the short, stubby fins *Tiktaalik* used for crawling along muddy stream bottoms and occasionally onto riverbanks became, with slight modifications, tetrapod limbs.

Before the end of the Devonian Period, tetrapodlike fish and their descendants, the true early tetrapods, had managed to spread throughout present-day Greenland and Europe and at least as far west as Pennsylvania in North America. Once on land, these animals became successful predators in a newly evolving ecosystem rich in insects and arachnids and, at least initially, with few predators or competitors. The rest is history.

A fossil of *Tiktaalik* (foreground) is shown in a display with an artist's model of what the creature may have looked like. It is shown raising itself up on its short, limblike fins.

Chronic Inflammation:
Pathway to Disease

Genetic researchers are exploring
the link between long-term
inflammation and serious—
even life-threatening—conditions.

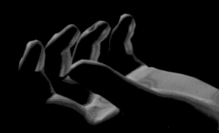

By Beverly Merz

The author:
Beverly Merz is a Boston, Massachusetts-based free-lance medical writer.

TERMS AND CONCEPTS

Antibodies: Molecules, produced by certain immune system cells, that attach to specific antigens and destroy them.

Antigens: Foreign substances in the body, including bacteria and viruses.

Autoimmune disorders: Diseases that occur when the immune system perceives the body's own tissues as foreign substances and attacks them.

Degenerative disorders: Illnesses associated with old age and the breakdown or failure of an organ or system.

Inflammation: The body's response to injury or infection; inflammation often involves heat, pain, redness, and swelling.

Inflammation, Acute: Short-lived inflammation that helps the body to destroy infectious organisms and dead or damaged tissues.

Inflammation, Chronic: Inflammation that persists for weeks, months, or years.

Leukocytes: Any of several types of white blood cells that play an important role in the body's immune system by destroying invading antigens.

Nonsteroidal anti-inflammatory drugs: Medications, such as aspirin, ibuprofen, indomethacin, and naproxen, that work by preventing the production of inflammatory molecules in the body.

What do a sliver, a heart attack, and some forms of cancer have in common? All three involve inflammation, the body's response to injury or infection. *Acute inflammation* (inflammation that is short lived) helps the body to destroy infectious organisms and dead or damaged tissues and to begin healing. In contrast, *chronic inflammation*—inflammation that persists for weeks, months, or even years—can damage healthy tissue. For decades, physicians have known that inflammation is the culprit behind such chronic diseases as asthma, multiple sclerosis, rheumatoid arthritis, and ulcerative colitis. By 2006, researchers also had found connections between inflammation and an increased risk of several major *degenerative disorders*—including heart disease, cancer, and Alzheimer's disease. (Degenerative disorders are illnesses associated with old age and the breakdown or failure of an organ or organ system.)

In 2005, the discovery of a gene that plays a pivotal role in controlling inflammation generated much excitement. Researchers had previously linked the gene with Alzheimer's disease, diabetes, infections, and obesity. In 2005, a team of scientists from the United States and Australia determined that the gene makes an anti-inflammatory protein called selenoprotein S that protects against cell damage and regulates immune cell function. The finding provided the first evidence of a biological link between inflammation and degenerative disorders.

In a sense, the discovery was another chapter in a 2,000-year-old quest to understand inflammation. Inflammation was first identified in ancient Rome, by the medical writer Aulus Cornelius Celsus. Celsus described in Latin what are still considered the condition's four classic signs—*calor* (heat), *dolor* (pain), *rubor* (redness), and *tumor* (swelling). Since Celsus's day, we have learned a great deal about acute inflammation—such as the swelling that results from a splinter in our finger. During the late 1900's, physicians began to recognize that inflammation can persist for too long and produce serious—even life-threatening—conditions. Physicians now believe that as people age, they can have constant, low levels of inflammation. Chronic inflammation, detectable only by laboratory tests, damages the tissues and organs of the body and sets the stage for degenerative disorders.

Many medical specialists focus their efforts on pinpointing the factors that trigger chronic inflammation and cause it to persist. *Molecular biologists,* for example, study the cellular and chemical players in the inflammatory drama and the systems and pathways that organize them. *Geneticists* probe the genetic origins of inflammation, and *epidemiologists* chart the lifestyle habits that encourage it. In addition, *pharmacologists* develop new anti-inflammatory therapies that target specific inflammation-causing mechanisms with greater preci-

THE CLASSIC SIGNS OF INFLAMMATION

Inflammation was first described in ancient Rome, by the medical writer Aulus Cornelius Celsus in his eight-volume work *De medicina (On Medicine),* published in about A.D. 30.

Celsus listed in Latin what remain the classic signs of the condition:

- *calor* (heat)
- *dolor* (pain)
- *rubor* (redness)
- *tumor* (swelling)

sion and fewer side effects than current medications do. In 2006, physicians could choose from among at least 30 anti-inflammatory medications, in addition to aspirin, a form of which has been available since Celsus's day. Genetic research promises new approaches to inhibiting inflammation, perhaps by developing drugs that can interfere with the process at an early stage.

The oldest surviving version of *De medicina* was published in Florence, Italy, in 1478.

Inflammation's role in the immune system

Inflammation is a complex response by the body's immune system that defends against diseases and harmful invaders. Physicians once assumed that the immune system went about this task in an orderly, step-by-step fashion. By the early 2000's, however, researchers understood that the structure of the immune system actually resembles the Internet, with numerous pathways and channels of communication among the cells, molecules, and involved tissues.

ACUTE INFLAMMATION

Acute inflammation, the body's immediate response to infection and injury, helps the body to destroy disease-causing organisms and remove foreign materials and poisons so that healing can begin.

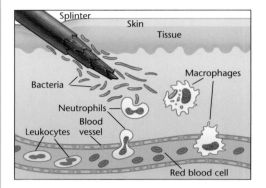

When a splinter pierces the skin, it damages body tissues and introduces bacteria into the wound. White blood cells called *leukocytes,* which circulate in the bloodstream, rush to attack the foreign invaders. Blood vessel walls become more porous, allowing leukocytes to pass through them to reach the site of the injury. Leukocytes called *neutrophils* engulf the bacteria and release chemicals to destroy them. Other leukocytes called *macrophages* engulf and digest the remains of the bacteria, dead neutrophils, and other cellular debris.

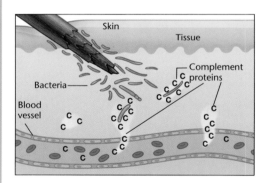

Immune system proteins called *complement,* which also circulate through the bloodstream, coat bacteria to make them easier to destroy. Complement proteins also emit chemical signals to attract inflammatory cells that stop the bacteria from multiplying.

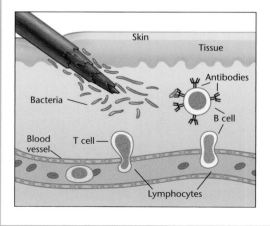

Some leukocytes, called *lymphocytes,* "remember" certain bacteria from previous encounters. *B lymphocytes* (B cells) produce antibodies that lock on to specific bacteria. *T lymphocytes* (T cells) signal B cells to produce antibodies. T cells also attack bacteria themselves and, when the bacteria have been destroyed, signal the immune system to stop the attack.

Critically reviewed by Joseph R. Podojil, Ph.D., Department of Microbiology-Immunology, Northwestern University, Chicago.

Types of immune systems

Immune-system processes are classified in several ways, which often overlap. Human beings have both *innate* and *adaptive* immune systems. Innate immunity is a general, automatic response to any foreign substance in the body. Human bodies also develop adaptive immunity over time, through exposure to many kinds of *antigens* (foreign substances in the body, including bacteria and viruses).

The appearance of an antigen triggers in both the innate and adaptive immune systems the release of several types of *leukocytes.* Leukocytes are made in the bone marrow and circulate through the bloodstream. Two types of leukocytes play a role in the innate immune system. Leukocytes called *granulocytes* contain microscopic sacs of *enzymes,* chemicals that digest *microbes* (germs) and other foreign substances. The most common type of granulocyte, called a *neutrophil,* attacks invading bacteria. Neutrophils travel to the site of the invasion, pass through blood vessel walls, engulf invading particles, and dissolve them.

Leukocytes called *monocytes* mature into cells called *macrophages* (from Greek root words meaning *big eaters*). Macrophages are the immune system's clean-up crew. They gobble up microbes, worn-out granulocytes, and tissue debris.

The innate immune system also includes the *complement system.* It consists of about 30 different proteins that circulate in the blood and are activated when they bind to invading microbes. The proteins coat the microbes to make them easier to destroy and emit chemical signals that attract inflammatory cells, which stop the microbes from multiplying.

The adaptive immune system targets specific microbes, such as cold viruses, that have previously invaded the body. Because the body "remembers" or "recognizes" such microbes, it can quickly mount a defense against

CHRONIC INFLAMMATION

Chronic inflammation contributes to diseases and conditions throughout the body. By the early 2000's, researchers had found a strong link between chronic inflammation and coronary artery disease (CAD), the most common type of heart disease.

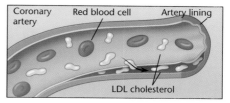

CAD begins when too much *low-density lipoprotein* (LDL) *cholesterol,* a fatty substance, builds up in the bloodstream. Some of the LDL cholesterol settles into the lining of the arteries, where it undergoes a chemical reaction that transforms it into an *antigen* (a substance foreign to the body).

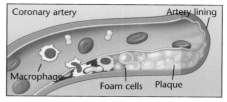

Immune system cells called *macrophages* attack and consume the LDL cholesterol. As they grow, macrophages turn into *foam cells,* which become trapped in the lining of the artery. Other immune system cells encase the foam cells in a shell of *fibrin* (a type of protein), forming *plaque.*

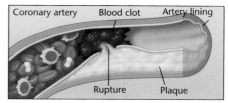

As the plaque grows, stressing its fibrin coat, the coat may rupture and break through the artery wall. Immune system cells that cause blood to clot rush to the scene. If a blood clot forms, it may block the artery, reducing the supply of oxygen to the heart. Heart tissue may die, causing a heart attack.

them. Two types of white blood cells called *lymphocytes* lead the charge. B lymphocytes, or B cells, mature in bone marrow; T lymphocytes, or T cells, mature in the *thymus,* a gland located near the base of the neck. B cells produce *antibodies,* molecules that attach to specific antigens. Antibodies bind to antigens, making them more vulnerable to macrophages, neutrophils, or T cells. Antibodies also activate the complement system, prompting it to destroy antigens or to attract other disease-fighting leukocytes. Finally, antibodies *neutralize* (render harmless) poisonous substances produced by microbes and other organisms and stop the progress of a disease.

T cells recognize foreign invaders and signal B cells to produce antibodies. They also attack and kill antigens themselves. Once the antigen has been destroyed, T cells signal the immune system to stop the attack.

The body's reaction

While the battle between the immune system and the invading cells rages, the body responds in a number of ways. Blood vessels *dilate* (expand) to quickly bring more blood to the site of the infection. The dilation makes the skin look red and allows *serum* (the clear liquid component of blood) to leak into the spaces surrounding the vessels, producing swelling. The blood vessels also become more porous, allowing neutrophils and other immune system cells to pass through their walls easily into the infection site. Some of the messenger chemicals in the blood act on the *central nervous system* (the brain and spinal cord) to raise body temperature—that is, produce a fever. The swelling in the tissues presses against nerves, causing affected parts of the body to ache.

Healing begins as the attack ends. *Fibroblasts,* cells that produce the proteins *collagen* and *fibrin,* migrate to the battle site. The fibroblasts build a connective framework on which new tissue cells are deposited. If the damage is severe, the body may not be able to produce enough new tissue cells to fill out the framework. Fibrin and collagen fill in the gaps, forming a scar.

Why does inflammation become chronic?

Chronic inflammation may occur for several reasons. The immune system may fail to completely rid itself of a microbe. Therefore, the system continues to respond to the microbe's presence. For example, persistent cases of *gingivitis* (gum disease) occur because the body's defense systems have difficulty destroying all the bacteria hiding in *plaque* (the sticky film deposited on teeth). In other cases, the immune system may remain in action even after it has eliminated the microbe. Researchers are still trying to learn why the immune response fails to shut down after the threat of danger has passed.

A third cause of chronic inflammation is mistaken identity, in which the immune system perceives the body's own tissues as foreign substances and wages an attack on them. Such conditions, including multiple sclerosis and rheumatoid arthritis, are known as *autoimmune disorders.* Just as an interminable war lays waste to the countryside, chronic inflammation wears away at the body. Because the immune system is so well equipped to handle potential assaults, it can devastate healthy tissue when it turns on itself.

The link between chronic inflammation and disease

Medical researchers began to establish a connection between chronic inflammation and disease as the result of several large *observational studies.* (In an observational study, researchers observe the effects of a particular risk factor, test, or treatment on a group of people without

DISEASES OR CONDITIONS ASSOCIATED WITH INFLAMMATION

DISEASE	ROLE OF INFLAMMATION
Asthma	Inflammation may cause permanent scarring and thickening of airway walls, leading to more frequent and more severe asthma attacks.
Atherosclerosis (heart disease)	Inflammation contributes to the build-up and disintegration of *plaque* (deposits of fat, calcium, and dead cells) in artery walls; plaque can block an artery, causing a heart attack.
Alzheimer's disease	Researchers have found inflammation in the brains of some people with Alzheimer's; however, its role in the disease is unclear.
Diabetes	Fat cells send out inflammatory chemicals that promote insulin resistance, which may lead to diabetes.
Helicobacter pylorus	Inflammatory bacterium causes stomach ulcers, which may lead to stomach cancer.
Hepatitis B or C	Chronic infection causes inflammation of the liver, which may lead to cancer.
Human papilloma virus (HPV)	Chronic infection causes genital warts, which may lead to cervical cancer.
Inflammatory bowel disease (ulcerative colitis, Crohn's disease)	Inflammation promotes the formation of colon cancer tumors.
Lung cancer	Smoking inflames the lining of the lungs, which may lead to the formation of tumors.
Rheumatoid arthritis	Inflammation causes irritation that wears away bone and *cartillage* (the spongy substance that covers the ends of bones).
Systemic lupus erythematosus	Malfunctioning immune system causes inflammation in various parts of the body, including heart, lungs, skin, other organs, and joints.

Critically reviewed by Joseph R. Podojil, Ph.D.

MEDICATIONS USED TO TREAT INFLAMMATION

TYPE OF DRUG	NAME	HOW IT STOPS INFLAMMATION	USE
Nonsteroidal anti-inflammatory drugs (NSAID's)			
▪ salicylates	aspirin	Block an enzyme called cyclooxygenase-2 (COX-2), which makes a chemical called prostaglandin; prostaglandins contribute to inflammation. Also block cyclooxygenase-1 (COX-1), an enzyme that helps protect the stomach lining from irritation.	Relieve pain and decrease fever and inflammation in many diseases and conditions, including rheumatoid arthritis, osteoarthritis, and back pain; commonly used to prevent heart attack and stroke.
▪ traditional NSAID's	ibuprofen (Motrin, Advil) naproxen (Naprosyn, Aleve) indomethacin (Indocin)	Block COX-2 enzymes (also block COX-1).	Relieve pain and decrease fever and inflammation in many diseases and conditions, including arthritis and menstrual cramps.
▪ COX-2 inhibitors	valdecoxib (Bextra) celecoxib (Celebrex) rofecoxib (Vioxx)	Block COX-2 enzymes with less effect on COX-1. (Vioxx and Bextra were pulled from the market in 2004 and 2005, respectively, because of an increased risk of heart attack and stroke.)	Relieve pain and swelling caused by arthritis, including osteoarthritis and rheumatoid arthritis.
Glucocorticoids (also called corticosteroids)			
	prednisone (Deltasone) prednisolone (Prednisolone)	Control inflammation by mimicking the action of cortisol, a corticosteroid that occurs naturally in the body.	Relieve swelling caused by arthritis, ulcerative colitis, asthma, and systemic lupus erythematosus.
Biologic response modifiers			
	etanercept (Enbrel) infliximab (Remicade) adalimumab (Humira)	Neutralize a protein called tumor necrosis factor-alpha (TNF-alpha) that is present in joints and blood and promotes inflammation.	Relieve pain and swelling of rheumatoid arthritis.
	anakinra (Kineret)	Prevents the protein interleukin-1 from prompting inflammatory cells to release enzymes that contribute to pain and inflammation.	
Other medications (traditionally used to treat cancer or other conditions)			
	azathioprine (Imuran)	Decreases inflammation by limiting the number and activity of inflammatory cells in the joints.	Relieve inflammation of severe rheumatoid arthritis.
	methotrexate (Rheumatrex)	Blocks the action of the vitamin folic acid; why this activity affects inflammation is not fully known.	

Critically reviewed by Joseph R. Podojil, Ph.D.

making any attempt to affect the outcome.) The Framingham Heart Study was one of the most important of these investigations. In 1948, researchers with the U.S. National Heart, Lung, and Blood Institute began to track a group of more than 5,000 residents of Framingham, Massachusetts, from early adulthood into old age, recording their health habits and the diseases they developed. The physicians hoped to determine which lifestyle habits are healthful and which lead to disease. Over time, the Framingham Heart Study and other studies yielded an intriguing connection. They revealed that men and women who take aspirin regularly—usually for several years—are less likely to have heart attacks. More recently, studies indicated that aspirin users have lower rates of several cancers and even Alzheimer's disease. These observations suggested that inflammation might be a common denominator in these degenerative disorders.

Eventually, the Framingham and other researchers began to track such physical characteristics as excess weight, elevated blood pressure, and high blood levels of glucose, cholesterol, and *C-reactive protein* (CRP). (CRP is a type of protein that is present only when there is inflammation somewhere in the body; researchers believe that CRP helps immune-system cells bind to foreign and damaged cells.) The data showed that the risk of heart disease rises as CRP levels in the blood increase. By the early 2000's, research also revealed a link between elevated CRP levels and an increased risk of heart disease, diabetes, several types of cancer, and Alzheimer's disease.

In the 1980's, research showed that inflammation, together with an excess of *low-density lipoprotein cholesterol* (LDL cholesterol, or "bad"

Willow bark is one of the earliest known treatments for the pain and fever of inflammation and was used by the ancient Greeks for that purpose. In 1828, scientists determined that a chemical called *salicin* in willow bark is responsible for its medicinal properties. A synthetic form of salicin is used today as an ingredient in aspirin and other nonsteroidal anti-inflammatory drugs.

cholesterol), leads to *atherosclerosis* (a form of heart disease, also called "hardening of the arteries"). By the 2000's, researchers better understood the process. When high levels of LDL cholesterol circulate in the blood, some of the LDL passes into the lining of the arteries. The LDL cholesterol also undergoes a chemical reaction that makes it a target for macrophages. Macrophages consume the LDL, growing in size until they turn into fat-filled cells known as *foam cells*. Fibrin accumulates over these cells, encasing them in a mesh cap to form arterial plaque. The foam cells in plaque continue to take in more LDL cholesterol. As they grow, they exert pressure on the fibrin cap, which may cause it to crack. As blood seeps into the fissures, clots form on the surface. If a clot blocks an artery supplying oxygen to the heart muscle, oxygen-deprived heart tissue may die. The result is a heart attack.

As plaque forms, CRP levels in the blood rise. Researchers found that people whose CRP levels ranked in the top third among study participants were twice as likely to have a heart attack as people in the bottom third, even if their cholesterol levels were low. Having both high cholesterol and high CRP raised the risk even more. Because CRP is easily identified in blood tests, it has become a particularly useful indicator of an increased risk of heart disease.

In recognition of the additional risk, the U.S. Centers for Disease Control and Prevention in Atlanta, Georgia, and the American Heart Association, headquartered in Dallas, in 2005 advised that people with an increased risk of cardiovascular disease may benefit from CRP screening. Physicians recommend that people with high levels of CRP exercise regularly, choose a healthful diet, refrain from smoking, and, in some cases, take drugs called *statins,* which lower the level of cholesterol in the blood.

How inflammation can lead to diabetes

High CRP levels can also increase the risk of *insulin resistance,* according to several large observational studies. Insulin resistance is a decline in the ability of tissues to use *insulin,* the hormone that regulates the body's use of food, particularly *glucose* (blood sugar). Insulin resistance often sets the stage for diabetes. Additional studies have found that blood levels of tumor necrosis factor alpha (TNF alpha) and other molecules of inflammation are also higher in people who ultimately develop diabetes. TNF alpha, which is produced by macrophages and T cells, has widespread effects on the immune system. It promotes blood clotting, which helps to contain infection, and the release of *prostaglandins* during acute inflammation. (Prostaglandins are chemical messengers that stimulate a variety of body functions.) During chronic inflammation, TNF alpha breaks down muscle tissue to provide fuel for the body, an action that may lead to the *wasting* (severe loss of weight and muscle mass) seen in people with cancer. The bodies of people with autoimmune diseases, such as rheumatoid arthritis, are not able to properly eliminate TNF alpha. It lingers in the tissues, where it continues to increase immune system activity.

Although researchers are still uncertain how inflammation contributes to diabetes, some evidence suggests that TNF alpha decreases the ability of insulin to move glucose from the bloodstream into the cells. In 2001, researchers discovered that insulin increases blood concentration of particles that help prevent inflammation and damage to cells in the arterial lining. When cells become insulin resistant, they lose the anti-inflammatory benefits of insulin, further increasing the risk of diabetes and heart disease.

A connection with cancer

Researchers also have discovered a connection between inflammation and cancer. "A sore that does not heal" has long been an important indicator of the possibility of cancer. In the 1860's, the German scientist Rudolf Virchow observed that cancerous tumors resemble wounds that the body is trying to heal with an influx of immune cells. Virchow proposed that cancer was somehow linked to chronic inflammation. However, his theory was ignored by scientists, who found it difficult to believe that the body's principal healing mechanism could have such a destructive effect.

Subsequent research proved that Virchow was right. By 2004, researchers had linked about 15 percent of all cancers to infectious diseases. Hepatitis B and human papillomavirus (HPV), for example, dramatically raise the risk of developing liver and cervical cancer, respectively. The bacterium *Helicobacter pylorus* is associated with stomach cancer. Ulcerative colitis and Crohn's disease are linked with colon cancer. Research also has shown that exposure to such inflammatory substances as cigarette smoke and the mineral asbestos are strong risk factors for lung cancer.

In addition to establishing a convincing connection between inflammation and cancer, scientists also probed for a physical connection. They found it in abundance. During the 1990's, medical researchers discovered that some of the immune system's most powerful tools may actually be helping cancerous tumors to survive and grow. They proved that leukocytes and other products of immune reactions damage cellular DNA (deoxyribonucleic acid, the molecule that transmits genetic information from one generation to the next). The damage creates *mutant* (changed) genes that allow cells to grow at an uncontrolled rate. Macrophages stimulate the formation of new blood vessels and produce numerous substances that help tumor cells grow. In a study published in 1994, cancer researchers found elevated levels of cyclooxygenase-2 (COX-2) in 80 percent of colon tumors. COX-2 produces prostaglandins that contribute to pain, fever, and inflammation. Researchers also have found increased levels of COX-2 in people with arthritis.

A possible connection to Alzheimer's disease

Observational studies also have led researchers to a possible connection between inflammation and Alzheimer's disease. In the early 1990's, researchers noted that people who regularly took aspirin and other medications called *nonsteroidal anti-inflammatory drugs* (NSAID's) for arthritis were less likely to develop Alzheimer's than were people who did not take such drugs. In addition, brain scans of people with Alzheimer's disease showed inflammation in regions affected by Alzheimer's. Such findings surprised many researchers, because they had long believed that the immune system has no influence over the brain. *Capillaries* (tiny blood vessels) in the brain have cells that are more tightly packed than those in the capillaries that serve other parts of the body. The brain capillaries create a so-called "blood-brain barrier" that most inflammatory cells and pathogens cannot cross. In addition, the brain has no pain receptors, making it difficult for physicians or patients to recognize the classic signs of inflammation in the brain.

In the late 1980's, scientists discovered that immune cells can enter the brain and that the brain contains its own population of immune cells. The major players in the brain's inflammatory response are cells called *microglia*. Like macrophages, microglia are phagocytes, or "eating cells," that absorb microbes and cellular

A bank of 1,500 parallel processors at the SBC Genomics Computing Center in San Antonio, Texas—the world's largest computing cluster devoted to analyzing genetic information—played a critical role in the 2005 discovery of the connection between the SEPS1 gene and inflammation. SEPS1 helps protect against cell damage.

debris. Microglia also release inflammatory compounds, such as TNF alpha. Just as fat-filled macrophages form the core of arterial plaque, microglia stuffed with *amyloid*—a protein fragment found between *neurons* (brain cells)—form the basis of plaque in the brains of people with Alzheimer's disease. An accumulation of plaque is one of several suspected causes of Alzheimer's.

Treating inflammation

Even before Celsus described the four signs of inflammation, physicians had sought ways to ease its symptoms. As early as the 400's B.C., the Greek physician Hippocrates proclaimed the value of willow bark for diminishing pain and fever. Willow-bark tea remained a mainstay of folk medicine for centuries. In 1828, scientists isolated the active agent in willow bark, salicin. When dissolved in water, salicin becomes salicylic acid. In 1899, Friedrich Bayer and Company of Germany began to market a synthetic form called acetylsalicylic acid, which was somewhat gentler on the digestive system. Popularly known as aspirin, the compound revolutionized pain control and gave rise, in the 1960's, to the NSAID class of drugs. The first generation of NSAID's—which included ibuprofen, naproxen, and indomethacin—worked by interfering with the COX enzyme, preventing the production of inflammatory prostaglandins. The drugs, which were effective in eliminating acute inflammation, also were used by people with such chronic inflammatory conditions as arthritis. However, the medications had a serious side effect: they raised the risk of stomach ulcers.

By the early 1990's, researchers had learned that the immune system produced several forms of the COX enzyme. By blocking all of them, early NSAID's prevented an enzyme called COX-1 from protecting the stomach lining. The discovery of COX-2 inspired pharmacologists to develop a drug that would target only that enzyme. The first such drug, celecoxib (sold under the brand name Celebrex), debuted in 1999. Celecoxib prevents COX-2 from making prostaglandins. Studies indicated that it was as effective in relieving pain and inflammation as the other NSAID's but caused fewer gastrointestinal side effects. Two other COX-2 inhibitors—rofecoxib (sold under the brand name Vioxx) and valdecoxib (sold under the brand name Bextra)—followed. However, several studies eventually indicated that people taking rofecoxib suffered heart attacks at a much higher rate than expected, especially if they already had heart disease. The U.S. Food and Drug Administration (FDA) requested that the manufacturers of Vioxx and Bextra withdraw those products from the market and called for the maker of Celebrex to include a *black box warning* on its packages indicating the possibility of serious side effects. A black box warning is the strictest drug warning issued by the FDA short of a recall.

Another class of medications called *glucocorticoids* (also known as corticosteroids)—including prednisone and prednisolone—is more potent than NSAID's. Glucocorticoids suppress the entire immune system rather than a single inflammatory pathway. They were first introduced in 1948 to treat rheumatoid arthritis. Glucocorticoids quickly slow and weaken the inflammatory process by inhibiting prostaglandins and certain other immune cells involved in inflammation. Glucocorticoid drugs may cause a number of serious side effects, including decreased muscle mass, reduced bone density, fluid retention, and insulin resistance. As a result, physicians weigh the risks of taking glucocorticoids against their benefits when prescribing these medications for long-term use.

By the early 2000's, researchers had developed several drugs called *biologic response modifiers.* Such medications halt the inflammatory process at a much earlier stage than NSAID's or glucocorticoids by targeting specific immune system molecules that trigger inflammation. For example, several drugs used to treat severe rheumatoid arthritis block TNF alpha. They include adalimumab (sold under the brand name Humira), etanercept (sold as Enbrel), and infliximab (sold as Remicade). Another drug, anakinra (sold under the brand name Kineret), blocks a molecule that prompts inflammatory cells to release enzymes that contribute to pain and inflammation.

Older medications called *immunosuppressants,* originally developed for use in cancer therapy or to prevent rejection of transplanted organs, are still used in treating patients with multiple sclerosis, rheumatoid arthritis, inflammatory bowel disease, and other autoimmune diseases. Also called *disease-modifying antirheumatic drugs,* these medications include azathioprine (sold under the brand name Imuran) and methotrexate (sold as Trexall and Rheumatrex). Researchers remain unsure how these drugs reduce chronic inflammation. Nevertheless, the medications often slow the progression of the disease and diminish symptoms.

Can diet help?

Drugs are not the only way to fight chronic inflammation. Because *saturated fatty acids* (components of fat derived primarily from animals and dairy products) also trigger the immune response, researchers have tested the effects of removing them from circulation in the blood. One way to accomplish this is to lower the intake of saturated fats and *trans fats* in the diet by substituting unsaturated fats, such as those found in vegetables. (Trans fats are chemically manufactured fats that are added to foods to improve their texture, flavor, and freshness. Trans fats also increase the amount of LDL cholesterol in the body.) Another way to reduce the effects of fatty acids is to lower the amount of body fat through exercise. Researchers have demonstrated that following a reduced-calorie diet and engaging in regular aerobic exercise can help many people to lower their levels of CRP, TNF alpha, and other markers of inflammation. (In a widely publicized study, researchers reported in February 2006 that a low-fat diet does not lower the risk for heart disease or cancer. However, many researchers pointed out that the dieters were unable to cut the fat in their diets to the recommended degree. They also noted that the study did not distinguish between the consumption of "good" fats and "bad" fats.)

Genetic research into inflammation

The identification in 2005 of the SEPS1 gene, which makes the anti-inflammatory selenoprotein S, may launch a new era of inflammation diagnosis and treatment. Individuals who carry slight variations of this gene may be predisposed to produce higher levels of inflammatory proteins. A test for those variations could be used to identify people at risk for chronic inflammation. Such people could then follow a low-fat diet and exercise program before they developed diabetes or heart disease. In addition, scientists are working to develop drugs that block the versions of selenoprotein S that raise levels of inflammatory proteins. People with those versions of the gene could take such drugs as a preventive measure.

Geneticists are looking for other genes responsible for specific inflammatory diseases as well. If they can identify a particular gene and the protein it *codes for* (carries the instructions for), they may be able to develop a drug to disable the gene or neutralize its protein.

Although the evidence of a link between inflammation and degenerative diseases continues to grow, scientists still have much to learn. The immune system has dozens of interrelated players. Such tasks as teasing out the molecules involved in different types of inflammation, learning which ones operate in a particular individual, and identifying genes in individuals that may predispose them to particular forms of chronic inflammation require specialized knowledge in a variety of newly developed disciplines. They include *bioinformatics* (the use of computers to organize large amounts of data in biological research); *genomics* (the study of genes and their function); and *proteomics* (the study of proteins and their function). Some biotechnology experts estimate that it will be economically feasible to decode the *genome* (the complete set of genetic information) for every human being within 10 years. If that prediction is accurate, it may also be possible to develop custom-designed anti-inflammatories. In another decade, aspirin may seem primitive, indeed.

THE EARLY EARTH

How can scientists learn about the Early Earth when Earth itself has erased nearly all traces of its first billion years?

By Steven I. Dutch

Descriptions such as alien, hostile, and violent only hint at the conditions you would encounter if you could travel back in time 4.6 billion years to the newly formed Earth. At first, you would see a waterless surface pockmarked with craters and littered with volcanic rocks. Over the next billion years, you would watch as rocky debris left over from the formation of the solar system—some of it the size of a small planet—pummeled Earth. You would see volcanoes spewing clouds of noxious gases and monstrous rivers of lava. But sooner than you might imagine in such an unstable world, you would notice a thick atmosphere enveloping the planet and water pooling in low places.

Then, about 3.5 billion years ago, you would discover that you were no longer the only life form on Earth.

Until the early 1970's, information about Early Earth was so sparse that geologists did not even have a name for the planet's first billion or so years. Many geologists now refer to this period as the Hadean Eon—after Hades, the ancient Greek underworld.

By 2006, scientists had developed a better understanding of the forces that had transformed Earth from an arid wasteland to a world whose oxygen-rich atmosphere nourishes an astonishing variety of life. They had been able to create a rough time line for the major events of

Meteorites and other debris left over from the formation of the solar system bombard Earth about 4 billion years ago, in an artist's illustration. The moon appears much larger than it does today because it was five times as close, only about 64,000 kilometers (40,000 miles) away.

The author:
Steven I. Dutch is a Professor of Natural and Applied Sciences at the University of Wisconsin in Green Bay.

the Hadean Eon, though many of the details remained unclear. They had discovered just how severe, violent, and long-lasting the bombardment period was. And they had found evidence that Earth's life forms and water had appeared much earlier than previously believed.

Most important, they had radically changed their theories about the origin of life. In the mid-1900's, scientists commonly assumed that life appeared in the warm, shallow seas of a primitive Earth whose temperatures were similar to those of today. By 2006, however, they had learned that some organisms, known as *extremophiles,* can thrive under conditions that would instantly kill human beings and most other life forms. At one time, scientists assumed that all life depended on the sun for energy, but they now know that some extremophiles get their energy from chemical reactions in hot springs bubbling up from the ocean floor. Other extremophiles live in oxygen-free environments deep underground. Under which of these—or other still unknown conditions—did the first life forms appear? Finally, scientists no longer wonder how Earth came to be so suitable for life. They know that from its first appearance, life has profoundly shaped the atmosphere, oceans, and other physical conditions on Earth.

The world in a grain of zircon

Trying to reconstruct Hadean-Eon conditions is challenging because our planet's geologic history is a story of unceasing change. Rock is continually destroyed by weathering, erosion, and other surface processes and recycled into new forms by heat, pressure, and the movement of the tectonic plates that make up Earth's outer shell. All this geologic activity has given Earth beautiful landscapes and a great variety of habitats for living things. However, this activity has also destroyed—and continues to destroy—the evidence of Earth's geologic past. Just as with human history, the further back in time scientists go, the more incomplete this record becomes. In fact, by 2006, geologists had only the tiniest bit of physical evidence from the early Hadean. The oldest known material of any kind on Earth consists of microscopic grains of the mineral zircon found in Australia; they date from about 4.4 billion years ago. The oldest known Earth rock is an *outcrop* (section of exposed bedrock) called the Acasta Gneiss in Canada's Northwest Territories. This coarse-grained rock formed about 3.9 billion years ago deep inside Earth from even older rock that was altered by heat and pressure. The discovery of the zircon has given geologists hope that they may find even older rocks. Rocks are far more complex than minerals and convey much more detail about the conditions under which they formed.

Scientists also look to outer space for clues about conditions on Early Earth. Certain types of meteorites, called *chondrites,* are the remains of rocks that formed early in the history of the solar system.

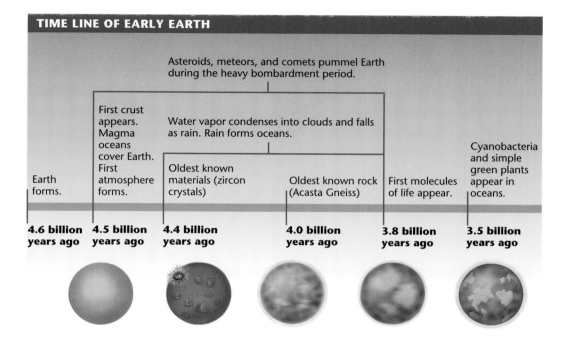

TIME LINE OF EARLY EARTH

Asteroids, meteors, and comets pummel Earth during the heavy bombardment period.

| First crust appears. Magma oceans cover Earth. First atmosphere forms. | Water vapor condenses into clouds and falls as rain. Rain forms oceans. | | | Cyanobacteria and simple green plants appear in oceans. |

Earth forms.

Oldest known materials (zircon crystals)

Oldest known rock (Acasta Gneiss)

First molecules of life appear.

| **4.6 billion years ago** | **4.5 billion years ago** | **4.4 billion years ago** | **4.0 billion years ago** | **3.8 billion years ago** | **3.5 billion years ago** |

Chondrites, which remained unchanged until they fell to Earth, are a valuable source of information about the geologic and chemical conditions on Early Earth.

In 1969, United States astronauts on the Apollo space missions began bringing back rock samples from the moon. Since then, scientists have discovered that some meteorites actually originated on the moon or even on Mars. They were blasted into space by giant meteor impacts, eventually landing on Earth. All these samples of extraterrestrial terrain were unaffected by weathering and other processes that alter rocks over time on Earth. As a result, they preserve an early record of conditions on the moon or Mars. Scientists presume the same sorts of events occurred on these celestial bodies as well as on Earth during the Hadean Eon.

Earth and the moon are formed

The formation of Earth probably began with the violent death of a star called a *supernova.* During this gigantic explosion, most of the dying star's matter was blasted into space. This expansion compressed molecules of gas and microscopic grains of dust nearby, forming a giant, rotating cloud called the *solar nebula.* The gravitational attraction between the molecules and grains caused the cloud to start contracting. As the nebula contracted, it spun faster and flattened into a disk. Most of the material in the solar nebula collapsed toward the center. Eventually, this central mass grew dense and hot enough for hydrogen atoms to

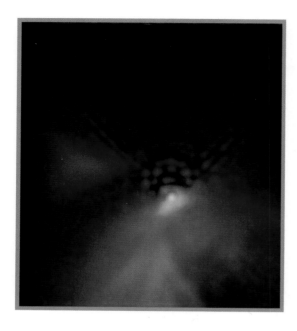

A young star (bright red) 450 light-years away in the constellation Taurus appears much the same as the sun did shortly after its birth 4.6 billion years ago. A dark band to the left of the star represents a rotating disk of condensed dust and gas from which the star formed. The sun, Earth, and the rest of our solar system formed from a similar disk.

begin fusing to helium atoms—the basis of nuclear fusion—and the central mass became a star, the sun.

The remaining material swirled around the sun at a faster and faster rate. As the dust grains collided, they stuck together to build larger objects—a process called *accretion*—and their gravity grew strong enough to sweep up more and more of the dust and gas molecules. Collisions between larger objects led to the formation of chunky rocks called *planetesimals*. Some planetesimals combined to form the planets, including Earth. Other planetesimals formed moons, asteroids, and comets.

Scientists believe that shortly after Earth formed, it collided with another planet-sized object in an event known as the Giant Impact. The impact shot a cloud of vaporized rock off Earth's surface and into orbit around Earth. Over time, the cloud cooled and condensed into a ring of small, solid bodies, which then stuck together, forming the moon. The Giant Impact would have melted most of or even all of Earth and the moon. According to some estimates, Earth would have been as hot as the sun for about 10,000 years afterward.

By about 4.5 billion years ago, Earth and the moon had cooled enough for light silicate rocks to rise to Earth's surface, forming the earliest crust. But that crust did not last long. In the interior, the *decay* (breakdown) of radioactive elements and increasing pressure generated great quantities of heat that melted the interior and rose to the surface. At the same time, frequent volcanic eruptions may have covered much of the planet's surface in red-hot flows of lava. Finally, asteroids, comets, and other chunks of space debris left over from the formation of the solar system continually bombarded the young planet. Each time an object hit the planet, the impact created a crater, melted some of the crust, and spread a blanket of debris around the crater. Scientists believe that the impacts occurred so frequently that debris from one impact buried the rocks heated by previous impacts. Erosion and the movement of the tectonic plates that make up Earth's outer shell have destroyed all evidence of these impacts. The heat from the interior and the bombardments became so intense that the crust melted, forming a magma ocean. Scientists have found evidence of a magma ocean in rocks brought back from the moon. They assume that the same process occurred on Earth, though no traces of a terrestrial magma ocean have been found.

THE "BIRTH" OF THE MOON

The moon formed as the result of a massive collision known as the Giant Impact, according to a widely accepted theory. About 4.6 billion years ago, Earth collided with a celestial object about the size of Mars (1). The force of the blow melted or vaporized most of the object as well as a considerable part of Earth, creating a fiery cloud (2). This vapor cooled and condensed into a ring of small, solid bodies orbiting Earth. The bodies slowly clumped together to form the moon (3).

A series of atmospheres

Our oxygen-rich atmosphere, so essential to life on Earth, is the latest in a series of atmospheres that have surrounded our planet since its formation 4.6 billion years ago. The newly formed Earth probably did not have a true atmosphere. Instead, gases escaping from the developing Earth may have formed a thin haze around the planet. But if this atmosphere existed, it was almost certainly stripped away by the impact that created the moon. Indeed, many of the larger impacts that occurred during the accretion of Earth would have effectively blasted any accumulating gas into space. As a result, Earth probably did not have a significant atmosphere until the end of the heavy bombardment period.

The development of Earth's magnetic field was another factor that helped in the creation of a dense atmosphere. After the sun ignited, eruptions began occurring on its surface, sending a continuous flow of electrically charged particles streaming into space. This *solar wind* swept away any haze that may have formed from gases escaping from Earth's interior. As iron and other heavy materials sank to Earth's center and began to melt, electric currents in the newly formed core generated magnetic fields, and a region of strong magnetic forces called the *magnetosphere* surrounded Earth. This magnetic field shielded the planet from the solar wind, allowing the formation of an atmosphere.

In 2006, scientists held two main theories explaining the formation of the first atmosphere. One theory involves a process called *volcanic outgassing*. According to this theory, as the interior of Earth became hotter, the heat caused certain elements to rise to the surface in volcanic

THE OLDEST KNOWN MINERALS AND METEORITES

A grain of the mineral zircon (left, embedded in plastic) is the oldest known material of any kind found on Earth. It was discovered in 1984 in Australia by Simon Wilde of Curtin University of Technology in Perth.

Chondrites (right), a kind of stony meteorite, are the remains of rocks that formed early in the history of the solar system. Because chondrites remained unchanged until they fell to Earth, they are a valuable source of information about the geologic and chemical conditions on Early Earth.

rock. Some of these chemicals—including ammonia, carbon dioxide, carbon monoxide, hydrogen, methane, nitrogen, sulfur dioxide, and water vapor—became the gases of the atmosphere. Additional gases, such as argon, were added by the decay of radioactive elements within Earth.

The presence of argon 40 in the atmosphere offers one of the strongest arguments for the outgassing theory. About 1 percent of the atmosphere consists of this *isotope* (form) of argon, a gas that does not readily combine with other elements. Argon 40 is created by the decay of a radioactive isotope of potassium, a relatively abundant element that makes up nearly 2.5 percent of Earth's crust. The abundance of argon 40 in the atmosphere implies that a large amount of this gas escaped from Earth's interior.

A second theory suggests that as comets slammed into Earth during the heavy bombardment phase, they released their cargo of frozen gases. These gases then formed Earth's first true atmosphere. Some scientists believe the zircon grains discovered in Australia provide some support for this theory. These grains show signs of having been weathered out of their original rock, transported by water, and then redeposited in newly formed *sedimentary rock* (rock composed of layers). In other words, the grains indicate that liquid water has existed on Earth for at least 4.4 billion years.

The presence of liquid water on Earth so long ago—perhaps only 150 million years after the planet's birth—raises difficult questions. According to most theories on how stars form, our infant sun 4.4 billion years ago was only about 70 percent as bright as it is today. As a result, the sun would have provided only about 70 percent of the heat that it does today. Accordingly, Earth would have been too cold for liquid water, which would have frozen.

Liquid water could have survived, however, if a dense atmosphere had surrounded Earth by that time. This atmosphere would have trapped enough heat to keep Earth's temperature above water's freezing point. Some scientists argue that volcanic outgassing could not have supplied enough gas to create such a dense atmosphere so fast. They suggest that the atmospheric gases, particularly water vapor, must have come from comets. Comets, which pummeled Earth in large numbers for millions of years, consist mainly of frozen gases.

Scientists were still exploring both the outgassing and comet theories in 2006. Both may well have played a role in the creation of the atmosphere. However it was created, Earth's atmosphere 4.4 billion years ago was much thicker than it is now. Most scientists agree that it contained little oxygen. Some scientists believe Earth's early atmosphere contained ammonia, helium, hydrogen, and methane, much like the present atmosphere of Jupiter. Others believe it may have contained a large amount of carbon dioxide, as does the atmosphere of Venus. If Early Earth resembled Venus, speculates geochemist Stephen Mojzsis of the University of Colorado at Boulder, the sky would have been reddish because of scattered sunlight, and any oceans would have been olive-green because of metals dissolved in them.

The oceans appear

For many years, some scientists believed that comets were also the sole source of the water that formed the oceans. By 2006, however, this theory had been largely discredited. One study that cast doubt on the comet theory involved the use of radio telescopes to analyze the chemical make-up of three comets as they came close to Earth in the 1980's and 1990's. These studies revealed that comets contain from 2 to 20 times as much *heavy water* as ocean water does. (Like normal water, heavy water contains one oxygen atom and two hydrogen atoms; however, it also contains at least one atom of deuterium, an isotope of hydrogen with twice the mass of ordinary hydrogen.) The higher proportion of deuterium in comet ice suggests that comets could not have been a major source of seawater. If comets had supplied water to the Earth's oceans, the oceans would be much richer in heavy water than they are.

Volcanic outgassing almost certainly created the oceans, most scientists believe. According to this theory, Earth's water originated from water molecules in the cloud of gas and dust that gave rise to the solar system. The water molecules were added to the accreting Earth with all the other materials that make up the planet. Later, the water vapor boiled out as steam. For hundreds of millions of years, volcanoes ejected the water vapor along with chlorine and other hot gases into the Earth's atmosphere.

Rain began to fall once the bombardment stopped and Earth cooled to a temperature below the boiling point of water (100 °C [212 °F] at sea level). Rain may have fallen in a deluge, drenching the planet. On the other hand, rain falling at the rate of 1 centimeter (0.39 inch) per year could have created an ocean 10 kilometers (6 miles) deep in only 1 million years. Many researchers believe that the first permanent ocean was likely in place from 4.3 billion to 3.8 billion years ago. This early sea may have covered the entire planet to an unknown depth as rainwater collected everywhere on the nearly featureless landscape.

The chemical composition of the earliest oceans was probably different from that of modern seawater. For example, the earliest oceans were probably not salty. Instead, they may have had high levels of other elements—including calcium, iron, and magnesium—that dissolved out of rocks by weathering. Studies by rovers on Mars have found *water-soluble* minerals (able to be dissolved in water) containing iron that may have formed in a similar but short-lived ocean on Mars.

If Earth's early atmosphere included high levels of carbon dioxide, the early oceans may have been highly acidic. The sulfur and hydrochloric acid that erupted from volcanoes also may have contributed to the

THE EVOLUTION OF EARTH'S ATMOSPHERE AND OCEANS

Gases expelled by volcanoes, including carbon dioxide and water vapor, may have formed Earth's first atmosphere. Some scientists believe that the numerous comets bombarding Early Earth may have released large amounts of gas, particularly carbon dioxide, to the atmosphere.

As atmospheric water vapor cooled and condensed and Earth's temperature fell below the boiling point of water—100 °C (212 °F)—rain began to fall, creating the oceans. Seawater may have become salty because volcanoes released large amounts of chlorine (Cl) that combined with water vapor (H_2O) to produce hydrochloric acid (HCl). As water flowed over the surface, the acid ate away the rocks, stripping them of sodium (Na). Some of the sodium atoms combined with chlorine atoms in the water to form sodium chloride, the primary salt in the sea.

Volcanic gases

Magma

acidity of the oceans. In this acidic and oxygen-poor setting, rocks would have weathered more quickly and many metals, such as iron, would have been much more soluble in seawater than they are now.

Seawater may have become salty because volcanoes released large amounts of chlorine in the form of hydrochloric acid, as well as water vapor, into the early atmosphere. The hydrochloric acid fell to the ground with rainwater. As the water flowed over the surface, the acid ate away the rocks, stripping them of their salts, particularly sodium. Some of the sodium atoms combined with chlorine atoms in the water to form sodium chloride—more commonly known as table salt—the primary salt in the sea.

Seawater is salty today because of billions of years of chemical processes in which some materials were added and others removed. For example, many processes remove such elements as calcium from seawater, but few remove sodium or chlorine. As a result, these elements tend to remain in seawater for a very long time. On the average, a sodium or chlorine atom remains in the ocean for about 100 million years.

The earliest continents

The discovery of the zircon grains in Australia indicates that crust had begun forming on Earth by 4.4 billion years ago. These grains were

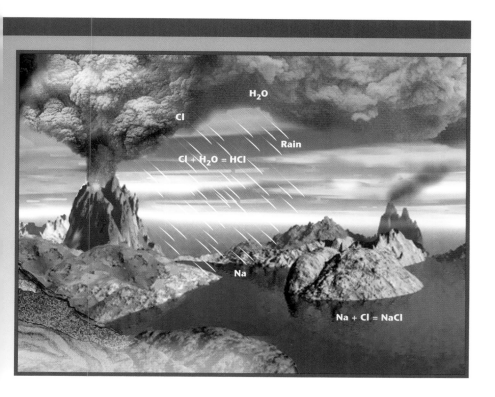

Organic (carbon-containing) molecules mixed with dust appear as wisps of green in a false-color image of the Christmas Tree Cluster taken by the Spitzer Space Telescope. Some scientists believe that the organic molecules that combined to form Earth's first organisms came to Earth in comets and asteroids created in space during the birth of the solar system.

not found in the rock in which they formed but in younger sedimentary rocks that are at least 3 billion years old. The Acasta Gneiss in Canada, the oldest known rock, is actually metamorphic rock—that is, rock that forms when heat or pressure or both cause changes in "parent" rock. That means that the Acasta Gneiss formed from even older rock, probably granite. Granite, in turn, usually forms by the slow cooling and crystallization of *magma* (molten rock) in which such light elements as potassium, silicon, and sodium become concentrated. Some crust also could have formed as the light minerals found in granite floated upward in Earth's magma ocean and in the pools of molten rock created by large meteor impacts. Geologists believe that, by about 3 billion years ago, at least 10 percent of the present continental crust had formed because rocks that old are found over about 5 percent of the present continents.

At first, Earth probably did not have large tectonic plates that, like modern plates, collided, moved apart, or slid past one another. Most of the earliest crust on Earth probably consisted of mosaics of small pieces of crust joined by collisions. In some places, magma from Earth's interior may have risen to the surface, where it cooled and was added to the crust. In other places, parts of the crust may have sunk into the interior. But probably none of these moving pieces of crust would have been wider than a few hundred kilometers. Because the inner Earth was more dynamic then than it is today, these mini-plates would also have moved much faster than modern plates, which travel at the rate of about 10 centimeters (4 inches) per year.

The appearance of life

The oldest known evidence of life on Earth consists of the 3.5-billion-year-old fossils of primitive, plantlike organisms called *cyanobacteria.* Sometimes called *blue-green algae,* cyanobacteria are simple, one-celled organisms that are related to bacteria. The organic compounds that combined to produce Earth's first life forms, however, are certainly much older than these primitive organisms—perhaps as old as 4.2 billion years.

Scientists have developed two main theories explaining the origin of life—the theory of chemical evolution and the theory of panspermia. The more widely accepted theory of chemical evolution was developed

PRIMORDIAL SOUP

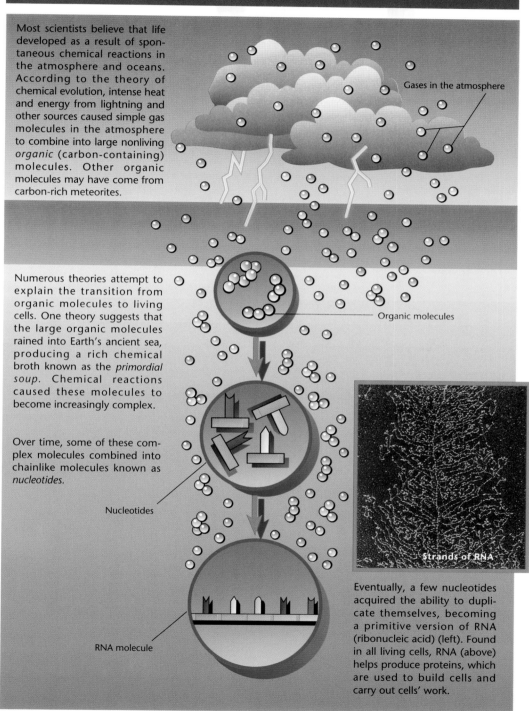

Most scientists believe that life developed as a result of spontaneous chemical reactions in the atmosphere and oceans. According to the theory of chemical evolution, intense heat and energy from lightning and other sources caused simple gas molecules in the atmosphere to combine into large nonliving *organic* (carbon-containing) molecules. Other organic molecules may have come from carbon-rich meteorites.

Gases in the atmosphere

Numerous theories attempt to explain the transition from organic molecules to living cells. One theory suggests that the large organic molecules rained into Earth's ancient sea, producing a rich chemical broth known as the *primordial soup.* Chemical reactions caused these molecules to become increasingly complex.

Organic molecules

Over time, some of these complex molecules combined into chainlike molecules known as *nucleotides.*

Nucleotides

Strands of RNA

RNA molecule

Eventually, a few nucleotides acquired the ability to duplicate themselves, becoming a primitive version of RNA (ribonucleic acid) (left). Found in all living cells, RNA (above) helps produce proteins, which are used to build cells and carry out cells' work.

independently during the 1920's by Soviet biochemist Alexander I. Oparin and by British biologist J. B. S. Haldane. Oparin and Haldane theorized that because hydrogen is the most abundant element in the universe, Earth's early atmosphere had large quantities of this gas. Under such conditions, the hydrogen-containing compounds ammonia, formaldehyde, hydrogen cyanide, methane, and water would also have been abundant. According to this theory, sunlight, lightning, and volcanoes powered reactions among these compounds that produced simple biological molecules, such as sugars and amino acids.

Two American chemists, Stanley L. Miller and Harold C. Urey, in 1953 provided the first experimental evidence in support of the theory of chemical evolution. They subjected a mixture of ammonia, hydrogen, methane, and water to the energy of high-voltage sparks for one week. After that time, amino acids and other simple biochemical compounds had formed. Scientists have repeated this experiment under various conditions. For example, some researchers have assumed that the early atmosphere contained little hydrogen but large quantities of carbon

LIFE IN ALL EXTREMES

Extremophiles, hardy organisms such as the bacteria *Nanoarcheum* and *Ignicoccus* (inset, stained red and green, respectively) that live in hostile environments, are providing clues about the possible nature of the earliest life forms on Earth. The bacteria live near hot springs (below) on the ocean floor and get their energy from chemical reactions.

Cyanobacteria (known also as blue-green algae), which live in layered colonies called *stromatolites,* may be the oldest life forms on Earth. Scientists have found fossils of these simple, one-celled organisms dating back to 3.5 billion years ago. Living colonies of cyanobacteria exist today in a number of places, including Shark Bay, Australia (left).

dioxide. They have replaced the hydrogen-rich "atmosphere" of the Miller-Urey experiment with various mixtures high in carbon dioxide and relatively low in hydrogen. These mixtures also have yielded biochemical compounds when exposed to sparks of energy. Other organic compounds on Early Earth may have come from carbon-rich meteorites known as carbonaceous chondrites. Scientists have found simple organic compound in several of this type of meteorite.

Most scientific research supports—or at least does not contradict—the idea that life arose through chemical evolution. For example, the surface of Earth experiences a continuous flow of energy as it gets light from the sun and radiates heat into outer space. Physics research has demonstrated that such an energy flow increases molecular organization. Thus, the evolution of complex biochemical molecules may be viewed as part of this natural process.

From molecule to cell

Scientists have developed three major theories to explain the transition from early organic molecules to living cells. All three theories are based on the idea that the simple organic compounds formed more complex ones, which then gave rise to the structures that make up cells. The oldest of these theories states that chemical reactions in the ocean or in lakes led to the formation of large molecules. These molecules then acted as *catalysts* (substances that speed up chemical reactions) to cause the formation of complex organic compounds. A second view holds that chemical reactions producing the first complex organic compound took place on the surfaces of clays or of minerals called *pyrites.* In this view, the clays or pyrites acted as catalysts.

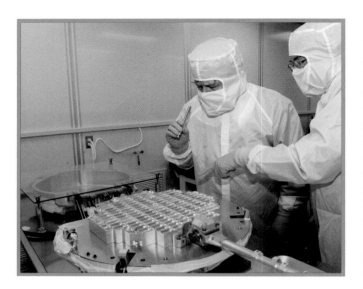

Planetary scientists examine blocks of a lightweight gel containing ancient comet dust brought back to Earth in 2006 by NASA's Stardust sample-return probe. The probe collected dust from the comet Wild-2. Scientists hope the particles, which date to the formation of the solar system about 4.6 billion years ago, will provide valuable information about the composition and formation of Early Earth.

A third theory is based on the knowledge that cell-like structures with membranes will form spontaneously in mixtures of certain *lipids* (fatlike substances) and water and that such structures fold into shells the size of small cells. Supporters of this theory argue that the chemical reactions leading to the formation of complex organic compounds took place inside and on the surface of these shells.

The earliest ancestors of living organisms were molecules that acquired the ability to make copies of themselves. Those self-replicating molecules may have been early forms of ribonucleic acid (RNA). Found in all living cells, RNA is important in making proteins and in the transmission of genetic information from one generation to the next. American chemist Julius Rebek of the Massachusetts Institute of Technology in Cambridge created the first self-replicating molecules in the laboratory in 1953. By 2006, scientists had found a number of these molecules. Very simple molecules are capable only of reproducing exact copies, and any variation in their structure prevents reproduction. More complex molecules, however, can reproduce even if the resulting copies are not exact. Scientists do not know which molecules gave rise to DNA, however. DNA (deoxyribonucleic acid) is a thin, chainlike molecule found in every living cell on Earth. It directs the formation, growth, and reproduction of cells and organisms.

The theory of panspermia

Because the leap from nonliving to living has not been replicated, some scientists suggest that the building blocks of life actually fell to Earth in meteorites and comets, a theory known as *panspermia*. For example, in 1969, investigators found *amino acids*, organic proteins found in living things, in the core of a meteorite that had drifted in space for 4.5 billion years. Scientists determined that the amino acids had been in the meteorite before it hit Earth. In 2004, geochemist Jennifer Blank of the Lawrence Livermore National Laboratory in Livermore, California, showed that amino acids could survive the extreme pressures and temperatures produced by a collision with Earth. Under these conditions, the amino acids also became more complex molecules called *peptides*. But protein is only one element found in living cells.

Scientists using the National Aeronautics and Space Agency's (NASA) Spitzer Space Telescope have found other evidence for the theory that life originated in space. In 2005, astronomer Fred Lahuis of Leiden Observatory in the Netherlands spotted forms of DNA and protein in the dust surrounding a young star called IRS 46. However, the theory of panspermia does not solve any of the important problems connected with the origin of life; it merely places the origin of life somewhere else. Wherever life appeared, it must have developed through chemical processes, something more likely to happen on Earth than in space.

Life in the extreme

Scientists can gain clues to the possible nature of the earliest life by looking at present-day life. The most primitive organisms on Earth, called *archaea*, look like conventional bacteria under the microscope. However, the way in which their DNA reproduces and the structure of their cell walls are different. Many archaea are extremophiles. Scientists have found these hardy organisms in boiling hot springs, in Antarctic lakes with permanently frozen surfaces, in desert lakes saturated with salt, and in extremely acidic runoff from mine wastes. One of the most interesting extremophile habitats is the highly polluted Rio Tinto in Spain. For many years, sulfide mineral ores in the rocks have been weathering out and releasing sulfuric acid into the river. Although the river is too acidic and toxic for most life forms, it is rich in extremophiles.

Extremophiles have shown researchers that life could have arisen on Earth even if conditions were different from those that exist today. Extremophiles could have survived an extremely hot or immensely cold Early Earth. They could have lived in highly acidic early oceans or flourished deep below Earth's surface.

In their efforts to understand Early Earth, geologists continue to look for older and older rocks. Although no Earth rocks older than 4 billion years have yet been found, geologists remain hopeful. Perhaps the original rocks that contained the zircons from Australia still exist.

■ FOR ADDITIONAL INFORMATION

Periodicals

Kotulak, Ronald. "The Cosmic Conversation." *Chicago Tribune Magazine,* January 8, 2006, pp. 10–16, 29.

Tyson, Neil deGrasse. "Living Space." *Natural History Magazine,* March 2005, pp. 31–33, 73.

Zimmer, Carl. "What Came Before DNA." *Discover,* June 2004, pp. 34–41.

Web sites

National Aeronautics and Space Administration—http://www.nasa.gov

NOVA—http://www.pbs.org/wgbh/nova/origins/

Turning Up the Volume on Hearing Loss

By Dennis Connaughton

Good hearing is an important health benefit, but excessive noise levels are causing unnecessary hearing loss among many people, especially children and teen-agers.

Young people with old ears. That is the phrase some physicians use to describe patients with a degree of hearing loss usually associated with much older people. While hearing loss is often a normal part of aging, hearing experts say that noise is accelerating the process. Loud music blasting at dance clubs and concerts, sound effects blaring from video games, booming sound systems in movie theaters, noise in the workplace, and the high-intensity sounds of everyday life can all cause a hearing decline over time. The soaring popularity of digital music players may be adding to the risk, according to many hearing experts. These devices often come with *earbuds*, earphones that fit snugly into the ear canal and sit close to the eardrum. Because these portable devices are so convenient to use and hold such large numbers of songs, people tend to listen to them more often and for longer periods than they do to radios and room stereo systems.

More than 31 million people in the United States—nearly 10 percent of the population in 2006—have experienced some degree of hearing loss, according to the Better Hearing Institute, an educational organization in Washington, D.C. Traditionally, the elderly make up the largest percentage of people with hearing loss. In 2006, age-related hearing loss affected 29 percent of the U.S. population over age 65. However, hearing loss among younger people is increasing. From 1971 to 1990, the number of children ages 3 to 17 with hearing problems rose by 12 percent, according to the National Center for Health Statistics. Among people ages 18 to 44, the increase was 21 percent. The Better Hearing Institute reported in 2006 that up to 15 percent of the U.S. population ages 40 to 65 had experienced some degree of hearing loss.

The author:
Dennis Connaughton is a free-lance medical writer in Oak Park, Illinois.

Types of hearing loss

Physicians categorize most hearing losses as either *sensorineural, conductive,* or *mixed.* Approximately 95 percent of all hearing loss in the United States is classified as sensorineural, according to the Better Hearing Institute. This type of hearing loss is produced by damage to the structures of the inner ear. These structures include the *cochlea* (an organ that changes sound vibrations to electric nerve impulses) and the sensory hair cells and nerve fibers, which transmit nerve impulses to the brain. Inner ear damage is most commonly the result of aging or prolonged exposure to loud noises. Age-related hearing loss is called *presbycusis,* a condition in which the sensory hair cells die.

Traumatic head injuries and sharp blows to the ear can also result in sensorineural hearing loss. Damage to the inner ear can be triggered by certain diseases and medical conditions, including such viral infections

HOW THE EAR WORKS

The ear consists of three sections—the outer, middle, and inner ear. The outer ear, which includes the pinna and ear canal, captures sound waves. The middle ear includes the eardrum and an air-filled chamber containing tiny bones, the malleus, incus, and stapes. The three bones amplify sound waves and carry them to the cochlea. The inner ear, which consists of the cochlea and the semicircular canals, converts the sound waves into nerve impulses that travel to the brain through the auditory nerve.

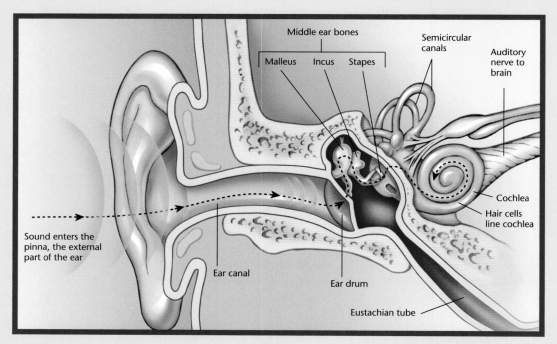

as mumps or measles; Ménière's disease, which increases fluid pressure in the inner ear; bacterial meningitis, a brain infection that can damage the acoustic nerve; and acoustic neuroma, a benign brain tumor. Certain antibiotics can result in sensorineural hearing loss. Aspirin also can cause hearing loss if taken in large quantities. Hearing loss is also associated with some birth defects and inherited abnormalities of the inner ear.

Conductive hearing loss is produced by a sound blockage located somewhere between the outer and inner ear. Sound travels in waves from the outer ear through the ear canal to the eardrum, a thin membrane that vibrates when hit by sound. Vibrations travel from the eardrum through the middle ear. The middle ear contains tiny bones that pass the vibrations to the inner ear membrane. In the inner ear, the vibrations are converted to electric impulses that are transmitted to the brain through the acoustic nerve.

A number of medical conditions produce conductive hearing loss. The most common, especially among children, is *otitis media*. This inflammation of the middle ear is most often caused by colds or other upper respiratory tract infections. Sticky fluids build up behind the eardrum, which can become infected.

Otosclerosis is another disorder that causes progressive conductive hearing loss. The condition produces an overgrowth of bone in the middle ear that eventually immobilizes the *stapes* (the innermost bone of the middle ear). Sound cannot pass to the inner ear if the stapes is unable to vibrate. Researchers are unsure about the cause of otosclerosis, though it may be inherited. A perforated eardrum, tumors, changes in air pressure, and a wax blockage or other foreign body in the middle ear are also common causes of conductive hearing loss.

Mixed hearing loss is a combination of sensorineural and conductive hearing loss. It can result from infections, tumors, genetic disorders, and head and ear injuries. A middle ear infection can damage the eardrum and the tiny bones that conduct sound waves and then move into the inner ear. Certain tumors can block the flow of sound waves in the ear canal or middle ear and then spread to the inner ear.

Preventing hearing loss

Most experts agree that hearing loss among people 40 to 65 years old is most often the result of exposure to excessive levels of noise. Such loss is almost entirely preventable. The National Institute on Deafness and Communication Disorders, a privately funded research foundation in Washington, D.C., reported in 2006 that nearly 30 million Americans annually were exposed to hazardous levels of noise or worked in dangerously noisy environments.

Research in the late 1990's also revealed that noise-induced hearing loss was affecting the nation's school-age population. The Centers for

A child's hearing is tested with a machine called an audiometer. Some audiometers use simple vibrations at different volumes to measure hearing ability. Others use spoken words or sentences.

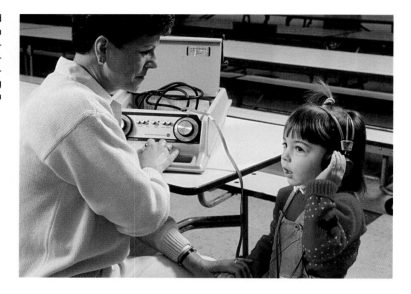

Disease Control and Prevention in Atlanta, Georgia, reported in 2001 that hearing loss among school-age children had become widespread in the United States. During the 1990's, 12.5 percent of all children between the ages of 6 and 19 (approximately 5.2 million people) had some degree of hearing loss. Hearing experts attribute much of this problem to avoidable noise, including prolonged loud music and video games.

Noise is measured in decibels. The higher the decibel level, the greater the effect on hearing. A whisper registers at 30 decibels. Normal conversation ranges from 50 to 65 decibels. A hairdryer or vacuum cleaner produces 70 decibels. These decibel levels do not damage hearing.

Damage occurs after prolonged exposure to sound beginning at 85 decibels, the level of busy city traffic. The noise from lawn mowers ranges from 85 to 90 decibels. Stereo volume set at the halfway mark registers 100 decibels. The higher the decibel level, the less exposure time needed to damage hearing.

The sound from the headphones of portable music devices played at full volume can be as loud as 125 decibels, according to the American Speech-Language Hearing Association, a nonprofit organization in Washington, D.C. The National Institute for Occupational Safety and Health, also in Washington, recommends that people limit to eight hours per day their exposure to sound higher than 85 decibels. The safe limit for listening to headphones is one hour per day with the volume at 60 percent of the maximum, according to a 2004 study by researchers at Children's Hospital Boston. Hearing experts also recommend using headphones that minimize background noise so that the volume can be kept to safe levels.

Physical effects of excessive noise levels

Prolonged exposure to noise at hazardous levels eventually injures the sensory hair cells in the inner ear. This injury causes swelling of the nerve endings in the cochlea, particularly at the base, where the nerve endings are extremely sensitive to high-intensity sound.

Excessive noise levels initially produce a short-term hearing loss, known as a *temporary threshold shift*. Sounds are muffled, usually for a few hours or a day. People who attend a rock concert, for example, often experience temporary threshold shift. They hear ringing or buzzing sounds and feel a sense of fullness in their ears. Normal hearing returns after damaged hair cells repair themselves, which usually begins the first day after exposure to the noise.

Hair-cell repair, however, can take as long as a month, depending on

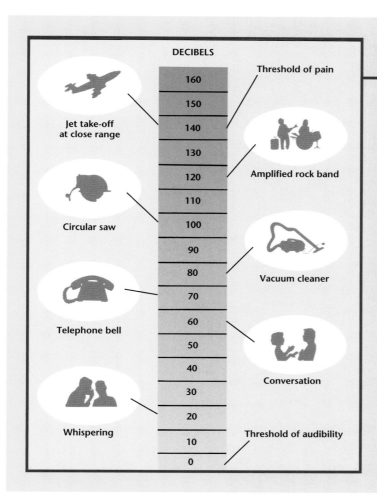

DECIBELS

160	Threshold of pain
150	
140	
130	
120	Amplified rock band
110	
100	
90	
80	Vacuum cleaner
70	
60	
50	
40	
30	Conversation
20	
10	Threshold of audibility
0	

Jet take-off at close range

Circular saw

Telephone bell

Whispering

MEASURING SOUND

Sound is measured in units called decibels. The softest sound that most people can hear has a level of about 0 decibels. Damage occurs after prolonged exposure to sounds registering 80 decibels or higher. The sound from the earphones of portable music devices played at full volume can be as loud as 125 decibels. The higher the decibel level, the less exposure time needed to damage human hearing.

the damage and an individual's sensitivity to noise. If exposure to loud noises is constant or frequent, as is often the case with digital music players, inner ear hair cells may lose the ability to bounce back. They die, producing irreversible hearing loss.

A second type of noise-induced hearing loss is acoustic trauma, which is produced by a single exposure to an explosion or other noise of high intensity. Such a sound can rupture the eardrum and blow apart the structures of the inner ear. A person who is experiencing acoustic trauma may become dizzy, feel pain, and hear ringing in the ear. The ear

REASONS FOR HEARING LOSS	TYPE	CAUSE
acoustic trauma	sensorineural	one-time exposure to explosion or loud noise
barotrauma	conductive	unequal air pressure on two sides of eardrum in air or under water
drugs	sensorineural	use of certain antibiotics, narcotics, diuretics, aspirin, quinine
eardrum, perforated	conductive	otitis media; sharp object; blow to ear; explosion; skull fracture; air pressure change; tumor
earwax	conductive	overproduction and build-up of cerumen (earwax)
foreign body in ear canal	conductive	children insert small objects in ears; flying or crawling insects
trauma	mixed	blunt blow to external ear, skull, or cochlea
mastoiditis, acute	conductive	spread of untreated otitis media to surrounding bone
Ménière's disease	sensorineural	unknown
cytomegalovirus (CMV)	mixed	prenatal exposure to cytomegalovirus
noise	sensorineural	prolonged exposure to noise, listening to music at high volume for prolonged periods
otitis media	conductive	upper respiratory infections, allergies, tonsil or sinus problems, blocked eustachian tube
otosclerosis	conductive	hereditary disease, primarily brought on by aging
presbycusis	sensorineural	changes in different parts of the ear that occur with aging
tumors	conductive	sebaceous cysts; bone tumors; and growths of excess scar tissue

closer to the explosion usually suffers more damage than the opposite ear. A person who experiences acoustic trauma may regain some or all of the lost hearing over time.

Recognizing hearing loss

How do people know they are losing their hearing? There are a number of warning signs. The first sign may be asking other people to repeat what they are saying or missing parts of conversations, especially when there is a lot of background noise, such as at a restaurant or party. Other

PHYSICAL CHARACTERISTICS	TREATMENT
damage or death of sensory hair cells	hearing aid; cochlear implant
rupture of blood vessels in middle ear or sinuses	antibiotic for possible infection; other treatments usually necessary
damage to cochlea and/or labyrinth	discontinue use of drugs; hearing aid; tinnitus therapy; cochlear implant
hole in eardrum	antibiotic drugs for possible infection; eardrum usually heals by itself
blockage of outer ear canal	removal by physician
blockage of outer ear canal	removal of foreign object by physician; mineral oil kills insects
bruising between cartilage and connective tissue	painkilling drugs; rest
bacterial infection of prominent bone behind the ear	antibiotics; surgical drain of bone abscess
increase in fluid in the canals in the inner ear; damage to cochlea and/or labyrinth	rest; drugs and therapy to relieve nausea and tinnitus; reduce dietary sodium
damage to middle and inner ear	hearing aid; speech and language therapy; medical care
damage to or death of sensory hair cells in inner ear	ear protection; hearing aid; tinnitus therapy; cochlear implant
pain; fluid; infection; fluid in the middle ear reduces sound traveling through ear	may resolve on own; antibiotics; pain relievers; fluid removal by physician
bone overgrowth immobilizes stapes in middle ear	hearing aid; stapedectomy operation to replace stapes
degeneration of hair cells/nerve fibers in inner ear	hearing aid; assistive and alertive devices; cochlear implant
growth blocks ear canal, or grows on nerve	surgical removal of tumor

An airport worker wears noise-reduction ear muffs to protect his hearing while a jet aircraft is readied for take-off.

warning signs of possible hearing loss include being unable to hear over the telephone; finding oneself unable to follow a conversation involving two or more people; turning the TV or radio to a high volume; being unable to determine where sounds are coming from; deciding that other people are not speaking clearly; responding incorrectly to a misunderstood question; or not hearing ringing doorbells or telephones.

Babies born with impaired hearing may fail to turn their head or move their eyes in response to voices or other sounds. An infant with impaired hearing may not form sounds or words at an age-appropriate time and may have difficulty learning to speak. A child with hearing loss may have trouble carrying on a conversation or may constantly turn one ear to hear a voice or other sounds.

Individuals who experience hearing loss may withdraw socially and experience confusion, depression, and feelings of inadequacy. They often deny or attempt to cover their loss. They may also exhibit embarrassment, frustration, impatience, and guilt.

Diagnosing a loss of hearing

While sensorineural and conductive hearing loss can be permanent, proper treatment helps many people with hearing loss improve their hearing. The first step is to see a physician. A family physician, in turn, may refer a patient with hearing problems to an *audiologist* (a specialist trained to evaluate hearing problems) or to an *otolaryngologist* (ear, nose, and throat doctor) or *otologist* (ear specialist) to determine the nature of the hearing loss.

The specialist will often order the patient to undergo a hearing test. During this test, an audiologist seats the patient in a soundproof booth and places a set of headphones over his or her ears. The audiologist broadcasts various tones and various noise levels through the headphones. An *audiometer,* a machine that measures the intensity of sounds, plots test results on an *audiogram,* a sheet on which hearing impairment is classified on a scale ranging from mild to profound.

A person with normal hearing can pick up sounds that range between 0 to 20 decibels. A person who is unable to hear sounds below 20 to 45 decibels is classified as having suffered a minor hearing loss. Moderate loss involves sounds between 45 and 60 decibels and moderately severe, between 60 and 75 decibels. Individuals with severe hearing loss are unable to hear sounds ranging from 75 to 90 decibels, and people with profound hearing loss do not hear sounds as loud as 90 decibels and above. With the audiogram results and other tests, the audiologist determines the degree of hearing loss and recommends appropriate treatment.

A hearing loss may cause people to feel socially inadequate and to withdraw from the company of others.

Treating hearing loss

Treatment for hearing loss depends on the cause and may include rest, ear protection, antibiotics, and even surgery. Otitis media, one of the most common reasons for children's visits to the doctor, is usually a temporary condition that may resolve itself within about three months without treatment. If there is a build-up of fluid or other complication, medications for pain may be used and antibiotics may be given to clear any infection. Surgical procedures for hearing loss include *tympanoplasty*, to repair perforated eardrums; and *stapedectomy*, to remove a bone damaged by otosclerosis and replace it with an artificial bone.

Hearing aids

Physicians also recommend the use of hearing aids in some cases of otosclerosis. Hearing aids are battery-operated electronic instruments that collect sound waves, amplify them electronically, and aim the sound into the ear canal. They are most often used to treat sensorineural hearing loss. Hearing aids do not cure the loss but increase the volume of the sound

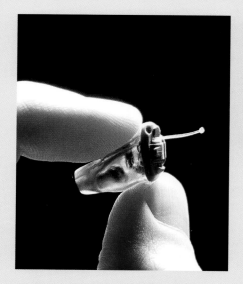

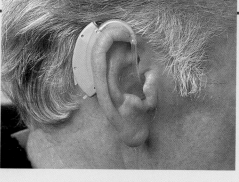

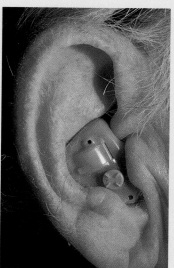

TYPES OF HEARING AIDS

There are three major types of hearing aids. The in-the-canal aid (above), which fits down into the lower portion of the ear canal, is the least visible of external hearing devices. The behind-the-ear aid (above, right) has an amplification device with a microphone that fits in back of the ear. The in-the-ear aid (right) has an amplifier, microphone, and speaker in a custom-fitted, shell-shaped container that sits on the opening to the ear canal.

that reaches the inner ear. The components of most hearing aids include an amplifier, receiver, microphone, earpiece, and batteries.

There are three major types of hearing aids—the behind-the-ear; in-the-ear; and in-the-canal. The behind-the-ear aid has an amplification device with a microphone that fits in back of the ear. Amplified sound travels through a flexible plastic tube to a small plastic *ear mold* (a specially fitted device that sits in the ear). Behind-the-ear hearing aids may also be mounted on eyeglasses.

The in-the-ear hearing aid is the most common type. This hearing aid has an amplifier, microphone, and speaker in a shell-shaped container that is custom fitted to the shape of an individual's ear. It sits on the opening to the ear canal.

The in-the-canal style of hearing aid, which fits down in the lower portion of the ear canal, is considerably smaller than the in-the-ear style.

One type of in-the-canal device can be placed deeply in the ear canal. Only a tiny portion of it shows at the opening to the ear canal. In-the-canal styles are popular because they are the least visible of all external hearing aids.

Hearing aids are equipped with a variety of electric circuitry. Most hearing aids are made with analog circuits, the same technology used to transmit television signals. When sound waves enter analog hearing aids, they are amplified across the entire spectrum of sound. Some hearing aids contain digital circuits, which convert incoming sound to binary numbers that are processed by a computer chip. Digital aids separate background noise from speech, eliminating some of the noise distortion that may accompany analog units. A third type of circuit is a programmable or combined digital/analog circuit. Aids with this type of circuit use digital circuits to adapt to a wide range of hearing loss but amplify the sound using analog circuitry. The duel technology in these aids allows the user to change settings in order to adjust to various environments.

Cochlear implants

A cochlear implant, which can be helpful to people with profound sensorineural hearing loss, is an electronic device that is surgically implanted into the inner ear. Cochlear implants compensate for the missing function of dead sensory hair cells in the inner ear. The implant consists of a small external microphone, worn behind the ear, that picks up sound. The sound travels through a wire to a speech or signal processor that converts speech into digital signals.

The best way to protect your hearing is to avoid situations that may inflict damage. Not all types of hearing loss can be prevented, but with a little caution, you can help preserve your hearing for a long time.

■ FOR ADDITIONAL INFORMATION

Books

Carmen, Richard. *The Consumer Handbook on Hearing Loss and Hearing Aids.* 2nd ed. Auricle Ink Publishers, 2004.

Christiansen, John B., and Leigh, Irene W. *Cochlear Implants in Children: Ethics and Choices.* Gallaudet University Press, 2002.

Dugan, Marcia B. *Living with Hearing Loss.* Gallaudet University Press, 2003.

Myers, David G. *A Quiet World: Living with Hearing Loss.* Yale University Press, 2000.

Olsen, Wayne, ed. *Mayo Clinic on Hearing.* Kensington Publishing, 2003.

Waldman, Debby, and Roush, Jackson. *Your Child's Hearing Loss: What Parents Need to Know.* Penguin Group, 2005.

An *organic* (carbon-containing) molecule called tetrafluorotetracyano-p-quinodimethane sits in a *nanotube* (hollow tube made of a single carbon molecule) in an image released in 2005 by the Oak Ridge National Laboratory in Tennessee.

All the materials and technology that we use in our daily lives, from our clothes to our cars, are made of at least trillions of atoms and molecules. Whether crafted by hand or assembled by machine, most of these objects were created without any thought for their tiniest parts. But what if, as the American physicist Richard P. Feynman asked, "we could arrange the atoms one by one the way we want them?" Would a shirt created with parts near the size of atoms and molecules be any different than a shirt made with a needle and thread? The answer, surprisingly, is yes—small objects and the materials made from them can have remarkably different properties compared with objects made from large-scale materials. Little things, simply because of their size, act differently than big things. Small objects can be harder and less brittle than large objects. Some tiny objects will even have different colors or become completely invisible depending on their size.

NANOTECHNOLOGY:
A REVOLUTION IN MINIATURE

Nanotechnology involves using the strange properties of tiny objects to create technology unlike anything we could build by large-scale manufacturing methods. Engineers have created hollow tubes made of single molecules and tiny machines invisible to the unaided eye. This technology brings benefits ranging from simple conveniences like stain-repellant shirts to powerful weapons against cancer and other deadly diseases. Some scientists believe that nanotechnology has the potential to revolutionize many industries and help people solve some of the world's biggest problems. Others think that the benefits of nanotechnology are outweighed by the potential health and environmental risks posed by these tiny objects. As the debate continues, nanotechnology is already appearing in many products that we use on a daily basis, from see-through sunscreens to high-speed computers. The technology made from tiny objects will likely have an enormous impact on our lives.

The authors:
Angela Berenstein is the Education Programs Coordinator for the National Nanotechnology Infrastructure Network at the University of California in Santa Barbara.

Albert H. Teich is the Director of Science and Policy Programs for the American Association for the Advancement of Science in Washington, D.C.

NANOSCALE BASICS

By Angela Berenstein

To understand nanotechnology, we must first look at the properties of matter near the atomic and molecular level—that is, on the *nanometer scale*, or *nanoscale*. We can also gain an appreciation of the amazing uses and potential of nanotechnology by studying some examples of nanoscale materials found in nature.

Nano is a prefix that comes from a Greek word meaning *dwarf*. The prefix is used to describe tiny objects or units of measurement that are "one-billionth of" something else. For example, a nanometer is one-billionth of a meter, or 0.000000001 meter (1/25,400,000 inch). To get an idea of how small this is, a human hair is about 100,000 nanometers thick. Nanoscale objects are from 1 to 100 nanometers long in at least one dimension and can be seen only with powerful microscopes.

Nanoscale science is the study of the behavior and properties of nanoscale objects. Nanotechnology uses nanoscale science to manipulate individual atoms, molecules, or nanoscale objects to create larger objects. Nanotechnology includes devices that measure the size of or take pictures of nanoscale objects, create other nanoscale objects, and fortify or improve *macroscopic objects* (objects visible to the unaided eye).

Nanoscale in nature

All living things are organized as systems of cells with nanoscale parts. In biological systems, nanostructures form naturally by a process called *self-assembly*. Atoms and molecules make chemical bonds and arrange themselves like puzzle pieces into larger, ordered nanostructures. Other nanostructures are created by nanoscale biological machines operating within cells. For example, tiny cell structures called *ribosomes* link amino acids together to form nanoscale proteins.

Nanostructures can give biological matter remarkable properties of strength and elasticity. For example, a tree trunk consists of vertical, tube-shaped cells. Each cell wall is composed of nanoscale fibers only 2.5 nanometers in diameter. Networks of these fibers make the wood strong enough to support large, heavy branches and—at the same time—flexible enough to bend in the wind.

The nanostructured surfaces of some plant leaves and insect wings give them remarkable *hydrophobic* (water repellant) properties. These surfaces do not repel water and other fluids because of their smoothness but because their nanoscale bumps prevent liquids from sticking. As a result, fluids form relatively large droplets when they

 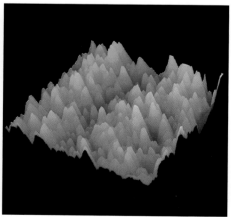

BENEATH THE SURFACE

A lotus leaf (left) appears to repel water and other liquids because it has such a smooth surface. Viewed on the nanoscale, however, the leaf actually has a rugged surface (right). The tiny bumps cause liquids to bead up and roll off the surface.

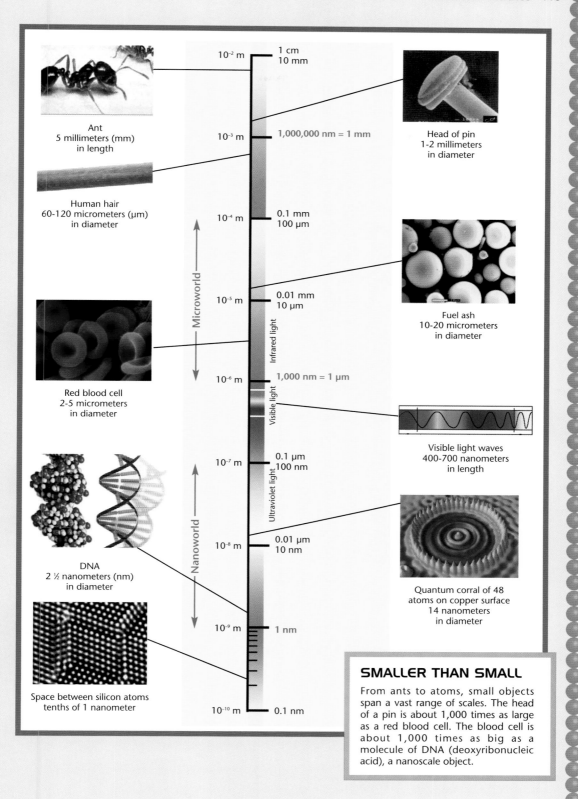

Ant
5 millimeters (mm)
in length

Human hair
60-120 micrometers (µm)
in diameter

Red blood cell
2-5 micrometers
in diameter

DNA
2 ½ nanometers (nm)
in diameter

Space between silicon atoms
tenths of 1 nanometer

10⁻² m — 1 cm / 10 mm
10⁻³ m — 1,000,000 nm = 1 mm
10⁻⁴ m — 0.1 mm / 100 µm
10⁻⁵ m — 0.01 mm / 10 µm
10⁻⁶ m — 1,000 nm = 1 µm
10⁻⁷ m — 0.1 µm / 100 nm
10⁻⁸ m — 0.01 µm / 10 nm
10⁻⁹ m — 1 nm
10⁻¹⁰ m — 0.1 nm

Microworld

Nanoworld

Infrared light

Visible light

Ultraviolet light

Head of pin
1-2 millimeters
in diameter

Fuel ash
10-20 micrometers
in diameter

Visible light waves
400-700 nanometers
in length

Quantum corral of 48
atoms on copper surface
14 nanometers
in diameter

SMALLER THAN SMALL

From ants to atoms, small objects span a vast range of scales. The head of a pin is about 1,000 times as large as a red blood cell. The blood cell is about 1,000 times as big as a molecule of DNA (deoxyribonucleic acid), a nanoscale object.

LESS IS MORE ON THE NANOSCALE

Nanoscale objects undergo faster reactions with chemicals in their environment than do objects outside the nanoworld because they have a higher surface area compared with their volume. To see why nanoscale reactions are faster, consider two objects of different sizes made of the same material.

The larger object consists of 90 atoms. It has 70 surface atoms (blue) and 20 interior atoms (yellow). The ratio of the object's surface atoms to its interior atoms is 70 to 20, or 3 ½ to 1.

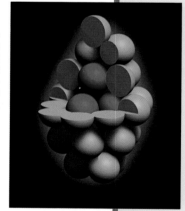

The smaller object consists of 35 atoms. It has 30 surface atoms (green) and 5 interior atoms (red). The ratio of this object's surface atoms to its interior atoms is 30 to 5, or 6 to 1.

Although the smaller object has fewer atoms overall, it has a higher ratio of surface atoms to interior atoms than the larger object above and so has more surface atoms available to react with chemicals.

come in contact with the tiny bumps and then roll across the uneven terrain like beach balls across a field of grass.

Fundamental properties of the nanoscale

Because nanoscale objects fall between the world of atoms and the macroscopic world, they have surprising properties. Some of these properties can be understood only with the theory of *quantum mechanics*—the branch of physics that describes the bizarre behavior of atoms, subatomic particles, and light. Other properties can be grasped with the more familiar physical concepts of our everyday experience.

The properties of all objects—from macroscopic to nanoscale—change systematically and smoothly with size, according to *scaling laws*. For example, one scaling law shows that if you were shrunk down to $\frac{1}{10}$ of your size, you would weigh only $\frac{1}{1,000}$ of your current weight. One consequence of these scaling laws is that the force of gravity, which has a strong influence on macroscopic objects, has only a small effect on tiny objects. As a result, a flea can jump many times its own height while an elephant cannot jump at all.

In the nanoscale world, in contrast, electromagnetic forces play a larger role. These forces are responsible for the attraction and repulsion of charged or magnetized objects—for example, the static electricity between two freshly dried socks. In the macroscopic world, if we lift the socks, the pull of Earth's gravity is often enough to separate them. However, the electromagnetic forces between tiny objects are far stronger than the gravitational force between them or between them and Earth. The stronger electromagnetic attraction causes nanoscale objects to stick to one another more easily than macroscopic objects do. This property may also create a major challenge for scientists trying to separate nanoscale objects to create nanotechnology.

Nanoscale objects also tend to be harder and less brittle than macroscopic objects. Both classes of objects have the same chemical bonds. Why, then, should nanoscale objects be stronger? One answer is that an object's strength is limited by any flaws in its structure. Small objects are less likely to have flaws than big objects and so are far less prone to cracking and breaking.

For example, glass is a material that will easily break when put under pressure. A pane of glass will first crack at its weak spots—tiny bubbles within the glass or small cuts on the surface. However, glass fibers—thin strands made of the same material—have fewer flaws and can support heavy loads without breaking.

In addition to their greater strength, nanoscale objects undergo faster reactions with chemicals in their environment than do macroscopic objects with the same total volume. This property derives from the fact that nanoscale objects have a high surface area compared with their volume. Imagine flattening a foam cube until it is as thin as a sheet of paper. Although the volume of the cube decreases by 99 percent, the cube's surface area drops by only 67 percent. Nanomaterials therefore have a higher percentage of their atoms on their surfaces than do macroscopic objects. Chemical reactions between an object and other materials typically occur on the object's surface, and so a larger surface area offers more atoms for chemical reactions.

Nanoscale materials also can have surprising properties because of *interference* effects. Interference is a phenomenon caused by two waves of the same kind—light waves, for example—passing through the same space. For instance, the rainbow of colors seen in soap bubbles results from interference. Light waves bouncing off the bottom of the soapy layer interfere with light waves reflecting off the top of the bubble. Depending on the thickness of the layer, some colors of light will be strengthened and others will cancel out, leaving a swirl of distinct colors.

Within nanoscale materials, electrons can interfere with one another. According to quantum mechanics, electrons behave like waves as well as like particles. As electron waves bounce around within nanoscale objects, they can be strengthened or cancelled by other electron waves. The distribution of electrons within matter determines certain fundamental properties of a material—such as its melting point or its ability to conduct electric current. By creating nanoscale features in matter, researchers can redistribute electrons and alter a material's properties without changing its chemical makeup.

Nanoscale objects also interact with light in different ways than do macroscopic objects. For example, in our everyday world, most objects appear colored because their chemical structure absorbs certain wavelengths of visible light and reflects others. As a result, the color of a macroscopic object depends on the material from which it is made.

In contrast, the color of a nanoscale object may depend on its size. Nanoscale objects do not reflect visible light. The wavelengths of visible light range from 400 to 700 nanometers—much larger than nanoscale objects. As a result, these objects are too small to affect the light waves and appear transparent. However, some nanoscale objects called *quantum dots* appear colored because as electrons move within a quantum dot, they give off particular wavelengths of visible light. Larger quantum dots emit light toward the red end of the visible spectrum, while smaller dots give off light toward the blue end of the spectrum.

SIZABLE PROPERTIES

Clumps of electrons (red) form by the process of *interference* within a nanoscale triangle only one atom high. Interference is an effect produced when two waves pass through the same space. According to quantum mechanics, electrons behave like waves as well as like particles. Interference increases the intensity of waves in some locations. The red bumps in the triangle represent areas where electron waves have strengthened one another. Each bump contains thousands of electrons. By confining atoms to nanoscale dimensions, scientists can redistribute electrons and so alter a material's fundamental properties without changing its chemical makeup.

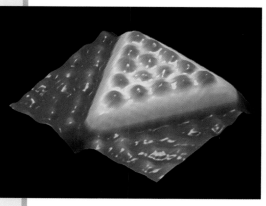

APPLICATIONS OF NANOTECHNOLOGY

By Angela Berenstein

Some of the smallest inventions ever created one day may help people solve some of the world's biggest problems. Nanotechnology is already revolutionizing many of the products found on store shelves and in our homes. From sunscreens to computer chips, engineers and designers are using nanotechnology to fortify existing products and to create new materials with remarkable properties. More than 200 consumer products used nanotechnology in 2006, according to the Project on Emerging Nanotechnologies, a research organization based in Washington, D.C.

When you wake up in the morning, you can slip on a pair of nanotechnology-based slippers that fight the mold, fungus, and bacteria that can cause unpleasant odors and skin irritation. The slippers are made of plastic foam with embedded *nanosilver* (nanoscale particles of silver). Silver is well known for its ability to kill microscopic organisms. Nanosilver has a high surface area and so it reacts strongly with these organisms, making the plastic foam a potent antimicrobial agent.

You may also encounter nanotechnology when getting dressed for the day. Researchers have developed shirts, pants, socks, and jackets that use nanotechnology to repel water, coffee, and other fluids. The fabrics are coated with nanoscale "whiskers" that repel liquids in the same way that the nanoscale waxy bumps on lotus leaves repel water. These fabric treatments have a rough structure on the nanoscale that causes fluids to ball up and roll off the fabric.

Scientists are also developing applications of nanotechnology that may have a dramatic impact on health. Researchers have created nanotechnology that shows promise as a powerful weapon in the battle against cancer. Nanoscale gold spheres injected into the body attach to cancerous cells. When *infrared light* (heat radiation) is shined on the cells, the metallic particles heat up, destroying the cancer cells and leaving nearby healthy cells unharmed.

Applications of nanotechnology fall into four broad categories: nanoparticles, nanostructures, nanostructured materials, and nanodevices. Nanoparticles are nanoscale versions of such macroscopic objects as metals and crystals. Nanostructures are arrangements of atoms or molecules that have at least one length dimension on the nanoscale. Nanostructured materials, in contrast to other applications of nanotechnology, themselves may not have any dimensions on the nanoscale. These materials may be macroscopic objects that include nanoscale objects. Nanodevices are tiny structures with at least one dimension on the nanoscale that are engineered to perform a particular machinelike function.

NANOTECHNOLOGY FOR SALE

The Nanotechnology Consumer Products Inventory cataloged 230 nanotechnology-based products on the world market in March 2006. The inventory was compiled by the Project on Emerging Nanotechnologies, a research organization based in Washington, D.C.

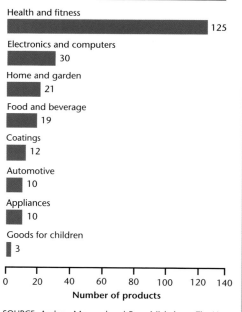

Nanotechnology-Based Products by Category

Category	Number of products
Health and fitness	125
Electronics and computers	30
Home and garden	21
Food and beverage	19
Coatings	12
Automotive	10
Appliances	10
Goods for children	3

SOURCE: Andrew Maynard and Evan Michelson, *The Nanotechnology Consumer Products Inventory*, March 2006. Project on Emerging Nanotechnologies, Woodrow Wilson International Center for Scholars, Washington, D.C.

NANOPARTICLES

Nanoparticles—tiny specks of matter measuring less than 100 nanometers in diameter—make up one of the most diverse categories of nanoscale objects. Nanoparticles include objects composed of carbon, naturally occurring metals, and *oxides* (chemical compounds made of oxygen and some other element). Scientists are using different types of nanoparticles to create materials that exploit many of the surprising properties of nanoscale objects.

For example, some nanoparticle-based materials make use of the unique interaction between light and nanoscale objects. Researchers have developed sunscreens based on nanoparticles of highly protective titanium dioxide (below). Titanium dioxide is a popular component of sunscreens because it reflects, absorbs, and scatters skin-damaging ultraviolet light. These sunscreens typically appear white (right, right half of woman's face) because titanium dioxide reflects visible light. Titanium dioxide nanoparticles have diameters of from 10's of nanometers to 100 nanometers—the same size as some wavelengths of ultraviolet light. Like larger particles of titanium dioxide, the nanoparticles absorb these harmful rays, preventing them from penetrating the skin. Nanoparticle sunscreens are transparent, however, (above, left half of woman's face) because the nanoparticles are much smaller than the wavelengths of visible light and so scatter almost no visible light.

Some materials made with metal nanoparticles interact rapidly with other materials because of their large surface area to volume ratio. When added to rocket propellants, for example, aluminum nanoparticles can make the fuel burn twice as fast, releasing significantly more energy.

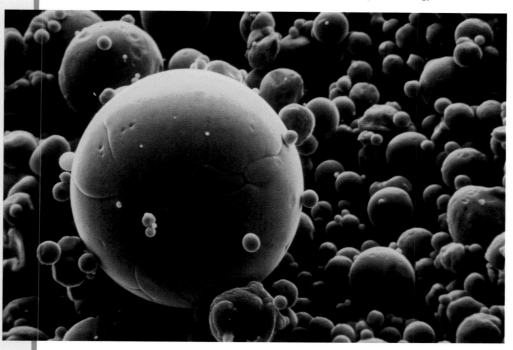

NANOCRYSTALS

Nanocrystals are miniature versions of macroscopic crystals—solids like snowflakes and salt that are composed of atoms arranged in orderly patterns. Nanocrystals vary in size from a few nanometers to hundreds of nanometers in diameter.

One application involves creating nanocrystals from *semiconductors*, materials that conduct electric current in some situations but not others. These objects, known as *quantum dots*, have a surprising visual property—their color depends on their size.

Within an atom, electrons are arranged in *shells* (energy levels) that surround the nucleus at varying distances. The distances between shells are determined by the type of material. When ultraviolet light shines on an atom, it can cause one of the atom's electrons to jump to a higher shell. After a short time, the electron spontaneously falls to a lower shell. In the process, it releases a *photon* (particle of light). The wavelength of the photon depends on the distance the electron falls. In the visible spectrum, a photon's wavelength also determines its color. Electrons that fall short distances *emit* (give off) light with long wavelengths—in colors that tend toward the red end of the spectrum. Electrons that fall long distances emit light with shorter wavelengths, and so their colors tend toward the blue end of the spectrum.

Biologists have used quantum dots to detect certain viruses in living cells. Different-sized quantum dots—shown in red and green—stick to structures that are unique to the virus's outer coat and to the surfaces of infected cells. When researchers shine ultraviolet light on these organisms, the dots become visible, highlighting the viral infection.

Like atoms, quantum dots have shells. For this reason, they are often called *artificial atoms*. The distance between the shells of quantum dots is determined by both the material type and the physical size of the dot. Large quantum dots have relatively less space between shells than smaller dots do. As a result, when their electrons are exposed to light, they fall a shorter distance and so emit redder colors. Smaller dots, in contrast, have relatively more space between shells and so their electrons drop farther. As a result, they emit bluer colors.

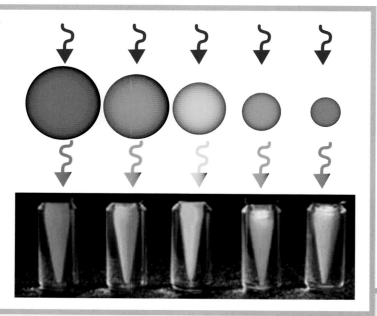

Quantum dots of different sizes exposed to ultraviolet light emit visible light in different colors. Larger quantum dots emit light in colors toward the red end of the visible spectrum. Smaller dots emit light in bluer colors.

NANOTUBES AND BUCKYBALLS

Nanotubes are nanostructures formed in the shape of hollow tubes made of a single carbon molecule (below). The tubes may be a few nanometers in diameter and several thousand nanometers in length. Because carbon atoms form extremely strong bonds with one another, nanotubes are about 100 times as strong as steel.

Sports equipment manufacturers have used carbon nanotubes to make stiff and lightweight tennis rackets and golf clubs. The rackets and clubs bend less than conventional equipment when they strike the ball, improving a player's ability to hit the ball to a particular spot. Lighter rackets and clubs are easier to swing and allow the player to hit faster and longer shots.

Carbon nanotubes also have unique electronic properties. Some carbon nanotubes conduct electric current the way metals do. The strong bonds between carbon atoms enable these nanotubes to conduct enormous amounts of current that would vaporize conventional metal wires.

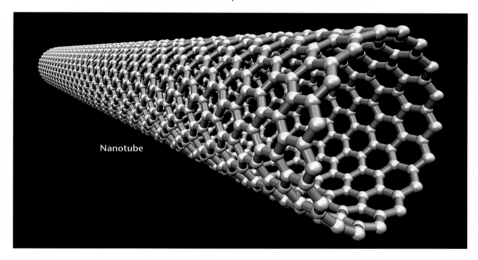

Nanotube

Researchers are also using nanotubes to create television and computer displays that are only a few millimeters thick. Within these devices, millions of nanotubes—only about one-millionth of a meter long—are packed together, with their open ends facing a screen. When an electric current flows through the nanotubes, they fire electrons toward the screen, causing chemicals on the screen to light up either as red, green, or blue spots. By mixing these colors, the television can create any color at any point on the screen.

Buckminsterfullerenes, also called *buckyballs*, are carbon molecules that are related to nanotubes. A buckyball, also known as C_{60}, consists of 60 carbon atoms bonded together in the shape of a soccer ball (right). Scientists named the C_{60} molecule buckminsterfullerene after the American engineer R. Buckminster Fuller (1895-1983) because its structure resembles a geodesic dome, a soccer-ball-shaped structure he designed. Researchers are exploring the use of C_{60} in a variety of fields. In medicine, for example, buckyballs may be used to fight the human immunodeficiency virus (HIV), which causes AIDS. Researchers have found that the buckyball disables HIV by fitting into a chemically active site on the surface of certain molecules that the virus uses to reproduce itself.

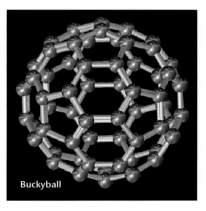

Buckyball

NANOCOMPOSITES

Nanocomposites are combinations of different types of nanoparticles or of nanoparticles and materials without nanoscale elements. Nanocomposites often have unusual mechanical and electronic properties. For example, plastics made using nanoscale fibers are typically lighter and less brittle than conventional plastics. The high surface area of the fibers enables them to bind tightly with other materials in the plastic, making the composite more resistant to stress. Exterior car parts made with plastic nanocomposites may be significantly more resistant to scratches and dents than conventional parts. In addition, car parts made with nanocomposites are significantly lighter than conventional parts because they provide the same strength and durability with less material.

Researchers have also created plastic nanocomposites (above) for use in disk drives and computer chip components (inset). These composites include nanotubes that easily conduct an electric charge. As a result, the composites prevent the build-up of damaging *static electricity* (electric charge that is not moving), a common problem affecting electronic devices.

NANOCLAYS

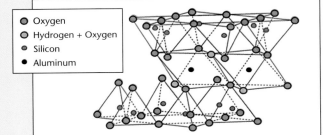

- ○ Oxygen
- ○ Hydrogen + Oxygen
- ○ Silicon
- ● Aluminum

Nanoclays consist of platelike particles made of *clay mineral* (left). (Clay mineral is a substance made of silicon, aluminum, hydrogen, and oxygen that is found in rocks and soils.) These plates are typically just one nanometer thick, though they can be up to 500 nanometers in diameter. Nanoclays can be combined with other chemical compounds to form nanocomposite coatings that function as strong barriers to liquids and gases. Within these materials, the stacked nanoclay plates form a maze that prevents liquid and gas molecules from escaping through a container's walls.

Nanoclay-coated plastic bottles, for example, contain pressure and carbonation better than noncoated bottles do and so reduce spoilage and retain flavor. As a result, the coatings add months to the shelf life of soft drinks.

Some tennis-ball manufacturers are using nanoclays to extend the life of their products (left). An extra nanoclay-coated rubber core inside the balls slows the escape of air from the core and helps prevent the balls from going flat as quickly.

NANOFILTERS

Nanofilters are structures with nanoscale pores that may be less than 10 nanometers in diameter. When gases and liquids are introduced into the filters, the gas and liquid molecules pass through the tiny pores. However, viruses and bacteria, many of which are larger than the pores, become trapped in the nanofilter.

Some nanofilters are used to purify water. They consist of cylinders whose walls are made of trillions of tightly packed nanotubes with nanoscale openings between them (below). One end of the cylinder is capped. When contaminated water flows into the cylinder, the water molecules pass through the walls. Meanwhile, such contaminants as *Escherichia coli*, which causes intestinal infections, and polioviruses are blocked by the nanotubes.

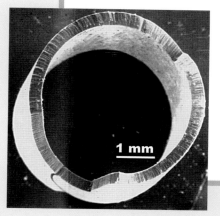

1 mm

Researchers have also created nanofilters using electrically charged nanoscale aluminum fibers. The fibers attract viruses; molecules of DNA (deoxyribonucleic acid), the substance that controls heredity; and other tiny objects by means of static electricity. These nanofilters could have applications in sterilizing drinking water and cleaning swimming pools. They could also filter out disease germs and other harmful organisms that could be used in biological warfare.

NANOELECTROMECHANICAL SYSTEMS

NanoElectroMechanical Systems (NEMS) are tiny machines with parts that can convert mechanical energy to electric energy—and vice versa. For example, one simple NEMS is a tiny scale. It consists of a paddle (below) that *oscillates* (vibrates) up to millions of times per second, depending on the *mass* (amount of matter) placed on it. These motions are measured electronically. Because the scale is so sensitive, researchers can weigh such incredibly tiny objects as cells and bacteria. Scientists have also weighed just a few hundred molecules by chemically binding them to a sample of gold only 50 nanometers in diameter. Future NEMS oscillators with even higher vibration frequencies could be used as scales for individual atoms and molecules.

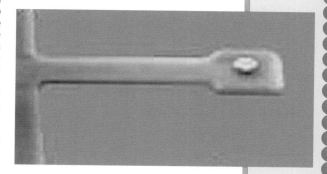

NEMS will also vibrate because of tiny electrical or mechanical disturbances, making them ideal sensors. Installed by the millions, for example, NEMS could monitor even subtle changes in the environmental conditions in hospitals or factories. In addition, because they require little energy to operate, NEMS could be used in computer chips, where they would consume much less power than the transistors currently in use.

MAKING AND SEEING NANOSCALE OBJECTS

By Angela Berenstein

Why is creating something small more complicated than creating something large? Engineers have created enormous structures that are thousands of times as large as a human being. In contrast, nanotechnology researchers must build structures that are 1 billion times as small as their creators. These objects are invisible to the unaided eye. In addition, they sometimes behave according to the laws of *quantum mechanics*—the branch of physics that describes the strange behavior of atoms, subatomic particles, and light.

Before scientists could create nanoscale objects, they needed to invent complicated techniques and tools for controlling matter on such a tiny scale. Despite these sophisticated techniques and tools, researchers have, so far, created only relatively simple nanoscale objects, such as tiny spheres and basic electronic circuit components. As scientists expand their understanding of the nanoscale, methods for creating more complex objects will likely develop.

Some nanoscale objects are created using "top-down" technology. In this method, engineers begin with a layered material from which they remove or add parts to create smaller features, the way a sculptor creates a statue with clay. Other nanostructures are created using "bottom-up" technology, also known as *molecular nanotechnology*. Just as workers assemble a skyscraper beam by beam, scientists create nanoscale objects using atoms and molecules as building blocks.

The top-down method

The top-down method consists of a multistep procedure called *selective layering*. Scientists begin with a base material, such as a silicon wafer, from which they will create a nanostructure. (A silicon wafer is a thin disc of pure silicon, one of the most common elements on Earth.) They may then add thin layers of metals, insulators, or other materials. Using a *patterning process*, scientists carve a pattern into a moldable substance spread out on the base material. They also can use one of several different *deposition processes* to add additional materials that will become part of the final object. Finally, scientists remove excess materials, leaving behind the desired nanostructure.

Photolithography, for example, is the process of using light to create a pattern on a silicon wafer. The term comes from Greek words meaning *light*, *stone*, and *writing*. The key to this process is *photoresist*, a light-sensitive chemical that is applied to the wafer in a thin coating. Ultraviolet light weakens the chemical bonds in positive photoresist and strengthens the bonds in negative photoresist. By shining light through a stencillike mask, scientists allow light to touch only the unshaded areas of the photoresist. The wafer is then placed in a chemical developer to wash away weakened photoresist. Left behind is a pattern that replicates the mask precisely—typically part of a basic circuit component. Because light spreads out as it travels and bends as it passes

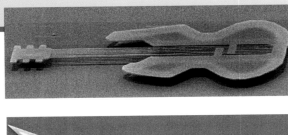

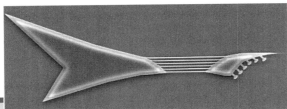

TINY TUNES

The first nanoguitar (left, top), made by physicists at Cornell University in Ithaca, New York, in 1997, was carved from silicon using electron beam lithography. The guitar is about 0.001 centimeter (0.0004 inch) long—about the size of a red blood cell. Each of its six strings is only about 100 atoms thick. In 2003, Cornell researchers created a nanoguitar that can be played (left, bottom) with a laser beam. The playable guitar is about five times as big as the original.

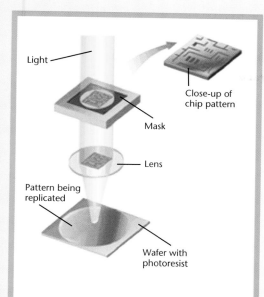

Light

Close-up of chip pattern

Mask

Lens

Pattern being replicated

Wafer with photoresist

CHIPPING AWAY WITH LIGHT

Photolithography is the most common technique for manufacturing nanoscale electronic circuits used in computer chips. A light-activated film called photoresist is applied to a wafer of silicon. Ultraviolet light passes through a stencil-like mask with a particular circuit pattern to a lens, which focuses the light on a small area of the wafer's surface. The light changes the chemical bonds in the photoresist in the areas not blocked by the mask. A chemical then removes this altered photoresist, leaving a circuit pattern on the wafer.

through the open areas of the mask, the size of the pattern is limited by the wavelength of light shone on the mask. As a result, photolithography can create patterns only as small as about 65 nanometers in length.

Another patterning technique, called *electron beam lithography*, or *e-beam lithography*, uses a narrow beam of electrons to create nanoscale features less than 10 nanometers in size. Instead of a photoresist, e-beam lithography uses electron resists that are sensitive to electron beams. While photolithography can create an entire pattern at once, e-beam lithography carves the pattern nanometer by nanometer by scanning the electron beam over the desired area. As a result, e-beam lithography is more expensive and time-consuming than photolithography.

Nanoimprint lithography can create the most precise features of any patterning process. This technique involves the use of a nanoimprinter to stamp a pattern into a layer of moldable material. Using this process, scientists can create surface features about 5 nanometers wide. Even simple nanoimprint molds must be created through complex procedures. Once made, however, they can be used to rapidly produce many copies of a nanostructure.

Many of the objects created with the top-down method are nanoscale components for tiny electronic circuits. These components are often made of metal or a *semiconductor* (substance that conducts electric current in some situations but not others). Scientists may use insulators to isolate conductive parts from one another. Several deposition processes can be used to produce nanoscale-thick layers of these materials.

In a technique called *sputtering*, researchers fire *ions* (charged atoms) at a target material, such as gold, knocking atoms off the target and onto a wafer surface. Scientists also can use the process of *evaporation*—heating metals or other materials until they become gases. When cooled, the material condenses on the wafer, much as water vapor condenses on a lid held over a steaming pot. In a third technique, tiny particles can be produced directly on a wafer through a process called *chemical vapor deposition* (CVD). In this technique, the wafer or a surrounding chamber heats one or more gases, such as oxygen and silane. When heated, silane can break apart into silicon and hydrogen. The silicon and oxygen can then react on the wafer's surface, forming a new material—silicon dioxide—also known as glass. With these deposition techniques, researchers can form layers with thicknesses of less than 1 nanometer.

In the final steps of the top-down method, excess resist and other materials not protected by the pattern are removed. Scientists can expose the wafer to such reactive gases as chlorine to chemically remove the materials. They also can bombard the surface with ions to physically knock off excess atoms. One cycle of the top-down method will create a simple nanoscale object or a basic circuit component, such as a tiny resistor. To create such complex devices as transistors and full circuits, this process must be repeated several times, so that the device is built up with many layers.

The bottom-up method

Scientists construct nanoscale objects from the bottom-up in the same way that workers on an assembly line manufacture everyday objects—piece by piece. At first, scientists created simple nano-

NO ASSEMBLY REQUIRED

A ring made of *nanorods* (nanoscale cylinders) of gold and chemical compounds forms by the process of self-assembly in a scanning electron microscope image. Chemists at Northwestern University in Evanston, Illinois, in 2004 created the nanorods using an aluminum template with rows of pores. After the template was dissolved, the nanorods attracted one another, clumping together into curved structures.

structures—such as a ring of atoms—by using the atoms' electric attraction to pull the atoms into place. However, more complicated bottom-up techniques use automated procedures to speed construction. A bottom-up approach called *molecular manufacturing* uses an ordered sequence of chemical reactions to guide molecules into such larger structures as molecular gears. In a related process called *self-assembly*, nanoscale objects form naturally because certain molecules join together due to their complementary shapes or tendencies to form chemical bonds with one another.

Some of the techniques used to manufacture nanoscale objects combine elements of both the top-down and bottom-up methods. Researchers may use templates created using lithography to guide self-assembly. When molecules are introduced into nanoscale holes, they self-assemble into nanostructures shaped like the holes. Scientists can also trap molecules between two contacts made of different types of metal, forcing the molecules to self-assemble into nanoparticles. As the space between the contacts increases, the nanoparticles can grow to become *nanowires*— long, thin wires used in tiny circuits.

Some of the deposition processes used in the top-down method are also used to create nanoparticles in the bottom-up method. By evaporating metals in a vacuum and rapidly cooling them with gas, scientists can condense nanoparticles made of atoms from both the metal and the gas. Researchers can also use the CVD process to heat gases and collect nanoparticles produced during the reactions.

Some nanoscale objects, such as carbon nanotubes, will stop growing when they reach a certain size determined by the temperature, pressure, or other properties of their environment. In other cases, growth must be stopped before a nanoparticle exceeds a desired size. There are two general

ways to control this growth: *volume restriction* and *arrested precipitation*. Volume restriction uses a material with nanosized pores as a template to physically prevent individual nanoparticles from increasing beyond a certain size. Arrested precipitation involves adding a chemical that, after a certain point, stops the reactions that produce nanoparticles. It may also involve adding a limited amount of a chemical so that the reaction will stop when the chemical is used up.

Seeing nanoscale objects

Nanoscale objects cannot be seen with the unaided eye or even with the *optical microscopes* commonly used in classrooms. These microscopes use lenses to bend light reflected by tiny objects and so magnify them for viewing. But because nanoscale objects are too small to reflect visible light waves, they appear transparent in even the most powerful optical microscopes. They can be viewed only with devices that interact with the object's surface.

The scanning electron microscope (SEM), for example, fires a beam of electrons through electromagnets and scans the beam across the object's surface. The beam knocks other electrons off the sample's surface, and a detector measures the electric current of the emitted electrons. The microscope then converts these measurements into a three-dimensional map of the surface, which can be viewed on a monitor.

With a scanning tunneling microscope (STM), a sharp metallic needle called a tip glides above an object's surface and measures the electric current produced. If the tip touches the surface, a large electric current flows through the needle. In contrast, without contact, there is no current. But if the tip is extremely close to the surface—from 0.5 to 1 nanometer—some electrons can "jump" across the gap, which produces a small current. This electron "jumping" is called *tunneling* in quantum mechanics.

Electrons are much more likely to jump across a narrow distance than a wide one, just as it is easier to jump across a narrow creek than a wide stream. When electrons tunnel, the STM measures the current. If the gap between the surface and the needle grows by even one atom, the current drops by a factor of 1,000. This sensitivity enables the STM to produce images in greater detail than the SEM can—down to the level of individual atoms. The measurements are converted into different colors, which correspond to the heights of surface features, and sent to a monitor for viewing. The STM can make pictures only of materials that conduct electric current.

Scanning probe microscopes (SPM's) use different types of tips to measure mechanical force, magnetic field, temperature, or other surface properties to visualize nanoscale objects. For example, some SPM's can take pictures of biological materials, including DNA (deoxyribonucleic acid), the substance that controls heredity, by measuring the mechanical force between the tip and these materials.

The atomic force microscope (AFM) is a type of SPM that works like a record player. It has a tiny, spring-loaded arm attached to a needlelike tip that drags across a surface. Instead of recording the tunneling current, the AFM records the small upward and downward deflections of the tip as it moves over the surface. Because the AFM does not measure an electric current, it can be used to study both conducting and nonconducting materials.

Challenges to working on the nanoscale

Computer models play a major role in scientists' efforts to work with nanoscale objects. Once scientists image nanomaterials, they can use computer models to understand the science behind their measurements. Nanoscale objects exist in an environment that is partly described by the physics of our *macroscopic world* (surroundings visible with the unaided eye) and partly by the laws of quantum mechanics. The mathematical equations that describe this overlapping region are complicated and so are more easily solved and visualized with computer models.

Scientists working on the nanoscale have thus far created only the most elementary nanoscale objects—often through time-consuming processes that yield only small quantities of nanomaterials. However, as researchers continue to improve their techniques and tools, they may be able to make nanoscale objects that are as complex as anything we can manufacture in the macroscopic world.

USING THE FORCE

A researcher uses an atomic force microscope (AFM) to study an object's surface with nanoscale resolution. The microscope drags a nanoscale tip along the object's surface and measures the small mechanical force between the tip and the object. The AFM then converts these data into a three-dimensional map of the surface. Scientists have created tips from nanotubes only 10's of nanometers wide (below). These tips can produce maps with ultrahigh resolution.

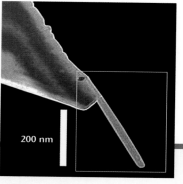

200 nm

A BRIEF HISTORY OF NANOTECHNOLOGY

By Angela Berenstein

The history of nanotechnology began about 3.8 billion years ago, with the development of the first living cells. Within these cells, nanoscale biological machines generated energy and assembled molecules into larger structures. The recorded history of nanoscale science began in Greece during the 400's B.C. with the idea that all matter is made up of a few simple parts. The Greek philosopher Democritus called the basic unit of matter the *atom*, which means *uncuttable*. However, more than 2,300 years passed before people began using atoms as building blocks to create technology.

Feynman inspires a revolution

The modern history of nanotechnology began on Dec. 29, 1959, at a yearly meeting of the American Physical Society at the California Institute of Technology (Caltech) in Pasadena, California. There the American physicist Richard P. Feynman delivered his now-classic lecture "There's Plenty of Room at the Bottom," in which he discussed the manipulation and control of objects on the atomic scale.

Feynman began with a thought-provoking question: "Why cannot we write the entire 24 volumes of the *Encyclopaedia Britannica* on the head of a pin?" If this were possible, he said, we would need to shrink the text by a factor equal to the area of all the pages of the encyclopedia divided by the area of the head of a pin. Feynman calculated that, because the head of a pin is about 0.16 centimeter (1/16 inch) in diameter, the encyclopedia would need to shrink by 25,000 times.

"How do we write it?" and "How would we read it?" Feynman asked. His answers outlined the principles of modern lithographic techniques, selective layering, and nanoimprinting. In addition, he argued that these methods could be used to cheaply produce nearly unlimited copies of the master.

Feynman then turned to the problem of how to miniaturize machines. At the time of his talk, computers were so big they filled entire rooms. In his presentation, Feynman argued that it should be possible to miniaturize electronic components to the point where wires would be as small as 10 atoms in diameter. Computers with such small, closely spaced parts would be able to process information more quickly than machines with components spread out across a room. Such parts would also give computers more components per unit volume and so enable them to carry out more complicated calculations.

Feynman proposed a novel technique for building nanoscale components from the bottom up. He envisioned tiny machine tools that could create even smaller tools until they reached the

MAJOR EVENTS IN NANOTECHNOLOGY

1959	1974	1977	1981
American physicist Richard P. Feynman sparks interest in what would later be known as nanotechnology with a lecture called "There's Plenty of Room at the Bottom," in which he describes the possibilities of manipulating matter at the atomic and molecular level.	Japanese engineer Norio Taniguchi writes a paper called "On the Basic Concept of 'NanoTechnology'," in which he defines nanotechnology as "production technology to get ... preciseness and fineness on the order of 1 [nanometer]."	K. Eric Drexler, an American engineer, originates the concepts of *assemblers* (nanoscale machines that can build objects atom by atom) and *molecular* (bottom-up) nanotechnology while an undergraduate at the Massachusetts Institute of Technology in Cambridge.	Gerd Binnig, a German physicist, and Heinrich Rohrer, a Swiss physicist, invent the scanning tunneling microscope (STM), which can produce images of individual atoms.

As soon as I mention this, people tell me about miniaturization, and how far it has progressed today. They tell me about electric motors that are the size of the nail on your small finger. And there is a device on the market, they tell me, by which you can write the Lord's Prayer on the head of a pin. But that's nothing! that's the most primitive, halting step in the direction I intend to discuss. It is a staggeringly small world that is below. In the year 2000, when they look back at this age, they will wonder why it was not until the year 1960 that anybody began seriously to move in this direction.

Richard P. Feynman, 1960

NANO WRIT LARGE

The American physicist Richard P. Feynman trumpeted the potential of the nanoscale in his classic lecture, "There's Plenty of Room at the Bottom," which he presented at the California Institute of Technology in Pasadena in 1959 and published in 1960. In 1999, researchers at Northwestern University in Evanston, Illinois, paid tribute to Feynman's vision by using an atomic force microscope to write one paragraph from his speech (inset) in an area one-thousandth the size of a pinhead.

nanoscale. These nanoscale tools, which he called "a hundred tiny hands," would then function as factories to manufacture nanoscale parts. Feynman speculated that engineers might eventually manipulate individual atoms one by one to create materials with unprecedented precision.

At the end of his talk, Feynman offered $1,000 prizes for two manufacturing competitions that he believed would lead to advances in miniaturization technology. One challenge—to create a rotating electric motor only 0.26 cubic centimeter (1/64 cubic inch) in volume—was met in 1960 through conventional technology. The other challenge—to reduce one page of a book to 1/25,000 scale—was not accomplished until 1985. In that year, Tom Newman, a graduate student at Stanford University in

California, used electron beam lithography to reduce the first paragraph of Charles Dickens's 1859 novel *A Tale of Two Cities* by a factor of 25,000.

Drexler's molecular machines

In 1976, 17 years after Feynman's inspirational speech, K. Eric Drexler—then an undergraduate at the Massachusetts Institute of Technology in Cambridge—wondered if machines could be built on the molecular level. He had observed that many biological molecules function as machines and began to think about creating custom-made molecules. In 1977, Drexler came up with the concept of an *assembler*, a nanomachine that could precisely manipulate chemically reactive molecules and control how they combine into more compli-

---

1985

Chemists Harold W. Kroto of the United Kingdom and Robert F. Curl, Jr., and Richard E. Smalley of the United States accidentally discover nanoscale hollow spheres of carbon now known as "buckyballs."

1986

Drexler writes the first book on nanotechnology, *Engines of Creation: The Coming Era of Nanotechnology*.

Binnig invents the atomic force microscope (AFM), which can image the surfaces of a variety of materials on the atomic level.

1989

American physicist Donald M. Eigler of the International Business Machines Corporation (IBM) uses an STM to move individual atoms to spell out the corporation's logo.

1990

Nanotechnology, the first journal dedicated to this field, begins publication.

ATOMIC INK

Xenon atoms spell out the logo of the International Business Machines Corporation (IBM) in an image taken by a scanning tunneling microscope (STM). In 1989, the American physicist Donald M. Eigler used an STM to move 35 atoms, one by one, into place on a nickel surface to create the IBM logo. Eigler became the first scientist to create a structure by manipulating individual atoms.

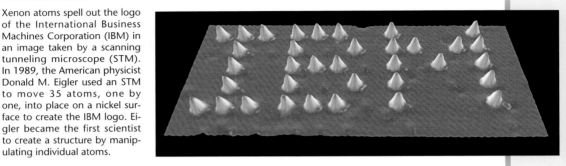

cated structures. He believed that assemblers could use atoms and molecules to manufacture virtually any object.

At first, Drexler kept the idea of assemblers to himself because he worried that the technology could lead to accidental environmental catastrophes or be used to intentionally harm people. Eventually, he decided that accidents could be prevented if strict safety rules, similar to those used to handle biological materials, were followed. Drexler also believed that because nanotechnology was certain to develop to some extent, the best way to prevent its abuse was to increase public understanding.

In 1981, Drexler published the first technical paper on *molecular* (bottom-up) nanotechnology. He began the article by paying tribute to Feynman's 1959 speech. However, Drexler argued that the nanotechnology of the early 1980's, which had created tiny objects from the top down, had fulfilled only part of Feynman's vision. Drexler called for the use of molecular engineering for "the structuring of matter to complex atomic specifications." The paper established Drexler as a leader in nanotechnology.

Since the 1980's, Drexler has written numerous papers and worked with other scientists to encourage research and development in molecular nanotechnology. His popular book *Engines of Creation: The Coming Era of Nanotechnology* (1986) dazzled the public with its portrayal of the benefits and risks of a nanotechnology revolution. In 1991, Drexler helped found the Institute for Molecular Manufacturing, a nonprofit organization based in Los Altos, California, dedicated to encouraging nanotechnology research. Today, Drexler contin-

MAJOR EVENTS IN NANOTECHNOLOGY continued

1991

Japanese physicist Sumio Iijima of technology manufacturer NEC Corporation in Tsukuba discovers carbon nanotubes.

1992

Drexler completes the first nanotechnology textbook, *Nanosystems: Molecular Machinery, Manufacturing, and Computation.*

1997

The first company dedicated to molecular nanotechnology— Zyvex Corporation of Richardson, Texas— is founded.

1999

American chemist James M. Tour and engineer Mark A. Reed create *switches*, one of the building blocks of electronic circuits, from single molecules, demonstrating the possible use of molecules to make ultrafast computer chips.

ues to investigate the potential social, political, and economic impact of nanotechnology.

Nanotechnology grows as an industry

During the 1990's, scientists succeeded in developing basic nanoscale devices. In 1998, for example, physicist Cees Dekker and his colleagues at the Delft University of Technology in the Netherlands used a single carbon nanotube to build a *transistor*, a device that controls the flow of electric current in most electronic equipment. In 1999, researchers at Hewlett-Packard Laboratories in Palo Alto, California, fulfilled one of Feynman's predictions by creating a wire only 10 atoms wide. While these types of nanoscale objects were primitive by Drexler's standards, they contributed to the formation of a practical nanotechnology industry.

In the early 2000's, the computer chip industry began using nanotechnology to build *microprocessors* (devices that do the actual computing in a computer). Since the 1960's, chip makers had worked to pack more transistors on chips to boost their processing power. To increase the number of transistors on a chip, manufacturers had to create smaller transistors, wires, and other electronic components. By 2002, these components had an average size of only 130 nanometers. In 2003, chip maker Intel Corporation of Santa Clara, California, broke into the nanoscale by manufacturing chips with an average component size of 90 nanometers. In 2006, Intel and other chip makers—including Advanced Micro Devices (AMD) Inc. of Sunnyvale, California, and International Business Machines (IBM) Corporation of Armonk, New York—were working to create chips with 65-nanometer components.

On Jan. 21, 2000, United States President Bill Clinton, in a speech at Caltech, unveiled the National Nanotechnology Initiative (NNI), a federal program dedicated to nanotechnology research, product development, and education. In his speech, Clinton also discussed the history of nanotechnology. "Over 40 years ago, Caltech's own Richard Feynman asked, what would happen if we could arrange the atoms one by one the way we want them?" Clinton said. "Just imagine, materials with 10 times the strength of steel and only a fraction of the weight; shrinking all the information at the Library of Congress into a device the size of a sugar cube; detecting cancerous tumors that are only a few cells in size. Some of these research goals will take 20 or more years to achieve. But that is why ... there is such a critical role for the federal government." Clinton's 2001 budget provided nearly $500 million to the NNI, almost doubling federal support for nanotechnology research.

In 2005, President George W. Bush substantially increased support for the NNI, allocating $1 billion, about 25 percent of all investment in nanotechnology worldwide. Several state-level nanotechnology initiatives were established in 2005, including programs in New York, Texas, South Dakota, and Oklahoma. Local organizations based in such cities as Chicago and Cleveland also promoted nanotechnology development. The United Kingdom Department of Trade and Industry estimated the global market for nanotechnology in 2005 at about $180 billion. By 2015, more than $1 trillion in all products produced around the world will include nanotechnology, according to the U.S. National Science Foundation.

2000

United States President Bill Clinton announces the National Nanotechnology Initiative (NNI), a federal program that encourages research and development in nanotechnology.

2002

The *Nanotechnology Opportunity Report*, the first business overview of the global nanotechnology industry, begins publication.

2004

The National Science Foundation funds the formation of the National Nanotechnology Infrastructure Network—a partnership of 13 universities that allow industry, academic, and government researchers to use their facilities for nanotechnology research.

2006

President George W. Bush's 2007 budget allocates about $1 billion for the NNI, nearly tripling the funding provided in the first year of the program.

NANOTECHNOLOGY: BENEFITS AND RISKS

By Albert H. Teich

The emerging nanotechnology industry suffered its first public setback on March 31, 2006. On that day, the German Federal Institute for Risk Assessment in Berlin issued a health warning about a bathroom cleaning spray called "Magic Nano." From March 27 to March 29, nearly 100 people experienced respiratory problems after using the product. Six people were hospitalized. The spray's manufacturer, Kleinmann GmbH, based in Sonnenbühl, Germany, quickly recalled the product.

The manufacturer said that "Magic Nano" contained nanoparticles of the element silicon as well as silicon dioxide. Media reports described the incident as the first recall of a product containing an engineered nanomaterial due to health or safety concerns. The recalled product was packaged in an aerosol can with a propellant. Tiny droplets sprayed from such cans may penetrate deep into lung tissue. Researchers in 2006 worked to determine if nanoparticles actually were responsible for the reported respiratory problems. The symptoms of most of the people exposed to the spray cleared up in less than a day. Nevertheless, the "Magic Nano" recall focused attention on the potential dangers of nanotechnology-based products.

Nanoscale particles exist in nature and have been produced as by-products of manufacturing processes for many years. However, in 2006, nanotechnology manufacturers could design and create nanoscale objects with specific shapes and sizes—including carbon nanotubes and buckyballs. Researchers do not know what effects these objects may have when released into the environment or inhaled into people's lungs.

Old lessons about new technology

Many of the scientists and businesspeople who were developing and promoting nanotechnology-based products were mindful of the importance of public opinion toward nanotechnology. They sought to avoid the sort of controversy that erupted over another product made using cutting-edge science—*genetically modified organisms* (GMO), which appeared in the late 1990's and early 2000's. (GMO foods come from plants whose ge-

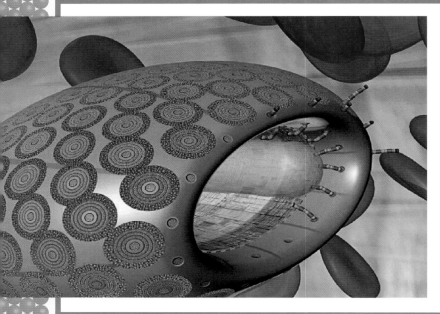

NANOMEDICINE

A nanoscale robot called a *microbivore* captures a bacterium while patrolling the bloodstream for foreign organisms, in an artist's illustration. Scientists envision using such robots in the future to fight disease. The disks on the robot's surface trap disease-causing agents. Telescoping "nanograpples" then transport the bacterium into the robot's mouth. The microbivore is designed to digest foreign organisms, breaking them down into simple, harmless molecules and then releasing these substances into the bloodstream.

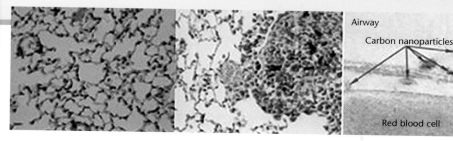

Airway

Carbon nanoparticles

Red blood cell

HIDDEN DANGERS

A mouse's healthy lung tissue (left) differs greatly from lung tissue scarred by inhaled carbon nanotubes (center). In 2005, scientists at the National Aeronautics and Space Administration's Johnson Space Center in Houston squirted nanotubes into the respiratory systems of mice and then observed their lung tissue. The nanotubes clumped in the animals' airways (right) and then passed into the bloodstream through nearby red blood cells.

netic makeup has been altered to give them more resistance to pests and disease.) Although the use of these products has significantly boosted food production, some politicians, environmental groups, and consumers opposed GMO foods because of their potential threat to human health and the environment. From 1998 to 2004, the European Union banned imports of GMO foods.

Nanotechnology advocates were working to address potential environmental, health, and safety problems before the use of nanotechnology became widespread. In 2003, for example, the United States Congress passed the 21st Century Nanotechnology Research and Development Act. The law called for studies of nanotechnology's potential impact on the environment and for research on preventing negative effects. The National Science Foundation (NSF) received most of the responsibility and funding for these projects as part of its role in the federal National Nanotechnology Initiative. In 2006, the NSF devoted about 17 percent of its $344-million nanotechnology budget to these studies. Among these were projects to anticipate how nanotechnology may affect the atmosphere, water systems, and agriculture.

Some firms looked to government regulation as a means of preventing problems related to nanotechnology. In April 2006, the U.S. Food and Drug Administration announced that it would hold a meeting later in the year to discuss the use of nanomaterials in the products it regulates—including food, drugs, cosmetics, and medical devices. Other companies were collaborating with public interest organizations to stimulate public discussions. For example, chemical giant DuPont Company, based in Wilmington, Delaware, in 2006 worked with New York City-based Environ-

mental Defense, a nonprofit organization formerly known as the Environmental Defense Fund, to develop a plan for responsible development, use, and disposal of nanomaterials.

Big benefits from little things

Scientists are also evaluating the potential benefits of nanotechnology. In the early 2000's, the NSF funded research on using nanomaterials to destroy airborne environmental pollutants, clean up groundwater, and develop environmentally friendly chemicals for use in industry. Because nanomaterials are the same size as the components of cells, nanotechnology has many potential medical applications. For example, in 2004, researchers at Oak Ridge National Laboratory in Tennessee developed a *nanobiosensor*, a probe that can study the inside of a cell without destroying it.

As scientists, businesspeople, and public interest groups weigh the benefits and risks of nanotechnology, these tiny objects are already appearing in a wide range of products. In March 2006, the Project on Emerging Nanotechnologies at the Woodrow Wilson International Center for Scholars in Washington, D.C., published an inventory of more than 200 nanotechnology-based products that were already on the market. These goods range from antiwrinkle skin creams to ultrafast computer chips. Nanotechnology has been incorporated in antibacterial socks, golf clubs with superstrong shafts, fast-drying shorts with built-in sunscreen, and a washing machine that sterilizes clothes as it washes them. Although such products offer benefits to consumers, they may also carry hidden dangers. While scientists continue to research the potential effects of nanomaterials, consumers may be the first to discover the real-world impacts of nanotechnology.

CONSUMER SCIENCE

Topics selected for their current interest provide information that the reader as a consumer can use in understanding everyday technology or in making decisions—from buying products to caring for personal health.

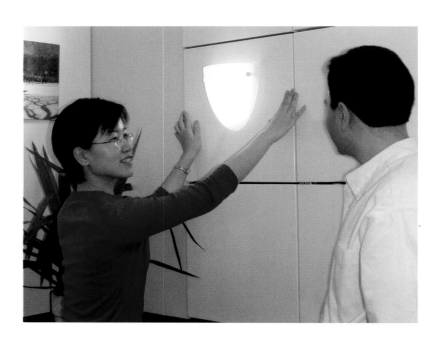

Tattooing began to enter the main-stream in the United States in the 2000's as a trendy form of body art. Long associated with carnival workers, gang members, military personnel, and prisoners, tattoos in 2006 appeared on a broader cross section of both women and men, who saw them as an expression of beauty or individuality. Approximately 24 percent of Americans ages 18 to 50 had at least one tattoo, according to a poll published in June 2006 on the Web site of the *Journal of the American Academy of Dermatology.*

The growing popularity of tattooing has heightened concerns among health experts about the risks associated with a procedure that involves needles and blood. Chief among these were infection with bloodborne *pathogens* (disease-causing microbes) and scarring, as well as allergic reactions to tattoo pigments. One of the most common complications of tattooing, however, may be dissatisfaction. As a person's lifestyle, skin, and tastes change over time, a tattoo's appeal may fade.

For at least 12,000 years, people around the world have used tattoos to ward off illness, indicate social standing, or mark such rites of passage as the transition to adulthood. Tattoos also have long been important cultural symbols among people in Africa, India, Indonesia, and various parts of North and South America.

Western societies have generally frowned on tattooing, in part because passages in the Christian and Hebrew Bibles seem to ban the practice. Likewise, Muslims have traditionally opposed tattooing because they view it as a mutilation of the body.

From the late 1800's through the mid-1900's, tattooing became increasingly popular among men serving in the military. The electric tattoo machine, patented in 1891 by New York City tattoo artist Samuel O'Reilly, made getting a tattoo faster and easier. Tattooing experienced another period of popularity in the 1960's and 1970's among hippies and other

The Skinny on Tattooing

Using a tattoo machine, a tattoo artist wearing disposable gloves injects ink into the skin of a customer's upper arm to create a tattoo of a rose.

members of the counterculture youth movement. In the 1990's and early 2000's, the growing numbers of tattoos on film stars, rock musicians, and sports figures fueled another surge in popularity.

The American Academy of Dermatology (ADA) in Schaumburg, Illinois, identifies two major types of decorative tattoos: *cosmetic* (also called permanent cosmetic makeup) and *professional,* which includes both cultural and modern tattoos. Cosmetic tattoos mimic the effects of eyebrow pencils, eyeliners, lipsticks, and other facial cosmetics. Cosmetic tattoos appeal to a wide range of people, according to the nonprofit

Skin deep

Tattooing involves injecting ink or another pigment below the surface of the skin. Professional tattooists use an electric, vibrating needles and tube to inject tattoo pigment at 50 to 30,000 times per minute into the *dermis,* or second layer of the skin, at a depth from 0.6 millimeter (0.02 inch) to 2.2 millimeters (0.09 inch).

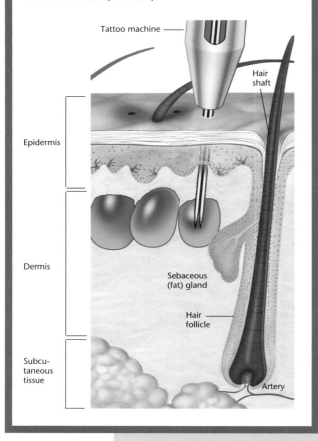

Tattoo machine

Hair shaft

Epidermis

Dermis

Sebaceous (fat) gland

Hair follicle

Subcutaneous tissue

Artery

tattoo is determined by tradition and may be done in a traditional way. For example, the artist may dip a tattooing comb in a mixture of soot and water, place it against the skin, and then tap it with a stick to drive the pigment into the skin.

Modern tattoos are created in tattoo salons by artists using an electric machine that is sometimes called a tattoo gun. The machine, which looks like a dental drill, consists of a group of at least three needles attached to a needle bar; a tube system for drawing ink through the machine; and an electric motor. A foot pedal controls the vertical movement of the needle bar. With each movement, the needles are drawn up into the tube holding the ink. When they move downward, they puncture the skin to a depth of from 0.6 millimeter (0.02 inch) to 2.2 millimeters (0.09 inch). With each puncture, the needles deposit a drop of insoluble ink into the *dermis,* the middle layer of the three layers of skin.

Anyone considering a tattoo should choose his or her tattoo artist carefully. Despite the growing interest in tattooing, the tattoo industry remained lightly regulated in 2006. The training requirements and hygiene standards for tattoo artists vary from state to state. For example, as of 2003 (the latest year for which statistics were available), 19 states required tattoo artists to take a class in infection control. Seven states required artists to obtain a vaccination for hepatitis B, a form of *viral hepatitis.* Viral hepatitis is a group of diseases that may cause serious liver damage or even death.

Some agencies of the U.S. government publish standards or guidelines on sterilization and infection control that are applicable to tattoo salons. For example, the Centers for Disease Control and Prevention (CDC) in Atlanta, Georgia, recommends that tattoo artists practice its Universal Precautions, procedures for preventing the transmission of bloodborne diseases during first aid or health care. These precautions, which

Society of Permanent Cosmetic Professionals in Des Plaines, Illinois. Among these are people with allergies to conventional cosmetics or those whose visual or motor problems interfere with their ability to apply facial cosmetics. Cosmetic tattooing may also be a convenience for people who are active in sports or simply do not want the bother of applying makeup each day.

Professional tattoos may be cultural or modern. The design of a cultural are mandatory for health care facilities, include the use of gloves, masks, and other protective equipment. The CDC also recommends that artists wash their hands thoroughly and inspect them for abrasions, cuts, and hangnails before working on a client. In addition, the CDC urges artists to wash all work surfaces thoroughly with a disinfectant approved by the Washington, D.C.-based Environmental Protection Agency (EPA) to kill both bacteria and viruses.

Many tattoo artists themselves work to reduce the potential risks associated with this invasive procedure, according to the Alliance of Professional Tattooists (APT), a nonprofit educational organization in Maitland, Florida, established in 1992. The APT developed and adopted guidelines for protecting both the artist and the client from bloodborne diseases.

The APT advises potential clients that the artist should be willing to explain the salon's sterilization practices and answer any questions. Before beginning the procedure, the artist should put on a new pair of disposable gloves. The dressings, inks, needles, and tube system that will be used should be "single-service" equipment. They should be removed from sealed, sterile packaging in front of the customer just before inking begins. The client should also be able to watch the artist pour a new supply of ink into the tube system. The area to be tattooed should be disinfected with water and antiseptic soap and, if necessary, shaved.

Health experts also recommended that the salon sterilize all reusable equipment in an *autoclave,* a closed vessel that produces superheated steam under pressure. Drawer handles, tables, and other equipment that cannot fit into the autoclave should be disinfected using an EPA-approved product. The artist should change gloves if he or she touches anything other than the client or sterile equipment during the procedure.

In 2003, officials with the CDC reported that the agency had no data linking tattooing with any U.S. cases of infection with HIV (human immunodeficiency virus), the virus that causes AIDS (acquired immunodeficiency syndrome). However, CDC officials also noted that a risk of transmitting HIV exists if instruments contaminated with blood are not sterilized or disinfected or are used inappropriately between clients.

There is also a possibility that viral hepatitis may be transmitted during tattooing. As of 2006, hepatitis was the most common bloodborne disease in the United States. The CDC has noted that fewer than 1 percent of people with newly acquired cases of hepatitis C, the most common cause of chronic hepatitis, had a history of being tattooed. Few studies have explored the health risks of tattooing, however. In 2006, the CDC was conducting a large study to evaluate tattooing as a potential risk factor for hepatitis C. Other diseases that could be transmitted by improperly cleaned tattooing equipment include syphilis, tuberculosis, and such skin infections as cellulitis and impetigo. Because of the risk of contracting a disease during tattooing, the American Red Cross and other blood banks require a one-year waiting period for donations from people who have gotten a tattoo in a state that does not regulate tattoo salons.

Tattooing involves five steps: stenciling, outlining (also called *black work*), shading, coloring, and bandaging. Professional tattooists may work from designs supplied by the client, create their own designs, or use *flash* (mass-produced designs). After the client chooses a tattoo design, the artist draws or stencils the design on the skin. This step prevents any distortion in the design, as the skin is stretched during the tattooing process.

The artist then outlines the image with a small group of needles—perhaps three—using ink with a thin consistency to create a permanent line in place of the stenciled image. The skin is then cleaned with soap and water.

During shading, the artist fills in the outline using thicker ink injected

Before you get a tattoo

- Make sure the tattoo studio appears clean and orderly, and employees have good personal hygiene.

- Ask to see the tattooist's portfolio. Look for sharp lines and bright colors in the designs.

- Ask for references and about professional memberships.

- Question the artist about specialties and length of career as a tattoo artist. Ask about the tattoo studio's business history. Observe the artist at work.

- Check to see if the studio has an *autoclave,* a vessel used to sterilize items with superheated steam under pressure. Ask employees how the work space is disinfected and about adherence to other sterilization standards established by public health agencies.

- Familiarize yourself with such possible adverse effects of tattooing as allergic reactions, dissatisfaction, infection, inflammation, and scarring.

- Determine the price for your tattoo before work begins. Get all agreements in writing or on a consent form.

- Proofread any text in your tattoo design.

- Review and approve the outline of the tattoo design and placement on your body before inking begins.

- Make certain that the tattoo artist washes his or her hands thoroughly and wears a fresh pair of single-use, disposable gloves that have no holes, and that the artist's hands have no exposed sores.

- Be certain that all needles are disposable and that a new needle and tube setup are opened in front of you just before work begins.

- Make sure that the artist pours a new ink supply into a disposable container.

- If you have doubts or discomfort, do not hesitate to leave.

Critically reviewed by David C. Breeding, R.S., R.P.E., C.S.P., C.H.M.M. Mr. Breeding is a director at the Environmental Health, Safety, and Security Services in College Station, Texas.

the punctures. Punctures that are too deep—entering the third layer of skin called the *subcutaneous tissue*—may cause excessive pain and bleeding. Punctures that are too shallow, penetrating just the first layer of skin called the *epidermis,* result in uneven lines and a poor-quality tattoo. Most people describe the pain of a tattoo inked into the dermis as similar to that of a bee sting. Tattoos placed in sensitive areas, such as places where the skin lies over a bone, are the most painful.

After completing the tattoo, many artists give their clients a pamphlet explaining how to care for the tattoo. Common advice includes keeping the area bandaged for a few hours and then gently washing it with a mild antibacterial soap and warm water. The area should be patted dry, not rubbed, with a soft, clean towel. A thin layer of antibiotic ointment should be spread on the tattoo. Except while washing, the client should keep the tattoo dry until it fully heals. Tattoo experts advise against exposing the tattoo to the sun, which may cause the tattooed area to swell and ooze, especially if yellow and green dyes have been used.

Tattooing may result in scarring, especially for people who tend to develop *keloids,* a mass of scar tissue that occurs at the site of a cut. Some clients develop *granuloma annulare,* a chronic skin condition that causes rings of reddish bumps. People also may experience allergic reactions to the tattoo pigments, most commonly to red inks.

Tattoos also have been blamed for causing complications in magnetic resonance imaging (MRI), which produce images of tissues inside the body. According to the FDA, a small number of people who have undergone an MRI have experienced swelling or a burning sensation in the tattooed area. Some FDA scientists believe that the metal particles in the tattoo pigments are disturbed by the machine's magnets. Some physicians have reported that the pigments interfere with the quality of the MRI images.

Tattoos also may become distort-

with a larger number of needles. The artist then cleans the area again and overlaps each line of color to make sure that no *holidays* (gaps in color) exist. The artist thoroughly cleans the tattoo and surrounding area a final time and covers the area with a sterile bandage. Bleeding usually stops within a few minutes.

The level of pain involved in tattooing depends on the skill of the artist, the client's tolerance for pain, and the location of the tattoo on the body. A skilled artist can control the depth of ed or blurred as the skin ages and loses its elasticity. In addition, people who lose or gain a significant amount of weight may find that their tattoos become distorted.

The most common problem with tattoos is dissatisfaction. As many as 50 percent of people who get tattoos later regret their decision, according to an article reported in 2003 at www.eMedicine.com, a physician-reviewed Internet database. A tattoo that delights a teen-ager, for example, may appear unprofessional in an office setting.

Dissatisfaction with tattoos is not a new phenomenon. In A.D. 543, the Greek physician Aetius described his attempts to remove a tattoo by scrubbing it with salt, a procedure called *salabrasion*. In 2006, people wanting to lighten or remove their tattoos had four main options: abrasion, cosmetic overtattooing, excision, and laser treatment.

In abrasion, a physician attempts to remove a tattoo by rubbing away the epidermis and dermis. The main forms of abrasion are salabrasion and *dermabrasion*. In dermabrasion, the tattoo is rubbed with a wire brush or a *diamond fraise* (a sanding disc). These processes may leave scars.

Cosmetic overtattooing involves camouflaging the tattoo by placing new pigments over the tattoo. The new pigments may form a new design, or they may be flesh-colored to hide the tattoo. Injected pigments may make the skin appear dull and unnatural, however.

In excision, a surgeon uses a scalpel to cut out the tattoo. The complete removal of a larger tattoo may require several surgeries. To reduce scarring, balloons called *tissue expanders* may be placed under the skin before the physician begins the procedure.

Lightening with lasers

A plastic surgeon removes a tattoo from a client using a laser (below, left). After four laser treatments, the tattoo is much lighter. Laser beams break up the pigments in tattoo inks, helping them to leave the body.

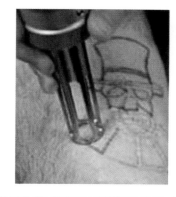

Many people choose laser treatments to lighten tattoos. Laser treatments should be done by a licensed dermatologist. In a laser treatment, a high-intensity beam breaks up the pigment into smaller pieces, which are then carried out of the skin by the body's lymph system.

Laser treatments require several visits, usually at least six, over a period of weeks or months. The cost may run to several thousand dollars, depending on the color, complexity, and size of the tattoo. Laser treatment is most effective on black and red ink. Bright colors do not respond as well and may turn to a less desirable shade. White lines turn black on contact with the laser. Laser treatment may also lighten the tattooed area too much, creating a condition called *hypopigmentation*. Other possible reactions include allergic reactions, scarring, and skin discoloration.

The decision to get a tattoo should be made carefully. Potential clients should carefully consider safety precautions and possible health risks. They should also think about whether they are going to want that pretty butterfly design to fly away in a few years' time. ■ Renée Despres

Removable Storage: Memory on the Move

Computer users in 2006 could carry around 500 songs, 10,000 digital photographs, or 1 million pages of text on a device no bigger than a pack of gum. Just 20 years earlier, it would have taken 100 bulky cartridges, each the size of a textbook, to hold that much data. Go back another century, and millions of *punch cards* would have been needed to store an equivalent amount of data. (Punch cards are paper cards on which information is recorded by means of holes punched according to a code.)

Punch cards were the first *digital* (numeric) portable storage media, devices for storing, transporting, and transferring data. Storage media originally were used to hold and process records of bank accounts, government data, and other information from large institutions. As storage devices shrank in size, individuals and businesses began to use them to store data files and distribute software. People also started

using storage media to store personal information in case of disaster.

In 1881, American inventor Herman Hollerith created a punch-card system that he later used to tabulate the results of the 1890 United States census. Hollerith's cards were about 18.7 centimeters (7 ⅜ inches) wide by 8.3 centimeters (3 ¼ inches) high by 0.02 centimeter (0.007 inch) thick. The number and arrangement of the holes represented a kind of code that a machine called a tabulator could read. Because each card held only a small amount of information, even what we consider a tiny computer file would have taken thousands of cards. In addition, if the cards fell out of order, their information was lost forever. Humidity could make the cards swell and jam inside the machine.

Computer users in 2006 could choose from three main classes of removable storage media. *Magnetic media*, including the floppy disk, store data in patterns of tiny magnetic particles. *Optical devices*, including the CD (Compact Disc), store information in tiny spots burned into thin plastic discs. *Solid-state devices*, often called *memory devices* or *flash memory*, store data electronically on computer chips.

Magnetic devices, the most common type of storage media, have

Removable storage media have shrunk from textbook-sized Iomega Bernoulli Box cartridges, released in the 1980's, to floppy disks measuring 13.34 centimeters (5 ¼ inches) or 8.9 (3 ½ inches) on a side, to compact discs with diameters of 12 centimeters (4.7 inches), to memory devices no bigger than a postage stamp, which were popular in 2006.

been widely available since the 1970's. These devices typically consist of plastic or metal disks coated with a material made of magnetic particles. All magnets have what is known as *polarity*— that is, they have a north and south pole. Computer data can also be thought of as having polarity. Computers store data in *binary* form, representing all numbers as either 0's or 1's. For example, the number eight is represented by the number 1000, while the number nine is 1001. Computers can store these numbers on magnetic media by orienting the magnetic particles to either north or south polarity, which represents 0's or 1's.

Most laptop and desktop computers use a hard drive to store data internally. A hard drive consists of several disks stacked inside a sealed metal box. As the disks spin, small metal arms called *read/write* heads glide over the disks. The heads read and write data by measuring and changing the polarity of the magnetic particles on each disk.

Hard drives can also be built into free-standing, portable cases that can be plugged into a computer at various input/output "ports." Each type of port has a uniquely shaped connector that corresponds to a particular data format. These include USB (Universal Serial Bus), the most common type of

connection; SCSI (Small Computer System Interface); and FireWire, also known as the IEEE-1394 format.

Many other removable media, including floppy disks, Zip disks, and tape cartridges, store data in magnetic form. While these devices are much smaller and lighter than an external hard drive, they can typically hold only a fraction of the amount of data. Computer information is measured in multiples of *bytes*. One byte equals a letter or other character. A *megabyte* equals 1,048,576 bytes, and a *gigabyte* equals 1,073,741,824 bytes. For simplicity, 1 megabyte and 1 gigabyte are often said to equal 1 million and 1 billion bytes, respectively. A typical floppy disk can hold 1.44 megabytes of data, or about 500 pages of text. In contrast, modern hard drives are measured in 10's to 100's of gigabytes. One gigabyte can hold about 500,000 typewritten pages, or the equivalent storage of nearly 1,000 floppy disks.

During the late 1990's and early 2000's, floppy disks became obsolete as the size of hard drives topped 100 gigabytes and CD's and other portable media grew in popularity. Indeed, most new computers in 2006 were not even equipped with a floppy drive. Instead, they relied on larger-capacity forms of removable storage, such as an external

How a DVD works

A DVD contains layers of digital data encoded in tiny pits. In a DVD player, a lens focuses a laser beam on the desired layer. As the disc rotates, the pits and the flat areas between them reflect patterns of light to a photodetector, which changes the patterns into electric signals.

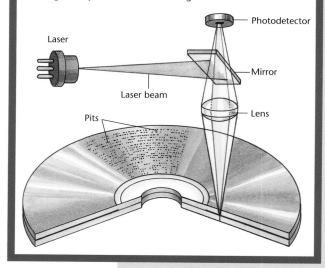

because software makers began distributing their programs on CD-ROM (Compact Disc-Read-Only Memory). A CD-ROM can hold about 650 megabytes, or about as much data as 450 floppy disks. As a result, CD-ROM's enabled software makers to release larger, more advanced programs.

Optical drives are named for their ability to write and read data using light. Such devices store information on CD's in a long, thin spiral of reflective and nonreflective dots, burned by a laser beam, on a metallic layer embedded within the plastic discs. If the spiral were unrolled, it would extend for about 5.6 kilometers (3.5 miles). To read the data, the drives fire a laser beam at the dots, and a sensor measures the amount of reflected light.

Computer users can create their own discs using either CD-R (Compact Disc-Recordable) or CD-RW (Compact Disc-ReWritable) media. CD-R's can be written on only once, while CD-RW's may be erased and rewritten on many times. These rewritable discs contain a metallic layer that can be made reflective or nonreflective, depending on the temperature of the laser beam used to burn the dots.

Optical media took a huge leap forward in 1996 with the introduction of the DVD format. Some sources say that DVD stands for Digital Versatile Disc or Digital VideoDisc. However, many people involved in the production of DVD's insist the letters have no specific meaning. Depending on the type of disc, DVD's can hold from 5 to 17 gigabytes of data. DVD's are the same size as CD's and store data in the same way. But DVD's can store data in two layers, one beneath the other. Like CD's, DVD's come in different formats, including DVD-R (Recordable) and DVD-RW (Read/Write).

Optical discs are popular because they are small, light, and inexpensive. Blank CD's and DVD's cost less than $1 in 2006. However, because the lasers that read the discs need a smooth, unblemished surface, a minor scratch can destroy data. In addition, DVD's come in a variety of formats, including DVD-ROM and DVD-RAM. As a result, discs

A typical USB (Universal Serial Bus) memory device connects to a computer through a USB port. Such devices allow users to easily share data between different computers.

hard drive. This trend was encouraged in part by the development of tiny external hard drives known as *microdrives*. These devices typically hold gigabytes of data, enabling users to sacrifice the large storage capacity of traditional hard drives for the ability to slip them easily into a shirt pocket.

Magnetic media have significant drawbacks, however. Data can be erased if a disk is placed too close to a magnet. In addition, magnetic media work slowly compared with other storage devices because the drives have moving parts—their read/write heads. Over time, these heads can wear down and make contact with the disks. As a result, magnetic media are prone to sudden failures that can wipe out most of the data stored on the disk.

In the 1990's, optical media succeeded magnetic media as the most popular type of removable storage created with one DVD drive may not work in another DVD drive.

Solid-state devices, or memory devices, which hit the market in the late 1990's, are extremely reliable and are compatible with many different computer systems. Many memory devices store data on removable cards no bigger than a postage stamp. These cards can also be used in such consumer electronic devices as digital cameras, digital music players, and video gaming devices. Other memory devices are self-contained units about the size of a finger. These devices are often referred to as *keychain devices* or *keychain drives*.

Solid-state devices get their name from the fact that they have no moving parts. They use a technology called flash memory to store data electronically on memory chips. These chips can typically be "written on" and erased at least 100,000 times. The devices are widely available in sizes of up to 2 gigabytes, with increasingly larger sizes appearing regularly on store shelves.

The most common type of memory device plugs into a USB port. Such devices are rapidly replacing the CD as a means of short-term storage. When connected to a USB port, the computer treats the devices like another disk drive, allowing the easy transfer of information even between environments that are not normally compatible. For example, USB memory devices typically work with both the Windows operating system made by Microsoft Corporation of Redmond, Washington, and with the Macintosh Operating System produced by Apple Computer, Inc.

In contrast to most *internal memory* (memory chips inside a computer), memory devices will retain their data even if the chips are disconnected from a power source. These media typically draw power from the computer or other device they are connected to.

In the future, computer experts predict, all types of storage technology—including magnetic, optical, and memory devices—will offer users higher performance at lower prices. External hard drives will continue to shrink in physical size and grow in storage capacity. They may also be more commonly

built with *wireless technology* that enables users to access the drives without physically attaching them to a computer.

Optical media will also continue to increase in capacity. Two successors to the DVD format, BD (Blu-ray Discs) and HD-DVD (High Definition-DVD), made their debut in 2006. While both formats use discs no bigger than a CD, they store data as even smaller dots, enabling them to hold more information than DVD's. The BD format can store up to 50 gigabytes of data, while the HD-DVD format can hold up to 30 gigabytes. An optical technology known as HVD (Holographic Versatile Disc) promises to store as much as 1,000 gigabytes of data and transfer the information at about 40 times the speed of DVD's. HVD's, which are the same size as CD's, store data within the volume of the disc instead of on a surface. In 2006, holographic technology was still in development.

In coming years, the capacity of memory devices is almost certain to increase, with their average capacity measured in gigabytes. The trend of building USB drives into utility objects—such as pens and pocketknives—will likely continue as the price of the technology falls.

Other trends may signal the end of removable storage, however. As Internet access spreads to computer users around the world, more people have begun storing their data online. For example, Internet users can store data files in their e-mail accounts or at sites on the World Wide Web. In addition, the widespread deployment of wireless technology and the growing computer sophistication of cell phones and other communication devices will allow users to transfer data wirelessly from one device to another. Like stacks of punch cards, portable storage devices may one day be seen as the quaint technology of a simpler time. ■ Dave Wilson

Flash memory cards, like the one shown above with its adapter, are as small as a United States quarter. This card holds 64 *megabytes* (millions of bytes) of data—enough storage space for at least 20,000 pages of text.

The Perks of Coffee

Cappucino is one of many popular beverages made using the beans of the coffee plant.

Java. Jamoke. Cuppa Joe. These nicknames for coffee suggest just how popular this beverage is in the United States. According to the Austin, Texas-based Coffee Research Institute, the average American coffee aficionado drinks three cups every day. People in many European countries drink two to three times that much. All told, people worldwide use nearly 7.2 billion kilograms (16 billion pounds) of coffee beans per year.

Coffee's rich flavor and aroma are major reasons for its popularity. Many coffee drinkers also crave the "buzz" imparted by the caffeine in coffee. In pure form, caffeine is a bitter white powder that is *toxic* (poisonous) to people. In the small quantities found in coffee, tea, chocolate, and some soft drinks, however, caffeine stimulates the nervous system. The growing popularity of lattes and other coffee-based drinks as well as expensive gourmet coffee has renewed the debate about the potential health benefits and risks of coffee and caffeine.

People have enjoyed coffee for centuries. The plant was originally cultivated in Ethiopia where, according to legend, goatherds noticed that their flocks were much livelier when the animals fed on the leaves and berries of the plant. Some 700 years ago, people learned to make a beverage from the seeds of the coffee berry.

The secret of coffee cultivation passed from Africa to Arabia to Turkey. By 1700, coffee had become popular in Amsterdam, London, Paris, Vienna, and other European cultural centers.

A mature coffee tree yields enough beans to produce about 0.7 kilogram (1.5 pounds) of roasted coffee annually. The process from tree to cup, however, is long. A cup of steaming coffee begins with the berry of a shrubby tree with glossy evergreen leaves. When the berry has ripened to a bright red, it is harvested and then processed by one of two methods.

In the first method, the berries are left to dry in the sun until their outer shell turns brown and the pulp inside evaporates, leaving the coffee beans (the seeds of the coffee berry) loose in the shell. The seeds are then extracted from the shell. In the second method, the pulp is removed quickly after harvesting, and the beans are run through fermenting and washing tanks. The beans are dried and left to cure for several weeks before they are separated from their shells.

The two species of coffee tree most important on the world coffee market are *Coffea arabica* and *Coffea robusta*. *C. arabica* flourishes in tropical areas, roughly between 1,100 and 2,400 meters (3,600 and 8,000 feet) above sea level. Beans from the *C. arabica* plant produce "arabica" varieties that are further classified as "Brazil" (produced in Brazil) or "mild" (produced outside Brazil). *C. robusta* (also known as *Coffea canephora*) thrives primarily in tropical lowlands of Africa and Southeast Asia. Beans from the *C. robusta* plant yield a type of coffee known as "robusta," which produces a bitter brew with 50 percent more caffeine than arabica.

Roasters heat the beans in rotating drums at high temperatures. As the coffee beans are heated, the moisture in the beans turns to steam. This heat causes reactions in the hundreds of chemical compounds in the beans, creating new compounds that give coffee its rich, complex flavors and aromas. The heat also causes sugars in the bean to caramelize, and the raw, green beans turn shades of brown or black, depending on the roasting time. (Longer roasting times produce darker, stronger-tasting coffee.)

Brewing coffee involves forcing hot water through ground beans. Although there are many ways to do this, the most popular is the drip method, in which hot water drips through the grounds and then passes through a porous filter.

Coffee makers also make several other coffee products, including decaffeinated coffee ("decaf")—coffee from which caffeine has been removed. Decaffeinating is achieved in one of two ways. In the first method, the beans are washed in cold water with chemical solvents that leach out the caffeine before roasting. The second method involves steaming the beans and scraping off the caffeine-rich outer layers. Instant coffee is produced by brewing coffee and then evaporating the water. Freeze-dried coffee is made by freezing freshly brewed coffee and then evaporating the frozen moisture. Adding hot water to the coffee powder or freeze-dried crystals produces a fresh-tasting cup of coffee.

A plantation worker plucks berries from a coffee plant. The colorful berries must be roasted before they can be used to make coffee.

A large roaster heats coffee beans, bringing out their flavors and aromas.

Support for coffee's health benefits is abundant. A number of studies indicate that substances called *antioxidants* found in coffee may help prevent disease. Antioxidants neutralize harmful substances called *oxidants,* which are by-products of energy production in the body's cells. Oxidants, also called *free radicals,* are chemically unstable molecules that grab *electrons* (negatively charged particles) from other molecules. Through this activity, antioxidants can damage cell structures. An antioxidant molecule, however, provides a free electron to an oxidant, thereby making it stable—and harmless. Many scientists believe that antioxidants prevent genetic damage to cells that can lead to cancer.

In August 2005, researchers with the University of Scranton in Pennsylvania released a study indicating that, of more than 100 kinds of foods—including fruits and vegetables—coffee is the number-one source of antioxidants in most Americans' diets. The researchers determined that the aver-

age adult daily consumed 1,299 milligrams (0.05 ounce) of antioxidants from coffee. Tea, the next highest source of antioxidants, contributed only 294 milligrams (0.01 ounce) of antioxidants to the average diet. In February 2005, a research team from Japan's National Cancer Center in Tokyo reported that daily coffee drinkers had half the risk of developing liver cancer of people who do not drink coffee. However, the researchers were unsure if it was the antioxidants or other chemical compounds in coffee that protected the coffee drinkers.

Studies have shown that coffee consumption might help prevent other chronic illnesses. A study published in 2001 by researchers at the Harvard School of Public Health at Harvard University in Cambridge, Massachusetts, found that coffee drinkers were half as likely to develop Parkinson disease as people who do not drink coffee. (Parkinson disease is a *degenerative* [gradually worsening] condition that destroys cells in certain regions of the brain involved in moving the body.) A 2004 study led by researchers with the Harvard School of Public Health found that coffee drinkers were 30- to 50-percent less likely to develop diabetes than those who do not drink coffee. In addition, researchers in June 2006 published a study showing that drinking one cup of coffee per day reduced the participants' risk of developing alcohol-related cirrhosis of the liver by 20 percent. Cirrhosis is a disease in which the liver becomes scarred.

Support for the claim that coffee is damaging to health is also plentiful. Most of this support points to the potentially adverse effects of caffeine on the body. A good deal of research has shown that overuse of stimulants, including caffeine, can cause *palpitations* (irregular heart rhythms). This has led to concerns that drinking caffeinated coffee might lead to heart disease.

Studies exploring the connection between coffee drinking and heart disease have been inconclusive, however. The majority of medical evidence suggests that moderate coffee drinking does not pose a significant risk of heart disease, and the American Heart Association does not list coffee drinking as a risk factor. However, a study presented at an American Heart Association conference in 2005 revealed that people who drank decaf coffee had higher blood levels of a particular type of fat that may increase levels of LDL cholesterol (known commonly as "bad cholesterol"), a risk factor for heart disease. The study was led by Atlanta-based physician H. Robert Superko and funded by the National Institutes of Health. Subjects who drank caffeinated coffee did not experience an increase in their levels of this fat. The researchers concluded that, for unknown reasons, decaf coffee increased the risk factors for *metabolic syndrome,* a set of symptoms that are related to the development of heart disease.

Caffeine consumption can also pose a problem for pregnant women. A 1999 study by researchers at the University of Utah in Salt Lake City found that pregnant women who drank more than five cups of coffee per day were twice as likely to suffer a miscarriage as women who drank less coffee. The researchers emphasized that moderate amounts of coffee (two cups per day) did not increase the risk of miscarriage.

Although coffee consumption was once thought to cause a variety of diseases, more recent studies have disproved many of those claims. For example, caffeine's ability to temporarily increase blood pressure raised concerns that drinking coffee might lead to *hypertension* (high blood pressure). However, a 12-year study of more than 150,000 U.S. women found no connection between coffee intake and an overall increase in blood pressure. In fact, the researchers discovered that among coffee drinkers, the more coffee the subjects drank, the less likely they were to develop hypertension. In addition, those women who drank four or more cups a day had a lower risk of developing the disease than those who did not drink any coffee. Results of the study, led by researchers at the Harvard School of Public Health, were published in November 2005.

Some foods and beverages relax the valve between the esophagus and the stomach, which can allow stomach acid to back up into the esophagus, causing a condition known as *acid reflux.* Some patients suffering from acid reflux report that coffee aggravates their condition. However, in 1999, researchers from the University Medical Center in Utrecht, the Netherlands, published a study showing no correlation between coffee drinking and acid reflux in healthy people.

Although scientific studies concerning coffee are somewhat contradictory, a growing body of research, including many of the studies cited above, suggests that drinking coffee in moderation is probably safe for most healthy people. This is good news for the millions of people worldwide who enjoy their daily cups of java.

◼ Rob Knight

A woman snaps a panel containing a light-emitting diode (LED) light fixture into a prototype modular lighting grid. Such grids, which were first demonstrated in 2005, will allow people to rearrange the lighting in their homes as easily as they rearrange their furniture.

Lighting: "LED"ing the Way

From a fire on the floor of a cave to a modern-day *torchère* (floor lamp), lighting has always been a key element for creating a safe and attractive home. In 2006, researchers were developing new ways to integrate the most modern form of lighting—the *light-emitting diode* (LED)—into the places where people live and work. LED's are tiny bulbs that fit into an electric circuit. These small light sources can fit into tiny spaces and will someday be integrated into walls and ceilings. Scientists predict that LED illumination systems will be a reality in the next 5 to 10 years.

In the 1800's, a device developed by American inventor Thomas A. Edison—the incandescent lamp—became the first commercially successful *lamp* (the term the lighting industry uses for what is commonly called a light bulb). Edison passed electric current through a *filament* (thin, coiled wire) to produce artificial light. In the early 2000's, the standard incandescent lamp remained the most commonly used lighting product. In fact, it has changed little over the last century. These bulbs will work in both antique fixtures and the latest *luminaires* (wall, ceiling, or free-standing light fixtures). Incandescents convert electric power into light by passing electric current through a *tungsten* wire filament. (Tungsten is an element found in certain minerals.) The filament *incandesces* (glows) when it becomes hot. The glass bulb housing the filament is filled with a mixture of chemically inactive gases—usually argon and nitrogen—that lengthen the life of the filament. A coating of a finely powdered white mineral called *silica* inside the bulb *diffuses* (scatters) the light and reduces glare. Over time, the tungsten evaporates, and the lamp eventually fails. (The evaporated tungsten forms the dark deposit often seen on the inside of a failed bulb.)

Many people prefer incandescent lamps to other types of lighting because of the "warm" (slightly yellow) light they emit. Incandescents also come in a variety of shapes, sizes, and *wattages* (a measure of electric power) and fit into most standard light fixtures.

Incandescent lamps have some disadvantages, however. Only 10 to 15 percent of the electric energy used in these lamps is emitted as light; the rest is wasted as heat. Incandescents emit infrared and ultraviolet energy, both of which can harm such delicate objects as paintings and fabrics. In addition, standard incandescents last less than a year under normal usage. (Installing a dimmer switch, which reduces the amount of power used, may extend the life of an incandescent lamp. Contrary to popular belief, though, frequently turning such lights on and off does not wear them out prematurely.)

Tungsten halogen lamps, a type of incandescent, emit "cooler" (whiter) light and are commonly used as accent lighting. In a halogen lamp, chemicals called *halogens* combine with the evaporated tungsten vapors in the bulb and then redeposit the element on the filament. This process increases the life of the lamp, which generally lasts longer than a standard incandescent.

Nevertheless, halogens operate at much higher temperatures than standard incandescent lamps and so carry a greater risk of causing a fire or burning anyone who touches the lamps. Also, dimming reduces the life of a halogen lamp. Lowering the *voltage* (the strength of electrical force) stops the cycle of evaporation and redeposition.

Fluorescent lamps—whether the traditional long, white tubes found in kitchens, baths, and workshops or the compact, spiral variety—produce light by chemical reactions in a system known as *gaseous discharge.* Such lamps contain a small amount of the element mercury. Electric current flows through the lamp between electrodes at each end of the tube and *vaporizes* the mercury (changes it from a liquid to a gas), producing ultraviolet rays. A *phosphor* coating inside the bulb absorbs the rays and *fluoresces,* or produces visible light. (Phosphors are substances that give off

light when exposed to energy, such as ultraviolet rays.) Fluorescent lamps may vary in color from "warm white" to "cool white," depending on the phosphor used for the coating.

Fluorescents require an additional electrical component called a *ballast.* The ballast provides an electric charge that begins the chemical reaction and then regulates the amount of electric current flowing through the bulb to protect the lamp from overheating and burning itself out.

A fluorescent lamp is more energy efficient than a standard incandescent because less of the electric current is given off as heat. Fluorescents last much longer than incandescents. They are most efficient when operated for more than three hours at a time. In addition, they work better in warm

A vest embedded with flashing LED lights gives police and fire department members—as well as construction workers, bicyclists, and joggers—greater visibility at night or in other low-light conditions.

Types of lamps

Lamp type	Life	Advantages	Disadvantages
Incandescent	750–1,000 hours	Inexpensive to purchase Fits most fixtures Has a warm color that is flattering to skin tones	85 to 90 percent of the energy produced is wasted as heat
Tungsten halogen	Over 2,000 hours	Emits more light and lasts longer than standard incandescent Small size	More expensive to purchase and more fragile than standard incandescent Hot to touch
Linear fluorescent	10,000–20,000 hours	Longer life, uses less electric power, and produces less heat than standard incandescent	More expensive to purchase than standard incandescent Requires an electronic or electromagnetic component called a ballast Requires special ballast for dimming
Compact fluorescent (CFL)	6,000–20,000 hours	Can be used in most standard fixtures Uses less electric power and lasts longer than standard incandescent	More expensive to purchase than standard incandescent Dimming requires special CFL lamps and ballasts Dimming and frequently switching on and off may shorten life
Light-emitting diode (LED)	30,000–100,000 hours	Uses less electric power than standard incandescent Does not radiate heat Can alter colors	Expensive, but price is dropping quickly as efficiency improves

Source: Lighting Research Center, Rensselaer Polytechnic Institute, Troy, New York.

Light and health

As the short days and long nights of winter drag on, many people begin to miss the sun. But for some individuals, light does more than cheer them up. It plays a vital role in their health.

A light box (center) provides high-intensity, bright artificial light for people with seasonal affective disorder (SAD), a condition similar to depression that often occurs in the fall and winter.

Physicians first began to use *phototherapy* (light therapy) in the 1980's, to treat a condition called *seasonal affective disorder* (SAD). People with SAD usually develop symptoms in the fall and winter. They may experience sadness, sleeping and eating disorders, fatigue, aches and pains, and anxiety. If these symptoms become severe enough to interfere with daily life, physicians may treat the patient with bright light from a special fluorescent lamp called a light box. People usually sit before a light box for a prescribed period of time one or more times a day.

Researchers have found that a hormone called *melatonin* may play a role in SAD. Melatonin is produced in the brain during darkness and helps regulate the body's *circadian* (internal) rhythms. Circadian rhythms control such functions as the sleep-wake cycle and the regulation of body temperature. In people with SAD, the body may produce too much melatonin, causing them to feel fatigued and depressed. Exposure to bright light helps to reduce the brain's production of melatonin.

Light may have a sinister effect as well. In 2005, researchers reported that exposing laboratory mice with human breast tumors to bright light at night caused the tumors to grow more rapidly than when the mice were exposed only during the day. The results helped explain previous findings that nurses who work the night shift appear to have a greater risk of developing breast cancer than their day-shift colleagues. Much more research must be conducted, however, before scientists can determine whether there is a direct relationship between prolonged exposure to artificial light and breast cancer and how much light at night is too much.

temperatures; the gases inside them are more difficult to excite at temperatures lower than 10 °C (50 °F).

Because mercury is a *toxic* (poisonous) substance, large quantities of burned-out fluorescent lamps must be taken to lamp recycling centers. However, homeowners, who usually have only one or two burned-out lamps per year, are allowed to throw them out with their regular household trash in many states.

Compact fluorescent lamps (CFL's), most of which have a built-in ballast, are more energy efficient than standard incandescents, last longer, and can be used in many basic luminaires designed for incandescent lamps. Nevertheless, several factors have slowed the acceptance of CFL's in the home market since their introduction in the late 1970's. CFL's are significantly more expensive than incandescents. In addition, some CFL's are larger than incandescents and so may not fit in all standard light fixtures. Finally, their color may range from yellowish to pinkish to bluish—all in the same room, depending on the manufacturer. Color among different CFL's is gradually becoming more consistent, however.

Although CFL's last longer than incandescents, frequently switching the lamps on and off can shorten life. And when used in enclosed luminaires, CFL's may burn out faster than normal because the excess heat build-up can weaken the ballast.

High-intensity discharge (HID) lamps also work on the gaseous discharge principle. The most common forms of HID lamps are metal halide, mercury vapor, and high pressure sodium. Such lamps require a ballast and generally take several minutes to reach their full light output as the elements within them grow sufficiently hot. HID's last longer than halogen bulbs and the color of their light is more similar to that of natural daylight than the light produced by halogens. HID lamps are particularly useful in such places as warehouses and parking lots, where high levels of light are needed to cover large areas. They are also used in residential settings, particularly outdoors.

Electrodeless fluorescent lamps—also called electronic lamps or E-lamps—produce light the same way a fluorescent lamp does but without the electrodes. In one type of E-lamp, a tiny antenna emits radio waves that collide with mercury atoms, which then release ultraviolet light. Phosphors inside the lamp absorb the radiation and produce visible light. In another type of E-lamp, microwaves excite sulfur vapor, causing it to emit light. Because they lack electrodes—the parts of a gas discharge lamp that have the shortest life—E-lamps last longer than incandescents.

The electrodeless lamp appeared in the early 1990's amid predictions that it would become more popular than the CFL for home use. However, E-lamps never caught on, because high production costs raised sale prices and the lamps interfered with the radio frequencies of other electronic devices. E-lamps also produced lower than expected light *efficacy* (the amount of light emitted relative to electric current used).

How do you choose the most efficient lamps for your home? The costs of purchasing, installing, and operating a lighting system are a major factor in deciding which lamp to purchase. Some lamps or luminaires may cost more to operate over time, so their lower purchase price may be cancelled out by higher operating costs. Operating costs include the cost of the energy consumed, as well as that of bulb replacements.

One way to compare the costs of operating different light fixtures is to calculate how many *watts* (units of electric power) the fixtures consume and how many *lumens* (units of light) they give out. Incandescents usually produce 12 to 15 lumens per watt, whereas CFL's produce about 50. LED's produce about 30 to 35, but that number is steadily rising. Luminaires come in a nearly endless variety of styles, so look for the ENERGY STAR® label. The label indicates that the fixture meets energy

Setting a mood

Special LED panels (below, left) can be paired with software designed to display a changing array of colors to create "mood lighting" for any room (below, right).

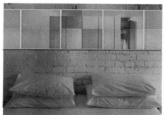

efficiency and quality guidelines set by the United States Department of Energy and the Environmental Protection Agency.

LED's are tiny chips made of *semiconductor* materials that create light without incandescence or fluorescence. (Semiconductors are materials that conduct electric current under some circumstances but not others.) The chips are made so that a semiconductor on one side has extra *electrons* and a semiconductor on the other is missing electrons. (Electrons are tiny particles that carry a negative charge.) The two sides are bonded together and have electrodes positioned at both ends. The entire chip is encased in a transparent plastic housing. When an electric current is applied to the LED, the extra electrons on the first semiconductor move to the semiconductor with missing electrons. In the process, they release *photons,* the elementary particles that make up light. LED's radiate only a fraction of the heat that other light sources do. The heat is usually vented out the back of the LED unit.

The chemical composition of the semiconductor material used to make the LED determines its color. The first LED's, introduced in the 1960's for use as indicator lights, were red or amber. The chemical element aluminum gallium arsenide produced the color red. Later, gallium nitride was used to make green and blue. As other colors became available, the use of LED's expanded to flashlights, digital watches, and traffic lights. With the development of white-light LED's and improvements in design and size, the devices moved into the home-lighting market as desk lamps and ceiling lights, among other uses.

LED's produce white light in two ways. In the mixed-color method, red, green, and blue LED's are grouped together to create white light. However, the light is not bright and may become unbalanced as the various LED's degrade at different rates. The other method is to coat the housing of a blue LED with phosphors, which convert some of the blue light into yellow light. The remaining blue light combines with the yellow light, producing a single, compact white-light source. However, a white LED made in this fashion is less efficient than a colored LED, because some of the light energy is lost in the conversion.

Lighting developers are particularly interested in LED technology because LED's are energy efficient, impact resistant, durable, and low maintenance. Nevertheless, because the devices are small, they must be grouped to produce significant amounts of illumination. It is the development of the *hardware* (the fixture that contains the LED cluster and connects the LED's to an electrical circuit) that has long kept LED's from being cost competitive with other lighting products.

One of the newest improvements, developed by the Lighting Research Center (LRC) at Rensselaer Polytechnic Institute in Troy, New York, is known as SPE™ (scattered photon extraction) technology. Typical commercial white LED's produce low overall light compared with traditional lighting technologies because the phosphors used to create the LED's light can scatter photons emitted by the semiconductor. Many of these photons return to the chip instead of being emitted as light. The SPE method helps the photons escape from the LED, significantly increasing light output and efficiency. The challenge lies in the LED itself. LED's are good at creating light, but not at getting the light out. Newer LED's are beginning to compete with incandescent and fluorescent lamps in terms of light output and efficiency.

Another lighting technology that may soon become available to consumers is the organic light-emitting diode (OLED). Most of the LED's on the market in 2006 were made from *inorganic compounds* (compounds that do not contain carbon atoms). Organic LED's can be made of a much wider variety of compounds, including *polymers* (large molecules formed by the chemical linking of many smaller molecules into a long chain). Polymers are flexible and can assume many different shapes, including thin films. Thus, OLED's can be made from materials that are much less expensive than regular LED's. Researchers in Japan, Germany, and the United States reported at a Materials Research Society meeting in late 2005 that the first white OLED's for general lighting use may enter the market as early as 2007.

In addition to improving LED's themselves, researchers have also been working with industry to create new ways to light spaces. To replace luminaires permanently mounted on ceilings and walls, developers are exploring more flexible, modular systems. The same team that developed the SPE technology also worked with researchers at the Alliance for Solid-State Illumination Systems and Technologies (ASSIST) to build a prototype interior lighting system that integrates LED's with other building materials such as ceiling panels, wallboard, stair risers, and even flooring. (ASSIST is a collaboration among researchers, manufacturers, and government organizations run by the LRC at Rensselaer.)

Many lighting researchers believe that *solid-state lighting* (lighting consisting of clusters of LED's) will revolutionize the way we light buildings. These researchers envision homes and offices with an electronic infrastructure that allows users to snap or mount various types of solid-state lighting anywhere in the ceiling or walls without the need for electricians or contractors. One day, changing your lighting may be as easy as rearranging your furniture.

■ Keith Toomey

SCIENCE NEWS UPDATE

Contributors report on the year's most significant developments in their respective fields. The articles in this section are arranged alphabetically.

AGRICULTURE

Drought, hurricanes, and floods affected various regions of the United States in 2005, leaving many agricultural producers in difficult circumstances at the start of the harvest season.

Hurricanes Katrina and Rita caused more than $1 billion in crop and livestock losses in the Gulf Coast region of the United States in September 2005, according to estimates from the U.S. Department of Agriculture (USDA). One of the hardest-hit crops was sugar cane. Louisiana is one of the largest sugar-cane-producing states in the United States. Thousands of acres of sugar cane ready for harvest were flattened by Katrina's winds, which blew up to 200 kilometers (125 miles) per hour.

Kurt Guidry, an economist at Louisiana State University AgCenter in Baton Rouge, estimated that 80 percent of the citrus acres in the state would not be suitable for replanting in 2006 because of soil contamination from saltwater deposited by the hurricanes. He also estimated that losses for *horticulture growers* (growers of flowers, fruits, vegetables, and other plants) in Louisiana would be about $19 million.

Power and fuel shortages plagued poultry and dairy producers in the Gulf Coast region following Hurricane Katrina. Mike Pepper, president of the Mississippi Poultry Association, estimated that 80 percent of that state's 9,000 poultry houses were damaged. Mississippi produces some 10 percent of the U.S. poultry supply.

The Gulf Coast's dairy industry was particularly hard hit by Katrina. Loss of electric power meant that dairy farmers were unable to refrigerate or process milk for days. Damage to Louisiana's roads and bridges prevented trucks from reaching farms to pick up the milk. Dairy economist Bill Herndon of Mississippi State University Extension Service in Mississippi State estimated dairy losses resulting from Katrina at about $6 million.

Drought. In the months before the hurricane season, severe drought gripped the nation's Corn Belt. The drought affected an area stretching across eastern Texas and Oklahoma, through Arkansas and Missouri, and into areas of Illinois and Indiana. Growers in Illinois, which is typically an ideal corn- and soybean-growing region, saw their worst drought conditions since the late 1980's, with many fields receiving less than half the normal amount of rainfall.

Remarkably, at grain harvest, U.S. farmers brought in the second largest corn and soybean crops on record, according to the USDA. Crop experts credited the bountiful harvest to improvements in farm management practices and crop genetics, particularly in the development of plants that can tolerate low rainfall conditions.

The nation's pumpkin crop was also affected by the drought. In Illinois, the largest pumpkin producer in the United States, the vegetables were about 20-percent smaller than usual.

Perennial crop goes biotech. The first *genetically engineered perennial crop* became a commercial reality in July 2005, when Roundup Ready alfalfa seed went on sale. (Perennial crops live for more than two years. Genetic engineering is a technique that alters the genes in an organism.)

The seed, which was approved by the USDA in June, allows growers to use the glyphosate herbicide Roundup on genetically altered alfalfa to control more than 200 species of weeds without damaging the alfalfa. Synthetic chemical herbicides such as glyphosate are frequently used in agriculture to control weeds, which can significantly reduce crop yields. The new biotech alfalfa is being produced jointly by agricultural biotechnology companies Monsanto of St. Louis, Missouri, and Forage Genetics International of Prior Lake, Minnesota.

Herbicides are classified by the kinds of plants they kill. *Broad-spectrum herbicides* kill all plants and, thus, cannot be applied directly to a field of crop plants. Broad-spectrum herbicides are most often applied *pre-emergence*—that is, before newly planted crops emerge. Until the 1990's, weeds that appeared after crop plants started growing were often controlled mechanically, either by cultivation or hoeing. Researchers realized that a crop plant genetically engineered to be resistant to a broad-spectrum herbicide would simplify weed management. In such cases, the application of a single herbicide would be sufficient to kill weeds without damaging the crop plant itself.

By the early 2000's, glyphosate-resistant crops had become a key part of U.S. agriculture. In 2004, 13 percent of corn acres, 85 percent of soybean acres, and 60 percent of cotton acres were planted with herbicide-resistant varieties, according to the Seed Biotechnology Center at the University of California at Davis. In 2005, alfalfa, the nation's third most important crop in economic value, was set to join the list.

New bird flu test. Concern about avian influenza (also called bird flu) heightened in October 2005, after the publication of an article by a team of scientists led by molecular pathologist Jeffrey K. Taubenberger of the Armed Forces Institute of Pathology in Rockville, Maryland. The article linked the 1918 Spanish flu *pandemic* (worldwide outbreak of a disease) to a *strain* (subtype) of bird flu. The 1918 flu epidemic killed as many as 50 million people after it *mutated* (underwent genetic changes) and spread to human beings.

Avian influenza causes a fatal disease in chickens, ducks, turkeys, and other domesticated birds. Poultry farmers worldwide grew increasingly concerned that the disease could spread to their birds. Some waterfowl—including seabirds, shore birds, and wild ducks—carry the virus but generally do not develop the disease themselves.

Bird flu viruses are classified by a combination of two groups of proteins found on the surface of the virus: hemagglutinin proteins (H), of which there are 16; and neuraminidase proteins (N), of which there are 9. From 2003 to May 2006, the bird flu strain classified as H5N1 had killed millions of birds and more than 100 people. The disease in avian populations had spread from Asia to a number of countries in Eastern and Central Europe as well as in Africa. A majority of the human deaths linked to the virus occurred in Asia, though several had also been reported in Africa and Europe.

MORE NUTRITIOUS MELONS

Plant physiologist Gene Lester (right) of the United States Agricultural Research Service (ARS) and farm manager David LaGrange of Starr Produce in Rio Grande City, Texas, examine cantaloupes grown with an experimental technique to make them more nutritious. In August 2005, researchers with the ARS announced that spraying potassium on melons as they grow increases their levels of such nutrients as beta carotene and vitamin C.

AGRICULTURE continued

An inexpensive test that can identify all subtypes of avian flu was being patented in 2005. Huaguang Lu, an avian virologist at Pennsylvania State University in University Park, developed the test, which provides results in a few hours instead of the usual three days needed for older testing procedures. Lu noted that same-day results could help slow an avian flu outbreak before it had the chance to become an epidemic. He also noted that it was especially important to get test results quickly in the cases of strains H5 and H7 influenza, which can be deadly in poultry.　■ Jeanne Bernick

See also **ATMOSPHERIC SCIENCE.**

ANTHROPOLOGY

The discovery of evidence suggesting that Neandertals and anatomically modern human beings coexisted for some time in what is now France was reported in November 2005 by archaeologists Brad Gravina and Paul Mellars of Cambridge University and *geochronologist* Christopher Bronk Ramsey of the Oxford Radiocarbon Accelerator Unit at Oxford University, both in the United Kingdom. (A geochronologist determines the time or length of existence of rocks and other geological formations and of geological periods.) The scientists announced the results of *radiocarbon dating* tests of animal bones found at three depths within a cave called the Grotte des Fées in Châtelperron, France. (Radiocarbon dating is a method of determining the age of an object by measuring its level of radioactive carbon.)

The upper and lower levels of the deposits excavated at the cave contain stone tools of a type called Châtelperronian, which scientists have long associated with Neandertals. The middle layer, however, contains a different kind of tools, called Aurignacian. Most anthropologists believe that Aurignacian tools were made by modern human beings, though the oldest known Aurignacian tools are 4,000 years older than the oldest known modern human fossils in Europe.

The deposits at Châtelperron had been dated before, but scientists found the results uncertain and contradictory. Archaeologists worried that the three layers of bones and tools had become mixed over time, blurring any distinction between Neandertal remains and those of modern human beings. The Oxford Radiocarbon Accelerator Unit employs a new process called *ultrafiltration* to ensure that samples are not contaminated in a way that skews the data. Dates obtained in this way are much more reliable than those produced even a few years ago.

Gravina, Mellars, and Ramsey showed that the lower layer (Neandertal) dates from about 41,000 to 39,000 years ago, while the upper layer (also Neandertal) dates from about 36,000 to 34,000 years ago. The middle layer, associated with modern human beings, dates from roughly 39,000 to about 36,000 years ago, between the times during which Neandertals occupied the cave. The scientists interpreted this finding to mean that modern human beings occupied the cave during a period when the Neandertals were living elsewhere.

The researchers were also able to compare the dates for the Châtelperron layers to new climate data from *cores* (cylindrical samples) of Greenland ice and of sediments from the eastern Atlantic Ocean. The data show that Europe's climate was fairly warm during the periods when Neandertals lived at Châtelperron but distinctly cooler when the modern human beings were there. The scientists speculated that the Neandertals may have had to retreat to warmer places when the climate became too cold, leaving northern France to modern human beings with their greater abilities to dress warmly and make efficient shelters.

Neandertals older? Another new set of dates from the Oxford Radiocarbon Accelerator Unit, reported in January 2006, suggests that Neandertals died out several thousand years earlier than anthropologists had believed. Scientists from Croatia, the United Kingdom, and the United States, led by geochronologists Tom Higham and Christopher Bronk Ramsey of the Oxford Radiocarbon Accelerator Unit, reported revised dates for Neandertal fossil remains from a site called Vindija Cave in modern Croatia. As recently as 1998, scientists believed these fossils dated to about 28,000 years ago, making them the remains of some of the last-surviving Neandertals. Applying ultrafiltration to the samples, the researchers concluded that a more accurate age for the fossils is around 33,000 to 32,000 to years ago. The new dates imply that the Neandertals disappeared from Europe earlier than most anthropologists had thought.

Anatomically modern human beings and Neandertals might have lived in Europe at the same time, though the scientists emphasized that they have no proof for this coexistence. Although Aurignacian tools in Europe go back at least 39,000 years, scientists have not found any fossils of anatomically modern human beings there that are older than about 35,000 years. The possibility remains, they stressed, that Neandertals themselves adopted Aurignacian technology from early modern human beings living in eastern Europe and passed it westward across Europe to other Neandertal groups. That hypothesis implies that Neandertals and early modern human beings interacted culturally—and perhaps more intimately—as the latter moved into eastern Europe.

Tracing ancient Americans. Researchers in 2005 provided evidence for the theory that two separate groups of early human beings colonized North and South America. In December, anthropologists Walter A. Neves and Mark Hubbe of the University of São Paulo in Brazil reported on their study of 81 ancient human skulls from the Lagoa Santa region of Brazil. After dating and measuring the skulls, the scientists compared them to other skulls from around the world. They found that the Lagoa Santa skulls are more similar to skulls from modern inhabitants of Melanesia and Australia than they are to the skulls of modern Native Americans or prehistoric Native American ancestors from northeast Asia.

Although human skeletons were first discovered in the Lagoa Santa region in 1842, they had never been dated using modern techniques. Using a variety of methods, Neves and Hubbe dated all the skulls from 11,000 to 7,500 years ago, with the majority of them dating from 8,500 to 7,500 years ago. The researchers noted that all these ancient Americans had been buried in the same way, which implies that they all belonged to the same human population.

FLEET FEET

The 26,000-year-old foot of an early modern human being shows signs of weakening in the bones of the lesser toes (those other than the big toe). Scientists in 2005 attributed this characteristic to the use of shoes. Anthropologist Erik Trinkaus of Washington University in St. Louis, Missouri, measured the strength in a number of fossil feet from ancient human beings and determined that people started wearing shoes from 40,000 to 26,000 years ago. When people walk barefoot, all five toes grip the ground, but, Trinkaus explained, footwear reduces the need for the toes to grip and so leads to a weakening of the smaller toes.

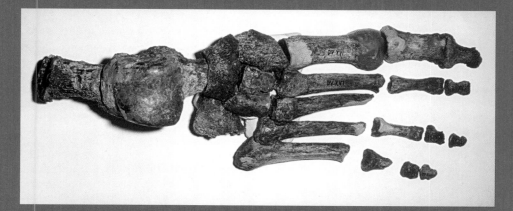

ANTHROPOLOGY continued

Neves and Hubbe compared the Brazilian skulls to skulls from elsewhere in the world. Modern Native American skulls are very similar to those of their ancestors from northeast Asia who migrated across the Bering Land Bridge into North America at least 10,000 years ago. However, the Brazilian skulls do not resemble those of northeast Asians or modern Native Americans. Instead of being short and broad, the Brazilian skulls are long and narrow. The closest matches for the Brazilian skulls were found in Melanesia, East Asia, and sub-Saharan Africa.

The results of this study suggest that the Americas were settled by two or more migrations of people from different parts of the world. The ancestors of modern Native Americans may have been among the last wave of immigrants and either replaced or absorbed the people who arrived before them.

Transitional species discovered? An international team of anthropologists reported in March 2006 the discovery of a hominid skull that may represent the transition from *archaic* (primitive) to modern human beings. Scientists with the Gona Paleoanthropological Research Project, led by paleoanthropologist Sileshi Semaw, discovered the unusually complete skull in the Gawis River drainage basin in northeastern Ethiopia.

Semaw estimated that the fossil skull dated to from 500,000 to 250,000 years ago. Genetic studies of modern human beings suggest that our species *(Homo sapiens)* first appeared in Africa during that time. The skull has some features linking it to earlier African *H. erectus* as well as other features similar to those of modern human beings, which suggests that the fossil represents a transitional species. The oldest known early modern human fossils also have been found in Ethiopia.

More "hobbit" findings. Scientists working on the Indonesian island of Flores reported in 2005 the discovery of additional fossils identified as belonging to *Homo floresiensis.* The announcement of the discovery of *H. floresiensis,* a tiny human species that existed from about 90,000 to 13,000 years ago, amazed the scientific community in 2004. Archaeologist Michael J. Morwood of Australia's University of New England in Armidale and colleagues from Australia and Indonesia reported finding more fossils of prehistoric human beings nicknamed "hobbits" after the little creatures in *The Lord of the Rings,* a series of novels by British author J. R. R. Tolkien. The scientists believe that the

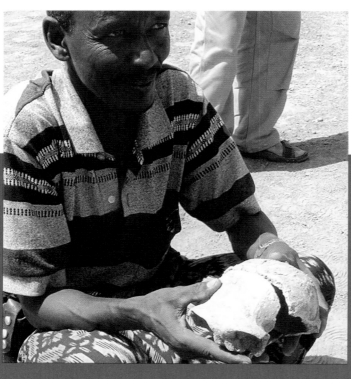

MIDDLE MAN?

Two halves of a skull found in Ethiopia in 2006 may represent a missing link between *Homo erectus* and anatomically modern human beings. Researchers dated the skull to from 500,000 to 250,000 years ago, a period represented by few human fossils. *H. erectus,* which many scientists believe was an ancestor of modern human beings, is thought to have died out about 400,000 years ago. The scientists noted that the skull, though different from that of modern people, shows some similar characteristics.

new fossils represent the remains of at least nine individuals and may help to resolve the controversy over whether the first specimen is a new species of *hominid* (the scientific family of human beings and prehistoric humanlike species) or an anatomically modern person suffering from *microcephaly*. (Microcephaly is a genetic disorder that results in an abnormally small skull and brain.) Like the original remains, the new fossils were found in a cave called Liang Bua on the Indonesian island of Flores.

Many anthropologists were skeptical of the idea that a diminutive human being with a brain the size of a chimpanzee's made stone tools and butchered dwarfed elephants in Indonesia as recently as 12,000 years ago. An alternative explanation, championed by Indonesian paleoanthropologist Teuku Jacob and others, is that the specimen was a modern human being with microcephaly.

The new fossils include a *mandible* (lower jaw) that is so similar to that of the original specimen that, Morwood said, it leaves little doubt that the fossils represent an authentic population of diminutive hominids. Both mandibles lack protruding chins, a characteristic of all modern human beings, even those with microcephaly.

Moreover, newly discovered limb bones show that though the Flores hominids were small (about 1 meter [3 feet] tall), they had unusually long arms in proportion to their body size. These arm proportions do not match those of any known members of the genus *Homo,* to which modern human beings belong. In fact, both the brain size and the limb

proportions of *H. floresiensis* recall African australopithecines, specifically the 3.2-million-year-old *Australopithecus afarensis,* nicknamed "Lucy," that was discovered in Ethiopia in 1974.

The resemblance between *H. floresiensis* and African australopithecines led archaeologists Robin Dennell of the University of Sheffield in the United Kingdom and Wil Roebroeks of Leiden University in the Netherlands to propose a reevaluation of the theory that the genus *Homo* originated in Africa. Writing in December 2005, Dennell and Roebroeks stressed that the oldest known Asian fossils belonging to the genus *Homo* are almost as old as the earliest *Homo* fossils from Africa. The scientists speculated that hominids may have begun migrating out of Africa soon after they adapted to life on the savanna, as early as 3.5 million years ago. It is even possible, the archaeologists argued, that the genus *Homo* originated in south or southwest Asia, with members of the group later migrating into Africa.

If this hypothesis is correct, the Flores hominids may have descended directly from a small-bodied australopithecinelike ancestor and not from a later hominid, such as *H. erectus.* More sophisticated analyses of *H. floresiensis* may help to resolve this issue, but the most useful evidence would be additional fossils, including specimens older than those discovered so far. By 2006, archaeologists were combing Flores and neighboring islands for just that kind of evidence.

■ Richard G. Milo
See also **ARCHAEOLOGY.**

ARCHAEOLOGY

The discovery of evidence indicating that early human beings lived in northern Europe about 700,000 years ago—200,000 years earlier than previously thought—was reported in December 2005. Archaeologist Simon Parfitt of the Natural History Museum in London announced that an international team of researchers had found 32 flint tools among fossils exposed by natural weathering processes at the bottom of a cliff face near Pakefield in eastern England. The tools were mostly *flakes* (small flint chips) used for cutting and scrap-

SEE ALSO THE SPECIAL REPORT, **FIRST CIVILIZATION IN THE AMERICAS,** PAGE 12.

ing. Among the Ice Age plant, animal, and insect remains discovered in the same layer were the bones of a species of water vole that became extinct about 700,000 years ago.

Dating ancient human activities by associating them with fossils of a known age is a common archaeological technique. The ar-

ARCHAEOLOGY continued

chaeologists at Pakefield also relied on a new procedure called *amino acid dating,* which examines the rate at which the proteins in plant and animal remains deteriorate. (Amino acids are organic acids that make up all the proteins in living things.)

In the tool layer, archaeologists also found the remains of a land snail from which they were able to recover amino acids. After determining that the protein sample had not been damaged by leaks, which would have affected the rate at which the protein decomposed, the scientists determined that the snail had lived at roughly the same time as the extinct water vole—about 700,000 years ago.

The fossils found with the Pakefield stone tools also provided a treasure trove of information on the climate and landscape of what is now England during this early period. These fossils included the skeletal remains of mammoths, rhinoceroses, hippopotamuses,

and saber-toothed cats. The scientists also analyzed samples of soil found in the fossil layer. They determined that the soil formed during a period when summer temperatures reached 23 °C (73 °F), winters were mild, and relatively little rain fell, all of which are characteristics of the Mediterranean climate of modern southern Europe.

Although the presence of the flakes shows that human beings lived in northern Europe about 700,000 years ago, scientists are unsure about when people first moved into the area or how large their population was. Fossilized human remains found in Italy show that people lived there about 800,000 years ago. Scientists assume that as human populations grew, they expanded slowly northward. The moderate climate at that time would have aided such movements. In addition, geographers have found evidence that about 700,000 years ago, what is now England was linked to mainland Europe by

MAYA DISCOVERIES

The creation of the world according to Maya mythology is one element on the earliest known Maya mural, revealed by archaeologists in Guatemala in December 2005. Scientists believe the 9-meter- (30-foot-) long wall painting dates to about 100 B.C., several hundred years earlier than any other example of elaborate Maya art. The mural indicates that Maya mythology, writing, and complex art existed much earlier than believed, the archaeologists said.

a land bridge. Although scientists still have much work to do before they will be able to understand how ancient people settled Europe, the Pakefield flints have provided valuable clues.

Origins of Maya art and writing. The discovery of the earliest known evidence of hieroglyphic writing in the Maya civilization of Central America was reported in January 2006. Archaeologist William Saturno of the University of New Hampshire in Durham announced that a panel of hieroglyphs found at the archaeological site of San Bartolo in Guatemala dates to about 250 B.C., about 300 to 400 years before the oldest previously known examples of Maya writing.

San Bartolo is located in the northeast corner of Guatemala, north of the well-known Maya site of Tikal. Archaeologists working at San Bartolo have recorded at least 100 structures, including a ballcourt (a feature typical of many Maya sites) and a palace. Dominating the site is a large central plaza with several temple-pyramids, the tallest rising 30 meters (100 feet). The site appears to have been occupied from about 800 B.C. through the Late Preclassic Period (about 400 B.C. to A.D. 250).

San Bartolo's golden age occurred during the Late Preclassic Period. One of the site's most important structures, the Las Pinturas temple-pyramid, was constructed during this time. The building consists of a series of layers or overlying structures—a new structure built over the previous one—a style typical of Maya pyramids.

While exploring a looters' tunnel dug into one of the earlier structures, archaeologists found a mural dating to 100 B.C. that showed Maya mythological themes. As archaeologists studied the mural, they tunneled deeper into Las Pinturas and found an even earlier building

The stone portrait of a woman dating to the A.D. 300's (right and in inset above) may be the oldest known example of Maya portraiture, archaeologists announced in December 2005. The portrait, which may be that of a queen or other high-ranking woman, may illustrate the role women played in Maya politics.

ARCHAEOLOGY continued

consisting of three masonry rooms. In the central room, they found several murals, including a painting of the Maya maize god. They also discovered a plastered stone block bearing a column of 10 hieroglyphs, painted in black. These hieroglyphs and the murals in the deeply buried three-room complex were radiocarbon-dated to from about 300 to 200 B.C.

Although some of the 10 glyphs appear typical of later Maya texts, experts have so far been unable to decipher the newly discovered glyphs. One glyph appears similar to a later glyph meaning *ruler*, *lord*, or *noble*. The researchers speculated that the rest of the glyphs might comprise the name of this apparently important person.

Medieval graveyard. The discovery of an enormous medieval cemetery in the British city of Leicester has given archaeologists an unparalleled opportunity to examine many details of medieval life in England. Excavations elsewhere in England in recent years have provided archaeologists with artifacts representing many aspects of medieval daily life, including food remains, types of pottery, and architecture. However, the cemetery at St. Peter's will give archaeologists their first opportunity to closely examine a large sample of medieval skeletons.

Archaeologist Richard Buckley of the University of Leicester Archaeological Services announced the discovery, which was made while preparing the area for the construction of a shopping center, in January 2006. The cemetery, along with the remains of the parish church of St. Peter's, to which it was linked, had been lost for centuries.

The first burials at the cemetery were made in the 1100's, with interments continuing until the mid-1500's. Historical records indicated that the church had been demolished in 1573. Buckley noted that the graveyard is the largest medieval English cemetery ever found outside London. Initial research documented 1,300 graves at the site, including some within the church itself.

Physical anthropologists specializing in *osteology* (the analysis of bones) hope to obtain information on the age, sex, diet, diseases, injuries, and social status of the people buried in the cemetery. For example, the teeth of the people buried inside the church are generally in better shape than those of the people buried in the cemetery outside. The difference likely stems from diet, with the people of lower status having eaten a diet of coarse bread that contributed to heavy tooth wear, decay, and *abscesses* (infection). The large number of infant and juvenile graves found at the site reflects the high mortality rate for children common during that period. Some graves hold a number of individuals. Archaeologists speculated that the graveyard may have become overcrowded or that people who died during epidemics were placed into mass graves.

Ancient warfare. Scientists in 2005 discovered the earliest evidence for organized warfare in the Middle East at a 5,500-year-old site that had once been a large settlement. Archaeologist Clemens Reichel of the Oriental Institute of the University of Chicago reported in December that excavations at the site, called Hamoukar, revealed a "war zone."

Hamoukar is located in northeastern Syria near the border with Iraq. The area lies between the upper reaches of the Tigris and Euphrates rivers in the birthplace of the first Mesopotamian civilizations. Hamoukar is a large mound rising about 15 meters (50 feet) that contains the remains of a series of villages and larger settlements.

Fieldwork at Hamoukar exposed one settlement that encompassed an area of about 32 hectares (80 acres). Researchers discovered two large building complexes constructed around courtyards. One of the buildings held a large kitchen in which researchers found grinding stones and a baking oven.

In another building, archaeologists discovered clay stamps generally used to close and mark baskets that contained food. Such seals have stamped motifs that indicate either the contents of the basket or its owner. The archaeologists believe that this building was not used as a residence but for communal (perhaps religious or administrative) storage or food distribution.

Researchers were especially intrigued by the remains of fortifications and the many artifacts associated with large-scale warfare found at Hamoukar. For example, surrounding the settlement was a 3-meter- (10-foot-) high mudbrick wall, almost certainly built for protection.

Researchers discovered evidence that walls within the settlement had collapsed during a major fire resulting from a military attack. They also found at least 1,200 round "sling bullets" fired by the attackers that had pockmarked the wall and the buildings in the settlement. These clay projectiles, which were 2.5 centimeters (1 inch) long, were hurled at the city and its defenders. Archaeologists also recovered 120 larger round clay ball "bullets" that ranged from 6 to 10 centimeters (2.5 to 4 inches) in diameter. Used with devastating effect to batter the settlement during the attack, the bullets show scratches and impressions caused by hitting buildings. The sling was an important weapon at this time and likely caused havoc among the defenders of Hamoukar. The archaeologists made replica slings, which they used to hurl replica bullets to demonstrate the devastating effect such an attack would have had.

The archaeologists also discovered large quantities of pottery characteristic of the Uruk culture of southern Mesopotamia in the layers of the settlement above those containing the evidence of destruction. The researchers believe that after the settlement lay in ruins, Uruk populations moved in to occupy the site.

Pompeii of the East. The remains of a little-known culture buried by the largest volcanic eruption in recorded history were found on the island of Sumbawa in what is now Indonesia, archaeologists reported in February 2006. The eruption of Mount Tambora on April 10, 1815, destroyed the Tambora culture on Sumbawa. At the time, the island was populated by about 92,000 inhabitants, all of whom were killed by volcanic gases, ash, and rock released by the eruption.

The eruption ejected 360 million metric tons (400 million tons) of sulfuric acids to a height of 43 kilometers (27 miles), creating atmospheric conditions that led to what is known as the "Year Without a Summer" in 1816. Global temperatures became unusually low, and colder-than-normal weather in France, Germany, and the northeastern United States destroyed many crops.

Haraldur Sigurdsson, a *volcanologist* (scientist who studies volcanoes) at the University of Rhode Island in Narragansett reported that fieldwork on Sumbawa had revealed deeply buried and well-preserved remains dating to the time of the eruption. Modern villagers on the island had previously discovered artifacts from the time of the eruption in a gully that had developed in the thick layer of volcanic ash and rock that covers much of the island. The villagers aided Sigurdsson's team in locating the remains.

ANCIENT CHURCH

An elaborate mosaic featuring the image of two fish adorns the floor of what may be the oldest Christian church in the Holy Land. Prisoners working on a new prison in the Israeli village of Megiddo discovered the mosaic. The fish is a Christian symbol that predates the cross. Experts believe the church dates from the 200's.

ARCHAEOLOGY continued

The researchers used ground-penetrating radar to search the gully for signs of the Tambora culture. (Ground-penetrating radar can detect underground features and objects within about 15 meters [50 feet] of the surface.) The team's initial excavations revealed a house burned by the eruption. Within the house were the remains of two people, along those of household goods, including pottery fragments, burned wood, and fragments of animal bone. The researchers also discovered porcelain cups and bronze artifacts made in a style common in Vietnam and Cambodia at that time. Early historic documents indicate that the Tamborans traded horses, honey, and sandalwood used for incense with other cultures in southern Asia.

The excavations revealed that Tambora suffered as much destruction from the eruption of Mount Tambora as ancient Pompeii did during the A.D. 79 eruption of Mount Vesuvius. Little had been recorded about the Tambora culture before the eruption. Further excavation will almost certainly shed light on this virtually unknown culture.

Peruvian observatory. A 4,200-year-old temple uncovered near the Peruvian capital of Lima is the oldest known observatory ever found in the Americas. In April 2006, scientists led by archaeologist Robert Benfer of the University of Missouri at Columbia reported finding the observatory at the top of a 10-meter- (33-foot-) tall pyramid at a site called Buena Vista. Benfer and his team believe that the celestial calculations made there were probably used to guide the planting of crops in ancient times.

Buena Vista, which lies in the fertile Rio Chillon Valley, covers about 8 hectares (20 acres) of a rock-strewn hillside. The pyramid and temple cover about 0.8 hectares (2 acres) of the site's central area. These buildings are the oldest at the site and have been linked to the Kotosh culture of the Peruvian Preceramic Period. (The Preceramic Period is the name scientists have given to the era of Andean civilization before the development of pottery.) The Buena Vista buildings predate other early Andean civilizations in the area by at least 800 years and the Inca civilization by about 3,000 years.

Much of the rest of the Buena Vista site is occupied by buildings of later cultures, including a ceremonial center with stepped pyramids. A great deal of this part of the site has been looted.

Excavations of the Kotosh pyramid and its temples revealed evidence of relatively minor looting. One large looters' pit, a hole about 2 meters (6 ½ feet) deep and 6 meters (20 feet) wide in the layers above the site, narrowly missed the observatory. Had the looters discovered the temple, important features of the site would almost certainly have been lost.

Originally, the rock surfaces of the observatory had been plastered over and then painted in red and white. Beside the doorways into the temple, the scientists discovered a painted picture of an animal, thought to be a llama. Carved into this painting is the image of a fox. In Andean mythology, the fox taught people how to farm.

During the excavation, Benfer noticed that a window over the temple's altar aligned with a notch on a distant hill. While surveying that area, the scientists found a small head carved into the rocks in the notch. A line from the temple window to this small head pointed southeast, at 114 degrees from true north. This alignment, an astronomer told the excavators, points to the sunrise on December 21, the summer solstice in the Southern Hemisphere.

The summer solstice represented the time that cotton and other crops were planted and then watered by the annual flooding of the Rio Chillon. The temple window was also aligned with the constellation of the fox, seen in the skies of the Southern Hemisphere at the time of the summer solstice. The combination of astronomical data found in the excavations led scientists to conclude that the temple (which Benfer named the Temple of the Fox) served as an observatory.

Additional finds in the temple included a large clay disk made of mud, grass, and plaster that was covered with a fine clay surface. The disk appears to show a frowning face, with upturned lines representing eyes and a downturned line representing the mouth. Eroded clay animals were found on both sides of the sculpture. Benfer believes the face may be that of the Andean god Pacha Mamma, who may have frowned on the sunset of the winter solstice, which was also the last day of the harvest. ■ Thomas R. Hester

See also **ANTHROPOLOGY.**

KENNEWICK MAN DETAILED

In July 1996, two college students wading along the shoreline of a lake near Kennewick, Washington, found scattered human bones along a bank. The bones, many of which were stained a light brown, had been underground for some time. The local coroner asked an archaeologist to work with police in recovering the skeleton. During an examination of the skeleton, the archaeologist discovered a stone spearpoint, like those used thousands of years ago, embedded in the right hipbone. The coroner sacrificed a fragment of one of the bones to the *radiocarbon dating* process to determine how long ago he had died. (Radiocarbon dating is a method of determining a sample's age by measuring its level of radioactive carbon.) The test revealed that Kennewick Man, as he was named, had died about 9,300 years ago. The discovery of the remains excited the scientific community because they represented the oldest and best-preserved skeleton found in the northwestern United States and one of the oldest found in the country.

In 1996, the U.S. Army Corps of Engineers—which had jurisdiction over the bones because they were found along a waterway under the Corps's control—determined that Kennewick Man was a Native American and, thus, would be returned to the Native American community for reburial according to their customs without further study. A brief study of the skeleton arranged by the U. S. Department of the Interior determined that Kennewick Man stood about 1.75 meters (5 feet 9 inches) tall, had muscular arms and legs, and was in his 40's when he died. More detailed studies of the skeleton, however, were delayed by a court battle between the Corps and a group of eight scientists who wanted to analyze the bones and sued to prevent the reburial. The Corps sided with Native American groups, who considered Kennewick Man an ancestor. In 2004, a U.S. court ruled that available information about the skeleton had failed to prove that Kennewick Man was a Native American and that the scientists could conduct their studies.

In July 2005 and again in February 2006, a team of 20 scientists, including forensic anthropologists, dental anthropologists, geologists, archaeologists, and a scientific photographer, studied and documented the remains. The studies, which are ongoing, have so far revealed fascinating information about Kennewick Man's skeletal and dental health, injuries, diet, activities, physical size, and biological connections to prehistoric populations in North America and Asia.

In February 2006, the team reported on their findings about the original location of Kennewick Man's burial and his position in the ground. Had Kennewick Man been purposely buried by others, or had he come to the

THE MAN IN QUESTION

Scientists examine Kennewick Man, a 9,300-year-old skeleton found in Washington state. The nearly complete skeleton is one of the oldest and best-preserved skeletons ever found in the United States.

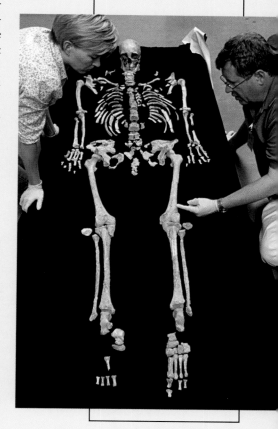

A PAINFUL POINT

The broken point of a projectile can be partially seen on the right side of Kennewick Man's hip. Although scientists initially thought that this wound might have contributed to his death, further research showed that Kennewick Man recovered from the injury and died many years later.

site by accident, perhaps carried there by a flood? To answer that question, scientists examined the location of the soil stains on the bones and the degree to which the bones had been bleached after their exposure to the sun. They also studied how the bones had fractured as they fell from the bank and how waves lapping against them had eroded their surfaces. In this way, they determined which bone surfaces had faced skyward after their exposure and which had remained buried and protected from weathering. To determine how deeply the skeleton had lain in the ground, the scientists examined mineral deposits on the bones. These minerals form in the upper layers of soil and are carried farther down by water before adhering to the bone surfaces. All these clues led the scientists to conclude that Kennewick Man had been buried intentionally, placed on his back, parallel to a nearby river, with his hands palm-down at his sides.

The scientists found that Kennewick Man had been injured several times in his life. His head had two depression fractures. Both injuries probably resulted from blows that had fully healed by the time of his death. The scientists studied the injury to his hipbone using computed tomography (CT), a scanning technique that produces three-dimensional images, to obtain images of the bone and spearpoint. They then used these data to create accurate plastic replicas of the two. The analysis of these reconstructions revealed that the tip of the spearpoint had broken off when it struck Kennewick Man's hip. The scientists used findings about the angle, depth, and location of the point in the hip to re-create the path of the spear. They concluded that the projectile had hit him with great force while hurtling downward and had been thrown from a distance. The spear likely had been hurled using an *atlatl*, a wooden or bone device that can increase the range and force of a thrown spear.

Damage to the base of the spearpoint indicates that someone tried to remove the point from Kennewick Man's hip by twisting the spear foreshaft to which it was attached. However, the point itself was too deeply embedded to pull out and broke free of the shaft. New bone had grown around the spearpoint, indicating that Kennewick Man had lived for many years after receiving this injury. So far, scientists have not been able to determine the cause of his death.

To help determine Kennewick Man's relation to other ethnic groups, the scientists measured his skull. They then compared these measurements with those from various populations of prehistoric and modern peoples. They found that the skull shows facial features characteristic of such East Asian groups as the Ainu of Japan and also of Polynesians, rather than modern Native Americans. Thus, scientists think that Kennewick Man's ancestors were more closely related to the ancestors of these modern Asian groups.

Scientists have discovered that Kennewick Man's teeth were worn down from a gritty diet and from using them to perform some kind of task. Currently, scientists are studying Kennewick Man's teeth as well as his bone chemistry to determine what he ate and for what kinds of tasks he might have used his teeth. In addition, further radiocarbon tests will pinpoint exactly when this mystery man lived. ■ Douglas Owsley

ASTRONOMY

Astronomers discovered three cometlike objects orbiting within the main asteroid belt, according to a report published in March 2006. The main asteroid belt, also known as the Main Belt, is a region lying between the orbits of Mars and Jupiter that contains most of the asteroids in the solar system. The discovery challenged the long-held belief that all comets originate in the distant reaches of the solar system.

Astronomers Henry H. Hsieh and David Jewitt of the University of Hawaii in Honolulu used telescopes in Chile, Hawaii, and Taiwan to study about 300 objects in the Main Belt. They found three objects, with diameters of from 2 to 5 kilometers (1 to 3 miles), moving in oval-shaped orbits located at a distance of about 3 astronomical units from the sun. (An astronomical unit equals about 150 million kilometers

SEE ALSO

THE SPECIAL REPORT, **EXPLORING THE SUBURBAN SOLAR SYSTEM,** PAGE 26.

THE SPECIAL REPORT, **THE EARLY EARTH,** PAGE 48.

[93 million miles], the average distance of Earth from the sun.) Hsieh and Jewitt found that these objects, like comets, *emitted* (gave off) dust when they passed through the portions of their orbits closest to the sun. This behavior led the astronomers to classify the three objects as comets in the Main Belt.

BLUE LAGOON ON THE RED PLANET

A "lake" of water ice lies in a Martian crater near the Red Planet's north pole, images from the Mars Express spacecraft revealed in July 2005. The ice, which is about 12 kilometers (7.5 miles) in diameter, rises about 200 meters (660 feet) above the floor of the unnamed crater. Mars Express photographed the ice during late summer in the northern hemisphere, which indicated that the lake is frozen year-round.

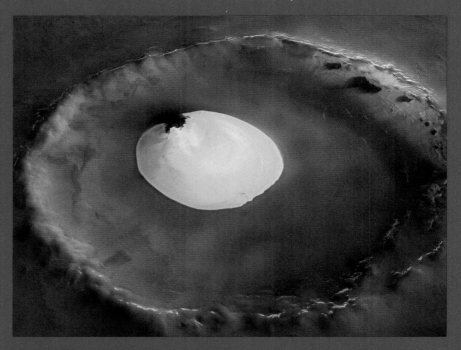

ASTRONOMY continued

Based on the amount of dust released by the cometlike asteroids and the speed at which the dust moved away from them, the astronomers concluded that the driving force behind the dust flow was *sublimation* of water ice. (Sublimation is the process by which a solid substance changes into a gas without first becoming a liquid.) As the three small bodies approach the sun, their surface temperature rises, and the expanding water vapor drives the dust from the surface. This process gives comets their cloudy atmosphere, known as a *coma*, and their spectacular dust tail.

The three asteroids were the first cometlike objects to be found outside the two known reservoirs of comets in the solar system—the Kuiper belt and the Oort cloud. The Kuiper belt is a band of objects made of ice and rock that circle the sun beyond Neptune's orbit. The Oort cloud is a cluster of comets and other objects in the outermost region of the solar system. A comet can leave the Oort cloud and enter the inner solar system when disturbed by a large gravitational force, such as the gravity of a passing star.

Hsieh and Jewitt considered the possibility that the cometlike asteroids had come from the Oort cloud or the Kuiper belt. However, computer simulations showed that it was unlikely that objects originally from these regions could end up in stable orbits within the Main Belt.

The astronomers concluded that the existence of icy, cometlike objects in the asteroid belt indicates that early in the history of the solar system, the Main Belt was cold enough for water to condense on some asteroids. Some scientists said that Hsieh and Jewitt's

discovery implied that large amounts of water were present in the Main Belt during the formation of the planets and that icy bodies in the belt could have been the source of Earth's water.

The discovery also provided the first direct evidence that asteroids contain water ice. Scientists had previously found evidence of water in some *meteorites*, rocks that have fallen onto Earth from space. Some meteorites are thought to be chunks of debris broken off Main Belt asteroids by collisions. Scientists had believed that the evidence of water in the meteorites suggested that asteroids also had water ice. However, astronomers had previously found only weak evidence for water ice—such as chemical signatures of water-bearing minerals—on the surfaces of some Main Belt asteroids.

Water geysers on Enceladus. Saturn's tiny moon Enceladus is erupting water vapor from what may be underground reservoirs of liquid water, scientists working on the National Aeronautics and Space Administration (NASA) Cassini mission reported in March 2006. The Cassini spacecraft, which arrived at Saturn in 2004, discovered the geysers.

In 2005, Cassini made three flybys of Enceladus—in February, March, and July. During

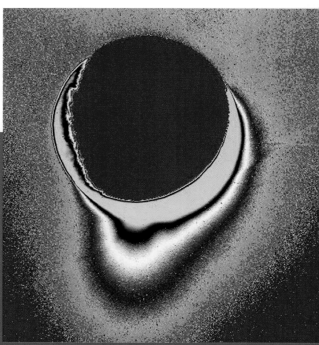

ENCELADUS ERUPTS

Sprays of icy material about 500 kilometers (310 miles) high shoot off the southern half of Enceladus, a tiny moon of Saturn, in an image (right) taken by the Cassini spacecraft in July 2005. The plumes were color-coded to enhance the contours in the spray. Scientists from the National Aeronautics and Space Administration (NASA) reported in March 2006 that the plumes may come from stores of pressurized liquid water that erupt through cracks (far right) in the surface of the moon's south pole.

Cassini's first two passes by the moon, on-board cameras took dramatic images of a series of *fractures* (breaks in the surface) near Enceladus's south pole. On the second flyby, Cassini's instruments found a strong flow of *ions* (charged particles) coming from the pole. During the third pass, the craft flew just 170 kilometers (105 miles) above the moon's surface and passed directly through a plume of gas emerging from the south pole.

Cassini examined Enceladus and its geysers using *spectrometers* (devices that break up light, heat, and other types of electromagnetic waves into a spectrum and display it for study). One spectrometer, which detects heat, found that the temperature in the fractures was –183 °C (–297 °F), warmer than their –203 °C (–333 °F) surroundings. A second instrument mapped the surface and found *organic* (carbon-containing) material in the cracks as well as water ice, which covers the rest of Enceladus's surface. A third spectrometer identified an atmosphere caused by plume gases spreading around the moon.

The spacecraft's instruments found that the plume caused fluctuations in Saturn's magnetic field and in the stream of electrically charged particles that orbits in the space around Saturn. These measurements extended over all three flybys and showed that the plume lasted for several months, though its intensity fluctuated.

Cassini also collected and analyzed the plume using a *mass spectrometer*, an instrument that can distinguish between different types of atoms and molecules. The mass spectrometer found that the plume consists of at least 90 percent water with small amounts of nitrogen, carbon dioxide, methane, and other *hydrocarbon gases* (gases made of carbon and hydrogen).

Based on the warmer temperatures at the fractures and the composition of the plume, scientists concluded that Enceladus must have a warm interior that is powering its geological activity. Computer models of the formation of the plume suggest that temperatures in the deep interior are high enough for liquid water to exist. Some scientists believe that liquid water could lie only tens of meters beneath the moon's surface.

Scientists in 2006 thought that Enceladus's interior heat could be generated by two separate mechanisms. The first mechanism is radioactive heating, the means by which Earth generates its internal heat. Cassini measurements indicate that Enceladus has a rocky interior made up of such elements as silicon, iron, and magnesium as well as some radioactive elements. Radioactive elements generate heat as they *decay* (break down). However, smaller bodies such as Enceladus tend to lose heat more quickly than larger bodies such as Earth. The ratio of their surface area—which determines a body's rate of cooling—to their interior volume—where heat is stored—declines with size. Therefore, some scientists doubt that all of Enceladus's heat is generated by radioactivity.

Tidal forces—physical strains caused by the uneven pull of gravity near a massive object—are another mechanism that could heat Enceladus. As Enceladus travels around Saturn in an oval-shaped orbit, the moon feels a varying force of gravity from the giant planet. This variation causes a periodic tugging and stretching of the moon, which generates frictional heating. Some scientists believe that the combined heating from tidal forces and radioactivity may give rise to the geologic activity seen on Enceladus.

Cassini's discovery of organic molecules and tantalizing hints of liquid water on Enceladus led many astron-

ASTRONOMY continued

omers to wonder if the moon might harbor life. However, Cassini is not equipped to detect the large molecules essential for life as we know it on Earth. Instead, astronomers sought to use Cassini to make more detailed observations to verify whether liquid water lies beneath the surface. Scientists called for a much closer flyby, perhaps within 50 kilometers (30 miles) of the surface of Enceladus. Further study of the data from the July 2005 flyby will allow NASA to decide whether to pursue this potentially exciting visit to Enceladus.

Bigger than Pluto. The object known as 2003 UB313 is larger than the planet Pluto, astronomers reported in February 2006. Scientists discovered 2003 UB313, informally known as Xena, orbiting the sun at about twice the distance of Pluto in July 2005, but its size was not measured until 2006.

German astronomer Frank Bertoldi and colleagues at the University of Bonn used the IRAM radio telescope in Granada, Spain, to measure the amount of heat radiated by 2003 UB313. Larger objects emit more heat than smaller objects of the same temperature. Based on their measurement, the astronomers determined that 2003 UB313 had a diameter of from 2,600 to 3,400 kilometers (1,620 to 2,110 miles). Even if the object had the lowest diameter in this range, it would still be larger than Pluto, which has a diameter of about 2,300 kilometers (1,430 miles).

In April 2006, the astronomers who discovered 2003 UB313—Michael Brown of the California Institute of Technology in Pasadena; Chad Trujillo of Yale University in New Haven, Connecticut; and David Rabinowitz of the Gemini Observatory in Hilo, Hawaii—used the Hubble Space Telescope to make a more accurate, visible-light measurement of Xena. The scientists found that Xena has a diameter of from 2,300 to 2,500 kilometers (1,430 to 1,550 miles), and so is about 5-percent larger than Pluto.

Making massive baby stars. In September 2005, two teams of astronomers reported the first evidence that massive stars, containing at least several times the mass of the sun, can form by *accretion*. Accretion is the process by which tiny particles of matter accumulate to form a larger object. Scientists had previously observed that lightweight stars, such as the sun, form by accretion, but theories indicated that massive stars could not form by this process. Massive stars are responsible for the creation of

elements heavier than carbon, including most of the materials that make up Earth. In addition, they end their lives as exploding stars called *supernovae*. Astronomers have long sought to understand how these stars are made.

The two groups of astronomers, led by Nimesh Patel of the Harvard-Smithsonian Center for Astrophysics in Cambridge, Massachusetts, and Zhibo Jiang of the Purple Mountain Observatory in Nanjing, China, reported that they had found evidence of accretion disks around massive young stars. When stars form by accretion, the gathering matter swirls around the central *protostar* (star in the process of formation) in the form of a flat disk.

Both teams located the disks by observing the sky at invisible wavelengths of the electromagnetic spectrum. Jiang's team used *infrared* (heat) radiation, which can travel through the dust and gas that usually surrounds young stars. The astronomers used a technique called *polarimetry*—the process of analyzing the *polarization* (direction of vibration) of light—to identify a disk around the Becklin-Neugebauer object, a young star in the constellation Orion, about 1,500 light-years from Earth. (A light-year is the distance light travels in one year, about 9.5 trillion kilometers [5.9 trillion miles].) Scientists estimated that the mass of the Becklin-Neugebauer object is at least seven times as great as the mass of the sun.

Patel's team discovered a disk surrounding a massive star in Cepheus-A, a "nursery" for young, heavy stars located about 2,400 light-years from Earth. The astronomers detected the disk with *submillimeter radiation*, ultrashort radio waves that can reveal structures hidden by gas and dust. They observed the shape of a disk centered on a protostar called HH2, which contains about 15 times as much mass as the sun.

The discoveries by Patel and Jiang's teams challenged established theories about the formation of massive stars. Scientists had believed that disks could not form around massive stars because the intense *radiation pressure* from the large protostars would disrupt the disks. (Radiation pressure is the physical force exerted by light on matter.) Instead, they had accepted the only known alternative to the accretion theory—that massive stars form from the merger of several smaller stars.

The new finding about the accretion of massive stars raised new questions. For example, why radiation pressure was not destroying the stars' surrounding disks.

Mystery of gamma ray bursts solved. Astronomers in October 2005 discovered the origin of the briefest, most intense bursts of energy in the universe. In that month, four teams of astronomers published papers in the journal *Nature* explaining the cause of three massive explosions known as short-hard *gamma ray bursts* (GRB's). GRB's are bright flashes of energy, most of which are emitted as *gamma rays*—electromagnetic radiation that has higher energies and shorter wavelengths than X rays do. GRB's were discovered in the 1960's and have since been observed daily, mainly in distant galaxies.

The bursts come in two distinct types: long-soft GRB's and short-hard GRB's. Astronomers had suspected that the two varieties were produced by different types of celestial events. Long-soft GRB's are long-duration bursts that last for at least 2 seconds. They release more total energy than soft-hard bursts, but each of their gamma rays typically has less energy and so is "softer" than those released by the short-hard bursts. Short-hard GRB's last less than 2 seconds—typically tens of *milliseconds* (millionths of a second)—making them difficult to locate and study. They release most of their energy in extremely powerful "hard" gamma rays. Both types of GRB's have *afterglows* (emissions of radiation from radio waves to X rays that last from days to years after an explosion).

The four teams of astronomers detected the short-hard bursts with two new satellites built specifically to observe GRB's. Gamma ray telescopes must orbit in space because Earth's atmosphere prevents the radiation from reaching the ground. The Swift satellite, a NASA mission, was launched in 2004. The High Energy Transient Explorer 2 (HETE-2) spacecraft, a collaboration between France, Italy, Japan, and the United States that is led by the Massachusetts Institute of Technology (MIT) in Cambridge, was launched in 2000. Both satellites were designed to detect GRB's and compute their locations on the sky. The satellites then notify Earth-based telescopes of the event. Within seconds, these telescopes, and the satellites themselves, can begin monitoring the afterglow of the explosion.

Three teams, led by Derek Fox of the California Institute of Technology in Pasadena, Jens Hjorth of the University of Copenhagen in Denmark, and Joel Villasenor of MIT, reported on a short-hard burst known as GRB 050709, which was detected by the HETE-2 satellite on July 9, 2005. Fox's team also studied GRB 050724, which the Swift satellite found on July 24, 2005. The teams used Swift and HETE-2 instruments together with space- and Earth-based telescopes to study the afterglow of the events in radio, optical, infrared, and X-ray wavelengths. The fourth team—led by Neil Gehrels of the NASA Goddard Space Flight Center in Greenbelt, Maryland—studied an event called GRB 050509b, which Swift found on May 9, 2005. Gehrels and his colleagues used the Chandra X-ray Observatory to monitor the X-ray afterglow of that event.

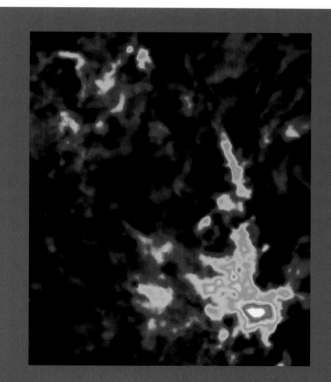

A STELLAR NURSERY

Part of a region of intense star formation called W51 is revealed in the clearest map ever produced of gas clouds that act as stellar birthing grounds in the Milky Way Galaxy. In March 2006, astronomers at Boston University in Massachusetts reported that they had used a radio telescope to image the clouds in fine detail. The scientists estimated that W51 would produce from 100 to 1,000 stars.

ASTRONOMY continued

The teams found that short-hard GRB's and long-soft GRB's are not caused by the same events. Scientists have linked long-soft GRB's with the collapse of massive stars at the end of their lives. The astronomers noted that the short-hard GRB's occurred in galaxies or regions of galaxies that were not forming new stars. In addition, the short-hard GRB's were not accompanied by supernovae.

All four of the teams concluded that the three short-hard GRB's most likely resulted from the collision of two compact stellar remnants, such as a neutron star and a *black hole* or a pair of *neutron stars*. A neutron star is the collapsed core of a massive star that has run out of nuclear fuel. Neutron stars measure only about 20 kilometers (12 miles) across, but they have a mass between 1.4 times and 3 times that of the sun.

THE FIRST STARLIGHT

Light from what may be the first stars to form in the universe (below) is unveiled in an image produced by NASA's Spitzer Space Telescope. NASA astronomers created the image by taking a picture of stars and galaxies in the constellation Draco (bottom). The image was captured as *infrared light* (heat radiation), wavelengths too long to be seen by the human eye. The scientists subtracted the light from the known stars and galaxies to reveal a background signal that they believe is the glow from stars that formed only hundreds of million of years after the beginning of the universe.

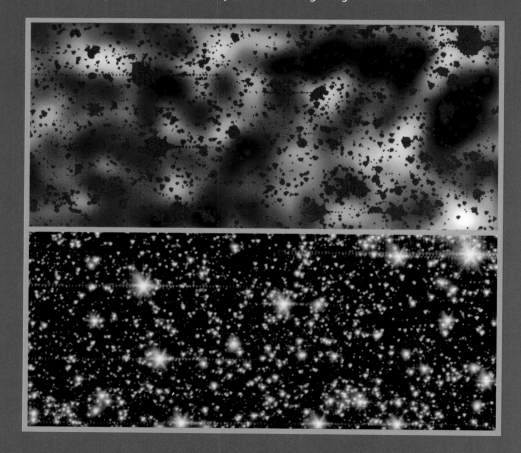

A black hole is a region of space whose gravitational force is so strong that nothing can escape from it. It can form when a massive star runs out of nuclear fuel and is crushed by its own gravitational force. When two of these objects collide, they form a single black hole. This process generates an enormous amount of energy, some of which is released as gamma rays.

Scientists believe that the long-soft GRB's occur when massive stars go directly to the black-hole stage without first becoming neutron stars. Some scientists noted that this explanation is consistent with the fact that long-soft bursts release more energy than short-hard bursts, because the long-soft bursts represent collapses that occur all at once, rather than in stages.

The first stars were born about 400 million years after the *big bang*, about 200 million years later than scientists had thought, astronomers announced in March 2006. The big bang was the cosmic explosion that scientists believe started the expansion of the universe.

A team of researchers led by physicist and astronomer Charles Bennett of Johns Hopkins University in Baltimore made the discovery using NASA's Wilkinson Microwave Anisotropy Probe (WMAP). Since its launch in 2001, the WMAP satellite has studied the cosmic microwave background (CMB) radiation, energy that formed in the heat of the early universe and cooled as the universe expanded. This energy now takes the form of electromagnetic waves called *microwaves*.

Scientists believe that the CMB radiation contains a record of the universe's first stars. During the period between the big bang and the formation of the first stars, the universe was filled with cold gas and dust. Over time, this material came together by the force of gravity until it formed huge stars with at least 100 times as much mass as the sun. The powerful light from these stars ripped electrons from the atoms of the remaining gas molecules, a process called *ionization*. The ancient light that became the CMB radiation then bounced off the free electrons, causing the radiation to become *polarized* (composed of waves that vibrate in a particular direction).

The WMAP team analyzed the polarization of the CMB radiation to determine when the first stars began to ionize matter in the early universe. They calculated an age of 400 million years after the big bang. Previous WMAP data released in 2003 had indicated that the first stars ignited about 200 million years after the big bang. However, many scientists contended that matter could not have gathered and formed stars within only 200 million years.

■ Jonathan Lunine and Theodore P. Snow
See also **PHYSICS; SPACE EXPLORATION.**

ATMOSPHERIC SCIENCE

The Great Plains of the central United States suffered from a persistent and wide-ranging drought as well as a continued warming trend in 2006. The warm, dry conditions, coupled with dry vegetation and strong winds, produced ideal conditions for wildfires that devastated parts of the region.

The drought affected a region extending from southern Texas to the Canadian border and from the eastern edge of the Rocky Mountains through Indiana. The Great Plains region normally experiences drought conditions about every 20 years, with dry periods lasting from 1 year to more than a decade.

The recent drought, which began in 1999, could be one of the most severe to hit the United States in the past 40 to 100 years, according to testimony given before the U.S. Senate in April 2006 by oceanographer Chet Koblinsky, the director of the National Climatic Data Center (NCDC) in Asheville, North Carolina, a part of the National Oceanic and Atmospheric Administration (NOAA). Koblinsky said that the current drought was, in some areas, comparable to the extreme droughts of the 1950's and the Dust Bowl years of the 1930's, when a series of destructive wind and dust storms struck the United States.

During droughts, the soil may lose nearly all moisture to a depth of more than 2 meters (6 ½ feet). As a result, the soil becomes loose and breakable on the surface, making it easy for even moderate winds to blow away the soil in clouds of dust.

In 2006, the United States also experienced its warmest January ever, according to a report issued by the NCDC. The United States had an average temperature of 4.2 °C (39 °F), which

ATMOSPHERIC SCIENCE continued

was 4.7 Celsius degrees (8.5 Fahrenheit degrees) above the average January temperature from 1895 to 2005.

In May 2006, the NCDC reported that the continental United States also had its warmest April on record. According to the report, none of the 48 contiguous states had cooler-than-average temperatures during April.

Higher-than-normal temperatures, a lack of precipitation, dry vegetation, and high winds promoted rapidly spreading wildfires in 2006. NOAA and the National Interagency Fire Center in Boise, Idaho, counted at least 13,000 wildfires from January 1 to March 13, more than twice the number of fires that have typically occurred during that period since 2001. By mid-2006, fires had scorched more than 375,000 hectares (930,000 acres) in the United States, mainly in Texas and Oklahoma, about 10 times the area that usually burns during that period.

Ozone recovering. The decline in the *ozone layer*, a layer of the atmosphere that protects life on Earth's surface from harmful ultraviolet radiation, has stopped, a team of climate researchers reported in August 2005. Since the late 1970's, scientists had observed a decrease in the concentration of ozone, a form of oxygen, in the layer—which is located from about 10 to 50

kilometers (6 to 30 miles) above Earth's surface. The main cause of ozone depletion is industrially produced chemicals called *chlorofluorocarbons* (CFC's), which were once widely used as refrigerants and as propellants in aerosol spray cans.

Climate scientists from NOAA, the Center for Integrating Statistical and Environmental Science at the University of Chicago, and three other U.S. universities used satellite data and surface measurements to examine global ozone concentrations. Their study was the first to focus on the *total ozone column layer*—that is, the amount of ozone in a column extending from Earth's surface to the edge of the atmosphere. Previous studies had focused on the topmost layers of the atmosphere.

The scientists found that the decline in the total ozone column layer had leveled off between 1996 and 2002. In addition, over some areas, the ozone layer had increased by a small amount, though concentrations were still well below levels measured in the 1970's. The scientists also found that ozone levels in polar regions were still below historical levels and that these areas might not return to normal levels for several decades.

Some scientists attributed the recovery of the ozone layer to the 1987 Montreal Protocol,

ICE HITS NEW LOW

The area of the Arctic Ocean covered by sea ice in September, when ice coverage reaches its annual minimum, in 2005 shrank to its lowest extent on record (right), according to a satellite image taken by the National Aeronautics and Space Administration (NASA). According to NASA scientists, the minimum Arctic sea ice coverage has been shrinking by about 10 percent each decade since 1979 (far right). The researchers attributed the decline to steadily increasing temperatures in the region from 1979 to 2005.

2005

an agreement between more than 180 nations that limited the production of CFC's. Other researchers cautioned that CFC's can take years to travel up to the ozone layer and may remain there for decades. As a result, they said, CFC's released in the past will continue to deplete the ozone layer for some time. Atmospheric ozone concentrations may not rebound for up to 40 years, the scientists said.

Greener, warmer Arctic. Higher temperatures in the Arctic and sub-Arctic due to *global warming* are transforming what was snowy terrain into shrub-covered land, climate researchers reported in September 2005. (Global warming is an increase in the average temperature of the Earth's surface that has occurred since the late 1800's.) The loss of snow is also accelerating the warming in these regions, which have experienced the greatest temperature increases from global warming.

Climate scientists from the U.S. Army Cold Regions Research and Engineering Laboratory in Hanover, New Hampshire, and Colorado State University in Fort Collins examined shrub and snow areas in western Alaska during the winters of 2000 to 2002. Previous studies had found that higher temperatures in the region had promoted shrub growth on the *tundra*—the cold, dry, treeless lands of the Arctic. The scientists observed that, in areas where shrubs were exposed, the dark vegetation reduced the amount

of sunlight reflected by the Earth's surface. As a result, more sunlight was absorbed by the ground. The scientists said that this increased absorption could accelerate local warming by up to 70 percent. They observed that from 2000 to 2002, spring melting began several weeks earlier in these areas than it did in snow-covered areas.

The team warned of a *feedback loop*, in which shrub growth causes increased warming, which, in turn, leads to more shrub growth. The scientists said that shrubs could take over more of the world's 4 million square kilometers (1.5 million square miles) of tundra, increasing regional temperatures by several degrees over the next few decades.

The scientists also cautioned that the feedback loop could lead to the melting of the sub-Arctic *permafrost*, a layer of permanently frozen soil. Permafrost traps carbon stored in dead plants within the soil. However, if thawed, the plants could decay, releasing their carbon into the atmosphere as carbon dioxide and methane—two *greenhouse gases* (heat-trapping atmospheric gases) that could further accelerate global warming.

The team also pointed to satellite data indicating greater plant growth across much of the Arctic. They warned that continued warming could lead to the growth of thicker types of brush that extend above the snow, changing

1979

ATMOSPHERIC SCIENCE continued

the white winter landscape of some Arctic areas into dark shrublands.

Hurricanes intensify. The number of severe hurricanes occurring worldwide each year has nearly doubled since the 1970's, a group of atmospheric researchers reported in September 2005. The trend had been accompanied by a rise in surface temperatures of the sea, the study found. Some scientists in 2005 contended that global warming had caused the increase in ocean temperatures and, in turn, the increase in the frequency of intense hurricanes. Other climate researchers argued that the increase was unrelated to global warming.

Atmospheric scientists Peter Webster, Judith Curry, and Hai-Ru Chang of the Georgia Institute of Technology in Atlanta and Greg Holland of the National Center for Atmospheric Research, based in Boulder, Colorado, studied worldwide hurricane data from 1970 to 2004. The researchers examined the number, duration, and intensity of hurricanes over that time. They found that, though the total number of hurricanes has decreased since the 1990's, the number of hurricanes reaching Category 4 or Category 5 status on the Saffir-Simpson Hurricane Scale has almost doubled since the 1970's. The scale ranks hurricanes in five categories, ranging from Category 1—the weakest—to Category 5—which can cause massive destruction. The 1970's experienced an average of 10 Category 4 and 5 hurricanes yearly. Since 1990, that average has risen to 18 hurricanes.

The scientists reported that since the 1970's, hurricane patterns have changed dramatically in the North Atlantic Ocean. This was the only region worldwide to experience more hurricanes overall. The North Atlantic has had an average of 8 to 9 hurricanes per year since 1995 but only 6 to 7 per year before then. In addition, the number of severe hurricanes increased. From 1975 to 1989, the North Atlantic saw a total of 16 Category 4 and 5 storms. From 1990 to 2004, the region was hit by 25 such storms.

The study by Webster and his team agreed with a report released in July 2005 by atmospheric scientist Kerry Emanuel of the Massachusetts Institute of Technology in Cambridge. Emanuel studied hurricanes in the North Atlantic and North Pacific oceans. He found that they had increased in average duration and intensity since the 1970's.

Webster and his colleagues said that increases in hurricane activity occurred while sea surface temperatures rose around the globe. Webster noted that temperature increases since the 1970's varied from 0.3 Celsius degree (0.5 Fahrenheit degree) to 0.6 Celsius degree (1.0 Fahrenheit degree), depending on the region.

In 2006, Webster and other climate researchers were working to understand how hurricanes affect Earth's climate. Webster noted that hurricanes evaporate surface waters, carrying tropical heat farther north or south. However, he said, climate scientists do not yet understand how hurricanes can evaporate water when winds are blowing at least 160 kilometers (100 miles) per hour, as they can during hurricanes. Webster acknowledged that his work supported a theory linking rising ocean temperatures to hurricane intensity. However, he called for further studies to answer the question of whether observed increases in the number of severe hurricanes are related to global warming.

Other scientists in 2005 disputed the idea that global warming was affecting hurricanes. Meteorologist William Gray of Colorado State University in Fort Collins said in September that the increase in hurricane intensity is due to natural, long-term cycles of water temperatures related to the speed of ocean currents. In November, NOAA issued a statement saying that its scientists had reached a consensus that global warming due to human activities was not responsible for an increase in either the number or severity of hurricanes.

In February 2006, researchers from both sides of the debate worked together to write a statement for release by the Tropical Meteorology Research Program of the World Meteorological Organization's Commission on Atmospheric Sciences in Geneva, Switzerland. The authors wrote: "The research issues discussed here are in a fluid state and are the subject of much current investigation. Given time the problem of causes and attribution of the events of 2004-2005 will be discussed and argued in the refereed scientific literature. Prior to this happening it is not possible to make any authoritative comment." Although the statement helped focus the discussion on the scientific questions to be addressed, the debate continued throughout 2005 and 2006 in the public media. ■ John Snow

See also AGRICULTURE; OCEANOGRAPHY.

Tracking Hurricanes and Improving Forecasts

Fifteen hurricanes tore through the Atlantic Basin during the 2005 Atlantic hurricane season, the most active season since record keeping began in 1851. The Atlantic Basin includes the Atlantic Ocean, the Caribbean Sea, and the Gulf of Mexico. The Atlantic hurricane season runs from June 1 to November 30. The 2005 hurricane count smashed the previous record of 12 hurricanes in one season, set in 1969.

The 2005 season saw a total of 27 *tropical storms* sweep through the Atlantic Basin, breaking the old record of 21 in 1933. Tropical storms are born when the winds in violent, swirling squalls called *tropical cyclones* race past speeds of 63 kilometers (39 miles) per hour. The new tropical storm then receives a name from a list managed by the World Meteorological Organization (WMO), a United Nations agency based in Geneva, Switzerland. Hurricanes are tropical cyclones whose winds have passed speeds of 119 kilometers (74 miles) per hour.

During the 2005 season, seven hurricanes became major hurricanes—Category 3 or higher on the Saffir-Simpson Hurricane Scale. The scale ranks hurricanes in five categories based on their *intensity* (maximum sustained wind speeds). Category 1 hurricanes, the weakest storms, have speeds from 119 kilometers (74 miles) per hour to 153 kilometers (95 miles) per hour. The most devastating hurricanes, Category 5 storms, have wind speeds of at least 251 kilometers (156 miles) per hour. Four hurricanes in 2005—Emily, Katrina, Rita, and Wilma—hit Category 5 status at some point. Based on estimates of historical storm strengths, scientists believe that 2005 may be the first year on record to produce four Category 5 storms.

Seven hurricanes, including four major hurricanes, made landfall in the United States in 2005, causing widespread death and loss of property from Florida to Texas. Hurricane Katrina, which struck the U.S. Gulf Coast in late August, devastated New Orleans, Louisiana. While the city escaped the full force of the hurricane's winds, a *storm surge* (rapid rise in sea level produced when winds drive ocean waters ashore) overwhelmed the city's flood defenses. Floodwaters covered about 80 percent of the city. Most of the more than 1,500 people who died because of the storm lived in New Orleans. Hurricane Katrina caused tens of billions of dollars in damage.

In September, the approach of Hurricane Rita triggered one of the largest evacuations in U.S. history. As the storm neared Houston, Texas, government officials warned at least 2 million Louisiana and Texas residents to evacuate the coast. Although the hurricane spared Houston, Rita's storm surge rushed over low-lying coastal communities in southwestern Louisiana and southeastern Texas. Rita caused about $10 billion in damage to the United States.

In April 2006, in recognition of the enormous loss of life and property caused by five hurricanes during the 2005 season, the WMO withdrew the names Dennis, Katrina, Rita, Stan, and Wilma from future use. More names were withdrawn in 2005 than in any other season.

Hurricanes occur in the Atlantic Basin and in the Northeast Pacific Ocean. Such storms are called *typhoons* if they occur in the Northwest Pacific Ocean, west of an imaginary line called the International Date Line. Near Australia and in the Indian Ocean, they are identified by the generic term tropical cyclone. A hurricane begins as a *tropical depression* (low-pressure area surrounded by winds that have begun to blow in a circular pattern), the weakest form of tropical cyclone, over warm ocean waters that intensifies by drawing up heat and moisture from the warm sea. Light winds in the upper atmosphere allow the storm to remain intact and grow.

Tropical cyclones are characterized by a central region of thunderstorms. The storms grow progressively stronger and more organized as pressure falls in the central region, which increases the speeds of the winds being sucked into the center. Hurricane winds spiral around a well-defined central "eye," while thunderstorms occur along inward-spiraling "rainbands."

A hurricane contains violent winds, torrential rains, and enormous waves. When a hurricane moves onto land, its winds can uproot trees and tear the roofs off houses. Debris can fly through the air at high speeds, causing injury or death. Heavy rains may cause flooding and mudslides. A hurricane produces a region of raised sea surface covered with enormous waves. As this region moves ashore, it produces a storm surge, driving ocean water far inland. Storm surges are one of the most deadly and destructive features of a hurricane because many coastal areas are densely populated and lie only a few meters above sea level.

Because of the threat posed by hurricanes to major population centers, the U.S. government, through the National Oceanic and Atmospheric Administration (NOAA), has developed tools and techniques for monitoring and forecasting hurricanes. Satellites, weather balloons, and radar monitor the atmosphere and the ocean. Such devices provide the first warning that a tropical cyclone is forming. Once a cyclone is detected, piloted "hurricane hunter" aircraft fly into the storm. Sensors on board and dropped from the planes record wind speeds, air pressure, and other measurements of the interior of the evolving storm.

Since the 1970's, forecasters have improved the accuracy of *track forecasts*, predictions of the course of a hurricane. Large-scale wind patterns determine a storm's course, and these data can be acquired easily by satellites and aircraft. Computer programs then use these observations to estimate the storm's future course. Improvements in computing technology have improved track forecasts.

Forecasts of storm intensity have not improved significantly since the 1970's, however. A hurricane's intensity is controlled by events happening in the core of the storm and in the local environment.

A RECORD SEASON

Five of the record 15 hurricanes that formed in the Atlantic Ocean during 2005 sweep through the Gulf of Mexico and the Caribbean Sea in a montage of images taken by the U.S. Geostationary Environmental Satellite system. Emily, Katrina, Rita, and Wilma reached Category 5 status—the most destructive level on the Saffir-Simpson Hurricane Scale.

Forecasting intensity is difficult, in part, because the core is so turbulent and hard to observe. In addition, unexpected changes in storm intensity may occur because a storm's intensity depends on humidity, energy transfer rate between the ocean and the air, and many other factors. Scientists in 2006 began to use unpiloted planes to gather detailed information about a storm to aid in improving intensity forecasts.

Hurricane forecasts helped to reduce the destruction from the season's most devastating storm, Hurricane Katrina. Using weather satellites, scientists watched the storm begin as a tropical depression on Aug. 23, 2005, about 324 kilometers (201 miles) southeast of Nassau, Bahamas. Based on wind measurements made by aircraft, the system reached tropical storm levels—and gained the name Katrina—on August 24. On the night of August 25, Katrina struck southern Florida as a Category 1 hurricane. The storm crossed the state from east to west, blowing down trees, damaging homes, and dumping at least 35 centimeters (14 inches) of rain in some places. In the morning, Katrina moved out over the Gulf of Mexico.

On the evening of August 26, the National Hurricane Center (NHC), based in Miami, Florida, used a three-day forecast of Katrina's course to predict a serious threat to the Alabama, southeastern Louisiana, and Mississippi coasts. On August 27, Max Mayfield, director of the NHC, called New Orleans Mayor Ray Nagin to warn him that Katrina was headed toward his city and could pose a major disaster. On August 28, a day before the storm hit, Nagin called for a mandatory evacuation of New Orleans.

Hurricane Katrina reached Category 5 status by noon on August 28. The storm hit its peak intensity, with wind speeds of about 280 kilometers (175 miles) per hour when it was about 315 kilometers (195 miles) southeast of the mouth of the Mississippi River. Katrina then turned north, heading toward New Orleans. The storm weakened slightly as it approached the coast.

On the morning of August 29, Katrina made a second landfall as a Category 3 hurricane with wind speeds of about 205 kilometers (125 miles) per hour near Buras, Louisiana, about 90 kilometers (55 miles) southeast of New Orleans. As the hurricane moved inland, its storm surge flowed into the channels and canals that surround and cut through New Orleans. The rising waters overtopped and breached the levee system built to protect the city, which lies about 2 meters (6 ½ feet) below sea level. Water from Lake Pontchartrain surged into the shallow bowl in which New Orleans sits, filling it nearly to the brim.

Hurricane Katrina killed at least 1,500 people and left hundreds of thousands of people homeless. The recovery from Katrina will take years, and its scars on New Orleans are likely to be permanent. Hurricane forecasters helped to prevent much damage and loss of life from the storm, but scientists hoped to improve hurricane forecasting to give people even more information about future hurricane threats. ■ John Snow

KATRINA UNVEILED

Hurricane Katrina roars toward Louisiana and Texas in an image taken by the Tropical Rainfall Measuring Mission (TRMM) satellite, a joint project of U.S. and Japanese scientists. TRMM can peer beneath the clouds to measure a storm's rainfall. Red regions represent areas of the highest rainfall—at least 5 centimeters (2 inches) of rain per hour.

BIOLOGY

Plants growing in normal conditions with abundant oxygen seem to produce a surprising amount of the methane gas entering the atmosphere. That discovery was reported in a January 2006 issue of the science journal *Nature* by geochemist Frank Keppler and his colleagues at the Max Planck Institute for Nuclear Physics in Heidelberg, Germany. The findings challenged the long-held belief that plants living in places with little or no oxygen accounted for most global emissions of methane, the second most important *greenhouse gas* (gas that traps heat within Earth's atmosphere). Many scientists believe that methane emissions are contributing to *global warming,* a gradual increase in the average temperature of Earth's surface that has been occurring since the 1800's.

Methane, a colorless, odorless compound that makes up a large part of natural gas, enters the atmosphere from both human-related and natural sources. According to the United States Environmental Protection Agency (EPA), landfills, natural-gas and petroleum production, and the digestive processes of cattle and other animals raised on livestock farms account for most human-related global methane emissions. Bogs and wetlands have widely been considered the chief sources of global methane emissions from natural sources. Scientists have believed that methane, also called *marsh gas,* is released as microbes living in oxygen-poor environments break down dead plants.

For their study, Keppler and his colleagues first exposed plants to radiation to kill any microbes that might be present and then placed the plants in a methane-free environment. Based on the levels of methane released by these as well as nonirradiated plants, the scientists calculated that plants living in normal oxygen-rich conditions throughout the world may account for from 10 to 30 percent of all global methane emissions.

The findings led some scientists to suggest that the large plumes of methane above tropical forests observed by Earth-orbiting satellites may be released by the trees themselves. However, Keppler and his colleagues emphasized that plants are not responsible for global warming. They also stressed the preliminary nature of their findings. Methane emissions from plants existed long before methane emissions from human-related sources began to affect the com-

SEE ALSO THE SPECIAL REPORT, **THE SECRET LIVES** OF JELLYFISH, PAGE 40.

position of the atmosphere, they noted. "It is the anthropogenic [human-related] emissions which are responsible for the well-documented increasing atmospheric concentrations of methane since pre-industrial times. Emissions from plants thus contribute to the *natural* greenhouse effect and not to the recent temperature increase known as 'global warming'," they insisted.

Dolphins' name game. Bottlenose dolphins create distinctive whistles that they and their relatives may use as "names," according to a study reported in the *Proceedings of the National Academy of Sciences* in May 2006. Birds called spectacled parrotlets show evidence of using labels, but, previously, human beings were the only animals clearly known to use names.

Previous research had revealed that infant dolphins create a signature sound that they use throughout their life. For example, a dolphin that becomes separated from its group may call out its signature whistle and alert others to its whereabouts. Scientists led by marine mammal researcher Vincent Janik of the University of St. Andrews in Scotland wanted to know if dolphins recognize the signature sounds themselves or the voices making them, the way a person might recognize the voices of friends before the friends say their name.

For their study, Janik and his colleagues temporarily captured seven female and seven male dolphins from a group of well-studied dolphins living in Sarasota Bay in Florida. Because of these studies, the researchers knew which of the dolphins were related. They recorded each dolphin's signature whistle and then created electronic versions of the whistles to disguise the dolphin voice. Finally, they played these electronically altered names along with other, random sounds to the dolphins through underwater speakers.

The scientists reported that nine of the dolphins turned toward the speakers when the scientists played the name of a close relative. This behavior suggested that the dolphins recognized the names themselves and not vocal cues from the dolphins speaking the names. The researchers also reported finding evidence that two dolphins may use a signature name to refer to a third dolphin.

Ant teachers. Ants of a certain species use techniques similar to those employed by human teachers to show each other how to find a food source, according to a study published in January 2006. Biologists Nigel R. Franks and Tom Richardson of the University of Bristol in the United Kingdom reported that their research has provided the first-known demonstration of formal teaching behavior in a nonhuman animal. The ants, which belong to the species *Temnothorax albipennis,* live in cracks in rocks along the southern coast of England.

For their study, the researchers outlined behaviors that must exist before an animal can be considered a teacher. During a lesson, the teacher must change the way it normally performs a task. For example, the teacher may perform the task slowly. For true teaching, the teacher must modify the task in some way that makes the teacher less efficient at the task but that allows the student to learn faster than it would on its own. The researchers also identified as true teachers those who keep track of how students are progressing and adjust their lesson accordingly.

To test the teaching abilities of the ants, the researchers took some of the ants from their rocky homes and placed them on a tabletop with food. The researchers watched as one ant found the food and returned to the colony. It then made another trip to the food, this time with a nestmate that did not know the way. Just how the food-finder attracts the nestmate is not clear, but Franks said a potential guide might signal with gestures or chemicals or both.

Leading a nestmate met the requirement for teaching, the researchers reported. The leader travelled slowly, in some cases, taking four times as long as a leader ant who knew the way. The leader also got feedback about the follower's progress, as the follower kept tapping its an-

BETWEEN TWO ROCKS

A young gorilla named Itebero smashes palm nuts between two rocks to get palm oil, in what biologists in Congo (Kinshasa) said was the first documented use of complex tools by apes other than chimpanzees. Researchers at the Dian Fossey Gorilla Fund International sanctuary in Goma reported the finding in October 2005. According to the researchers, scientists may have underestimated gorillas' understanding of their environment. They also said that the ability to use tools may have developed among primates earlier in their evolution than scientists had believed.

BIOLOGY continued

tennae against the leader's legs and abdomen as they ran. The leader ant slowed down when it got too far ahead of the follower and sped up when the follower got too close. Sometimes the follower stopped moving and looked around—as if checking for landmarks, the scientists suggested. The leader stopped then, too. When the follower started running and tapping again, the leader ant began running as well.

Some scientists disagreed with Franks and Richardson's conclusion. These scientists argued that the ants were merely transferring information to one another, a common behavior among many types of animals.

New species, new ecosystem. The discovery of eight previously unknown animal species in a "new and unique" *ecosystem* was reported in May 2006 by researchers from Hebrew University in Mt. Scopus, Israel. (An ecosystem consists of all the living and nonliving things in a particular area and the relationships among them.) The researchers found the animals in a cave, about 100 meters (330 feet) below the surface.

All the animals are *invertebrates* (animals without backbones). Four of the animals are aquatic creatures and four are land animals. The aquatic animals are *crustaceans* (a type of animal with an exterior skeleton), two of which are saltwater species and two of which live in fresh or *brackish* (slightly salty) water. All four aquatic species live in an underground lake. Bacteria living in the cave apparently provided the main source of nourishment for all the animals.

The scientists suggested that the eight species developed millions of years ago and have remained relatively unchanged because of their isolated environment. They reported that the cave lies beneath a layer of chalk that prevents water from seeping through from the surface.

Ultrasonic frogs. For the first time, scientists have found an amphibian that creates and responds to *ultrasonic calls* (sounds too high-pitched for human beings to hear). Auditory neuroscientist Albert S. Feng of the University of Illinois at Urbana-Champaign, auditory behaviorist Peter M. Narins, and their colleagues reported

SENSITIVE NATURE OF TUSKS

The mysterious spiral ivory tusks of male narwhals have millions of tiny tunnels throughout them that might make them sensitive to the pressure, *salinity* (saltiness), and temperature of the Arctic waters in which these unusual whales live, according to a report presented in December 2005 by dentist Martin Nweeia of the Harvard School of Dental Medicine in Boston. The 3-meter- (9-foot-) long tusk is the narwhal's only tooth that emerges beyond the skin.

in March 2006 on their finding about the concave-eared torrent frog *(Amolops tormotus),* found in China. Previously, only certain types of mammals, particularly bats, dolphins, and some rodents, were known to have the ability to communicate using ultrasonic sounds.

Ultrasound has a *frequency* higher than 20,000 *hertz.* Frequency is the number of cycles of vibration per second. Higher-pitched sounds have higher frequencies than do lower-pitched sounds. One hertz is one cycle per second.

The scientists became interested in the Chinese frog because of its unusual sunken ears. Unlike most frogs, the concave-eared torrent frog has an ear canal. In 2000, a research team headed by Feng reported that male concave-eared torrent frogs produce birdlike chirping sounds.

In 2006, Feng and his colleagues reported on studies of the frogs' hearing that were made using recording systems that produce a *sonogram* (diagram showing all the low- and high-pitched sounds produced). When the scientists looked at the sonogram of the frog's chirps, they saw a strong ultrasonic component. The researchers then recorded a male's call and played it to eight other male frogs. Five of the eight frogs soon called in return, suggesting that they could hear and respond to ultrasound. The male frogs seemed to be able to hear sounds up to about 20 kilohertz, higher than the maximum frequency most people can hear.

The scientists also tested the frogs' hearing ability by playing ultrasonic chirps to one frog while measuring the electrical activity in its midbrain, which is involved in processing sounds. The level of activity there indicated that the frog heard the sound. In contrast, when the scientists played the chirps after blocking the frog's ears, its brain showed no activity. From these results, the scientists concluded that the frog was detecting the sound through its ears and not through its skull.

The frogs live in an area of rushing water and waterfalls. The scientists theorized that the frogs may have evolved the ability to hear ultrasound as it gave them an advantage in communicating above the constant noise of their environment.

Big deal over smallest fish. Scientists disagreed in 2006 about which of two candidates should hold the title of the world's smallest fish. Ralf Britz, an *ichthyologist* (scientist who studies fish) from the Natural History Museum in London and his colleagues touched off the debate in January 2006 by suggesting that *Paedocypris pro-*

ALL AGLOW

Parts of a four-o'clock flower *fluoresce* (glow) in a photograph illustrating what may be the discovery of a number of plant species that can emit fluorescent light in a range visible to the human eye. The ability of some plants to fluoresce in the ultraviolet range is well known. The four-o'clock's color display results from a pigment that appears yellow in white light, researchers from the University of Murcia in Spain said in September 2005. The color may improve the plant's ability to attract insects for pollination.

genetica, which belongs to the same group of fish as carp and lives in peat swamps in Sumatra, was the smallest fish. The scientists first described the fish in the online edition of *Proceedings of the Royal Society B.* The researchers found a 7.9-millimeter- (0.3-inch-) long adult female *P. progenetica,* which they suggested was the world's smallest fish as well as the smallest *vertebrate* (animal with a backbone). In arguing for this honor, the scientists looked only at what were clearly adult fish.

Within a few days of this announcement, another researcher challenged both of *P. progenetica's* titles. Theodore W. Pietsch of the University of Washington in Seattle pointed out that in an August 2005 issue of *Ichthyological Research,* he had described a newly discovered species of fish from the Philippines, in which a mature male

BIOLOGY continued

measured only 6.2 millimeters (0.24 inch) long. The male of this type of anglerfish, named *Photocorynus spiniceps,* attaches itself to a female's body for life. The male takes its nutrients from the female's body in return for providing sperm for reproduction. In response, Britz contended that even if the male anglerfish is the smallest adult vertebrate, the peat-swamp fish could still be the smallest free-living fish.

New honeysuckles, new fly. Brushy honeysuckle, an Asian plant that invaded the northeastern United States at least 100 years ago, has contributed to the evolution of a new kind of fly, researchers reported in July 2005. The fly belongs to a group of *Rhagoletis* flies that lay their eggs on only one type of fruit. Bruce A. McPheron, an *entomologist* (scientist who studies insects) from Pennsylvania State University in University Park, said that the fly developed from the mating of a fly that lays its eggs on blueberries with a fly that lays its eggs on snowberries. The new species prefers brushy honeysuckle.

Scientists estimate that almost half of all species of plants are *hybrids* (crosses of two types of organisms). In general, new animal species evolve when one species splits off from another.

Scientists believed that hybridizations among animals rarely start new species but just interbreed—and blend in—with the parent species. Such animals often are not as well suited to the environment as either of their parent species. The scientists theorized that the honeysuckle fly was able to survive because it had a new habitat—the brushy honeysuckle—and so did not have to compete with its parent flies.

McPheron's group began to study the honeysuckle fly after an entomologist noticed an infestation of the insects on brushy honeysuckle on the Penn State campus. He found it curious that a plant introduced to the United States relatively recently would have a specialized pest.

To determine the parentage of the new fly, McPheron and his colleagues analyzed its DNA (deoxyribonucleic acid, the material that controls heredity). The scientists then compared this DNA to that of other *Rhagoletis* flies. They found that the hybrid fly had evolved from a mating of the blueberry and snowberry flies. The scientists also determined that the cross-species breeding that produced the new fly appeared about 100 generations earlier. ■ Susan Milius

See also **CONSERVATION; ECOLOGY.**

BOOKS ABOUT SCIENCE

Here are 12 important new science books suitable for the general reader. They have been selected from books published in 2005 and 2006.

Archaeology. *1491: New Revelations of the Americas Before Columbus* by Charles C. Mann focuses on the latest archaeological research about the Americas before the arrival of Europeans. When these explorers landed in the New World, they were not stepping onto a nearly empty continent inhabited by a few primitive nomads. Cities in Mexico and Peru rivaled European capitals in population density and commercial activity. Complex so-

1491

NEW REVELATIONS OF THE AMERICAS BEFORE COLUMBUS

CHARLES C. MANN

cieties had built giant pyramids in Central America and immense ceremonial mounds in the Mississippi Valley. Some native-born peoples had developed ingenious methods of counting and writing. Others practiced advanced forms of agriculture, with elaborate irrigation and crop breeding systems. Mann helps set the record straight on the remarkable achievements of pre-Columbian peoples that conquering armies—and the ravages of time—have almost erased. (Alfred A. Knopf, 2005, 465 pp., $30)

Astronomy. *The Rock from Mars: A True Detective Story on Two Planets* by Kathy Sawyer explores the controversy over a Martian meteorite found on the Antarctic icecap in 1984. In 1996, scientists at the United States National Aeronautics and Space Administration (NASA) investigating the rock reported finding internal structures that resembled the fossils of ancient microorganisms found on Earth. Sawyer, who covered the Mars rock story for *The Washington Post,* explains the basis for these claims and why some scientists doubt that the rock contains evidence of life on Mars. (Random House, 2006, 416 pp., $25.95)

The Planets by Dava Sobel, the best-selling author of *Longitude* and *Galileo's Daughter,* presents a series of evocative essays on the nine major planets and the sun and moon. Sobel describes such events as the violent collision of a comet with the massive planet Jupiter in 1994, contrasting it with the gentle probing by the Galileo spacecraft in 1995. She re-creates the Pacific voyage of Captain James Cook in the 1700's to view the passage of Venus across the face of the sun, an event that happens only twice in a hundred years. Each essay provides information in a different compositional form, including a personal letter and a biographical account, accompanied by striking images. (Viking, 2005, 288 pp., $24.95)

Biology. *Aglow in the Dark: The Revolutionary Science of Biofluorescence* by Vincent Pieribone and David F. Gruber describes remarkable discoveries in microbiology revealed by the cool and penetrating light of bioluminescence. Biochemists Pieribone and Gruber explore the world of organisms—from flashing fireflies to luminous jellyfish—that generate their own light. They describe clams that squirt glowing blue liquid to signal distress and deep-sea anglerfish that lure their prey with luminous stalks on their noses. The authors recount how biochemists isolated these creatures' light-emitting molecules, called *luciferins,* and learned to attach them to the biochemical

molecules of other creatures. In the process, they created a vital new way for biologists to study the workings of cells. (Harvard University Press, 2006, 260 pp., $24.95)

Our Inner Ape: A Leading Primatologist

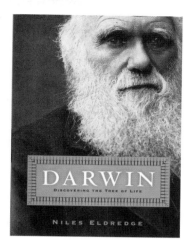

Explains Why We Are Who We Are by Frans de Waal draws on de Waal's extensive research with chimpanzees, bonobos, and other primates. De Waal, a professor in the psychology department at Emory University in Atlanta, Georgia, discusses such human issues as sex, violence, power, and conflict resolution from the perspective of related species. Although he cautions about drawing overly broad conclusions from animal behavior studies, de Waal shows that we can learn much about the human condition from other primates. We can even, perhaps, take some lessons from them on how to build a more productive and peaceful society. (Riverhead/Penguin, 2005, 288 pp., $25.95)

History of science. *Darwin: Discovering the Tree of Life* by Niles Eldredge draws on the notebooks and journals of Charles Darwin to recount how the noted British naturalist's observations of nature gradually turned him from a Biblical creationist into a pioneer of evolution. Nearly two hundred years after his birth, Darwin's work remains a focus of controversy. Scientists have long accepted his ideas about evolution by natural selection, finding them powerful tools for explaining the variety in nature. But Eldredge, a curator at the American Museum of Natural History in New York City, points out that Darwin's ideas continue to challenge traditions explaining the origin and meaning of human existence. (W.W. Norton and Company, 2005, 256 pp., $35)

Mathematics. *The Infinite Book: A Short Guide to the Boundless, Timeless, and Endless* by John D. Barrow explores some mind-boggling

BOOKS ABOUT SCIENCE continued

and paradoxical ideas about infinity. What is the biggest number? What happens when you add two infinite numbers together? What does it mean to say that there is no end to the universe? If the universe were infinite in space and time, would there be an infinite number of worlds with an infinite number of readers of this review? Some parts of the book strain the mind, but veteran science writer Barrow makes the paradoxical seem understandable. (Pantheon Books, 2005, 352 pp., $26)

Meteorology. *Sky in a Bottle* by Peter Pesic traces attempts to answer the simple question "Why is the sky blue?" from the speculations of

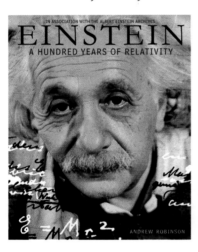

the ancient Greeks to the modern theory of quantum mechanics. Pesic, a physicist at St. John's College in Santa Fe, New Mexico, makes numerous stops along the way to discuss the views of artists, writers, and scientists as diverse as Leonardo da Vinci, Johann Wolfgang von Goethe, Vasily Kandinsky, and Albert Einstein. The book moves from Chinese myths about the sky to the chemistry of paint to the perception of color. Heavily footnoted, Pesic's work provides both a narrative of the history of science and a reference book on the optics of the atmosphere. (MIT Press, 2005, 262 pp., $24.95)

Physics. *Einstein: A Hundred Years of Relativity* by Andrew Robinson, in association with the Albert Einstein Archives, celebrates the 100th anniversary of the publication of Einstein's articles on relativity and quantum mechanics. Essays by a variety of authors reveal the many sides of the German-born physicist's genius. British physicist Stephen Hawking comments on Einstein's science, while other essayists describe such topics as Einstein's *pacifism* (antiwar stance), his passion for music, and his status as a pop-culture icon. Photographs from all stages of Einstein's life and reproductions of rare personal letters provide a record of

the man whom *Time* magazine named the "Person of the Century." (Harry N. Abrams, 2005, 256 pp., $29.95)

Warped Passages: Unraveling the Mysteries of the Universe's Hidden Dimensions by Lisa Randall traces scientists' search for an understanding of the building blocks of the universe. A highly respected particle physicist, Harvard University professor Randall explains such complex subjects as relativity, quantum mechanics, string theory, and hidden dimensions. Research conducted by Randall and her colleagues indicates that the fundamental forces and particles that make up everything are manifestations of an even more fundamental shadow world in which strange entities called D-branes determine the laws of physics. A new particle accelerator under construction in Europe, called the Large Hadron Collider, may confirm these remarkable conjectures within the next few years. (Ecco, 2005, 512 pp., $27.95)

Psychology. *Us and Them: Understanding Your Tribal Mind* by David Berreby cites findings from a wide range of animal and human studies that show how many human prejudices—including those about race, culture, or religion—are linked to instinctive animal behavior. According to these findings, both animals and human beings are hard wired by nature to distinguish between members of their own species and members of other species or social groups who may be dangerous predators or competitors for a limited food supply. Science writer Berreby explores why behavior that is acceptable in the wild may be labeled dysfunctional in a complex human society. (Little, Brown and Company, 2005, 384 pp., $26.95)

Technology. *Infrastructure: A Field Guide to the Industrial Landscape* by Brian Hayes uses a classic nature-guide format to explain the mysterious artificial world that modern societies have created. What are the structural elements of a cell-phone tower, power-station chimney, or hydroelectric dam? How does a wind generator, an oil refinery, a highway tunnel, or a grain elevator work? How does a truss bridge differ from a suspension bridge? Hayes, an experienced science writer, provides clear descriptions of the technology and design behind these and many other familiar sights of the modern landscape. (W.W. Norton Company, 2005, 512 pp. illus., $49.95)   ■ Laurence A. Marschall

■ BOOKS ABOUT SCIENCE FOR YOUNGER READERS

These seven new books about science are suitable for the younger reader. They have been selected from books published in 2005 and 2006.

Archaeology. *Mummies: The Newest, Coolest, and Creepiest from Around the World* by Shelley Tanaka provides graphic descriptions of mummies from various places throughout history. Tanaka emphasizes the scientific importance of mummies, discussing what these long-preserved bodies reveal about the health, diet, culture, and migrations of ancient societies. Each chapter introduces discoveries from a particular region, such as the remains of King Tutankhamun of Egypt. Modern-day mummies, including Vladimir Lenin of the former Soviet Union, get their due as well. (Abrams/Madison, 2005, 48 pp. illus., $16.95)

Astronomy. *Stars and Galaxies* by Ron Miller takes a comprehensive look at the sun, the stars, the Milky Way, and surrounding galaxies. Miller explains the various theories describing how the universe began and how it will end. His examples make even difficult concepts easier to understand, such as comparing the way a loaf of bread rises during baking to the way stars move apart in an expanding universe. Stunning photographs and Miller's original artwork illustrate this volume in the World Beyond series. (Lerner/Twenty-first Century, 2005, 96 pp. illus., $27.93)

Ecology. *The Prairie Builders* by Sneed B. Collard III captures the beauty of the tallgrass prairie that once spread across the Midwest. Bison, elk, butterflies, and birds once populated the area but, by the beginning of the 1900's, that wilderness had been tamed into farmland. Then, in 1989, a prairie restoration project in Iowa began to reintroduce native plants and animals to thousands of acres. Collard chronicles the work being done at Iowa's National Wildlife Refuge and encourages readers to become involved in similar efforts. (Houghton, 2005, 72 pp. illus., $17)

Fossil studies. *Encyclopedia Prehistorica: Dinosaurs* by Robert Sabuda and Matthew Reinhart presents the Mesozoic Era's major dinosaurs, each roaring into three-dimensional view in its own pop-up. Six spreads include paper sculptures of such favorites as the gargantuan *Brachiosaurus,* a spiked *Ankylosaurus,* and a ferocious, highly detailed *Tyrannosaurus rex.* Mini books on each spread as well as a brief paragraph of text cover other dinosaur-related topics. (Candlewick, 2005, 12 pp., $26.99)

Geography. *The North Pole Was Here: Puzzles and Perils at the Top of the World* by Andrew C. Revkin gives readers a sense of life in the Arctic. In 2003, Revkin, an environmental reporter for *The New York Times,* joined climatologists and oceanographers on an expedition to the North Pole for an examination of Arctic warming. History, adventure, scientific observation, and experiments are supplemented by Revkin's dispatches to the *Times* and a variety of stunning photographs. (Kingfisher/New York Times, 128 pp. illus., $15.95)

History of science. *With a Little Luck: Surprising Stories of Amazing Discoveries* by Dennis Brindell Fradin brings 11 scientific discoveries up close and personal. Whether explaining how Isaac Newton made the connection between gravity and a falling apple or Alexander Fleming grasped the importance of the mold growing in his lab, this book makes clear that attention to detail and concentration on a problem can lead to revelation. Fradin combines personal stories with science, technology, and history as he writes about discoveries in astronomy, archaeology, and nuclear technology. (Dutton, 2006, 144 pp. illus., $17.99)

Nutrition. *Chew on This: Everything You Don't Want to Know about Fast Food* by Eric Schlosser and Charles Wilson takes a look behind the scenes at the fast-food industry. Schlosser, author of *Fast Food Nation,* an adult best seller about the industry, chronicles how fast-food companies have achieved such spectacular international success. Readers are treated to tours that take them from feeding stations to processing plants and, finally, behind the counter. (Houghton/Graphia, 2006, 304 pp. illus., $16)

■ Ilene Cooper

CHEMISTRY

The creation of chemically powered artificial "muscles" made of metal or carbon that are up to 100 times as *efficient* as human muscles was reported in March 2006 by researchers at the University of Texas at Dallas in Richardson. (The muscles use 100 times less energy to move than human muscles do.) Unlike existing artificial muscles, which require an outside power source for movement, the new muscles generate their own energy. According to the Texas researchers, robots equipped with the new muscles could move in lifelike ways, rather than with the stiff movements of current robots.

Human muscles produce energy by chemically combining nutrients from food with the oxygen we breathe. The artificial muscles, developed by researchers led by chemist Von Howard Ebron, generate energy by chemically combining a fuel—such as hydrogen gas or alcohol—with oxygen from the air. They are able to do this because the muscles are also *fuel cells* (devices that convert chemical energy into electrical energy). In this case, the artificial muscles, which have been built only in the laboratory, convert the chemical energy of the fuel and oxygen into an electric current that causes them to move.

Ebron and his team created two models of the artificial muscles. One device consisted of a thin metal wire that shortens when heated and lengthens when cooled. This wire was made of an *alloy* (mixture) of nickel and titanium. It was coated with a layer of the metal platinum, which acted as a *catalyst* (a substance that speeds up a chemical reaction). When the researchers combined atmospheric oxygen with a fuel—hydrogen, *methanol* (a type of alcohol), or a chemical called formic acid—and passed the mixture over the coated wire, the resulting chemical reaction caused electrons to flow through the wire. The moving electrons, which passed from the fuel molecules to the oxygen atoms, caused the wire to heat up and contract, lifting a small weight tied to it. When the scientists turned off the fuel and oxygen supplies, the wire cooled off and relaxed.

In the second device, the moving part was a thin paperlike sheet composed of millions of carbon *nanotubes*—tiny tubes of carbon atoms whose diameters are about $1/100,000$ as wide as a human hair. The movement was caused by the build-up of positive charges in the nanotube sheet, which caused it to bend.

SEE ALSO SCIENCE STUDIES, NANOTECHNOLOGY: A REVOLUTION IN MINIATURE, PAGE 112.

Ebron and his group suggested that their artificial muscles could lead to the development of robots that are not tethered to wall outlets by electrical cables or weighed down by heavy and bulky batteries, as was the case in 2006. Artificial limbs powered by small chemical packs, as well as aircraft and submarine surfaces that change their shape on demand, are other potential uses of the artificial muscles, the Texas team said.

NMR analyses without the magnets. In 2006, researchers in Germany announced the development of a modified nuclear magnetic resonance (NMR) device powered by Earth's magnetic field instead of the field produced by the bulky magnets currently in use. Chemists rely heavily on NMR spectroscopy to identify molecules and determine their *chemical structures* (arrangement of atoms). However, NMR spectroscopy, which measures the magnetic signals given off by the *nuclei* (centers) of atoms in a molecule, requires powerful magnets and instruments that can fill a room and cost at least $1 million. In January, researchers led by chemist Stephan Appelt of the Central Institute for Electronics in Jülich, Germany, announced that they had developed a simple, lightweight, low-cost NMR instrument that relies on Earth's natural electromagnetism.

NMR instruments bombard samples with magnetic fields emitted by powerful electromagnets and with energy from radio waves. In response, some atomic nuclei in a molecule change the way they spin. During this change, the nuclei absorb energy and emit a signal. The signal provides information about the nature and location of other atoms, which enables scientists to determine the structures of nearby molecules.

Normally, the stronger the external magnetic field, the easier it is to detect a signal from the atomic nuclei. However, the atoms in a molecule emit another type of magnetic signal, known as "J-coupling," that does not depend on the strength of the outside field. This signal is caused by interaction of the spins of different nuclei in a molecule. Measuring J-coupling requires exposing a sample to an outside magnetic field that is free from fluctuations, however, which is difficult to achieve with the large magnets currently used in NMR.

Appelt and his team observed that Earth's magnetic field, which is approximately 100,000 times as weak as magnets typically used in NMR instruments, is strong enough and sufficiently uniform to detect the J-coupling. To use this magnetic field, however, the scientists had to block out human-made magnetic fields that would overshadow Earth's magnetic field. They did this by first taking their equipment into the countryside, where there are almost no artificial magnetic fields. Then they placed all the samples being analyzed inside a magnetized cylinder called a Halbach magnet, which shielded the samples from any magnetic fields except Earth's. Using the shielded equipment, they were able to easily observe the J-coupling signals and identify test samples.

The researchers estimated that shielded devices using Earth's magnetic field might some day cost only thousands of dollars, significantly less than current instruments. Portable NMR machines could enable scientists to quickly and rapidly monitor chemical structures during manufacturing and track their movements in the environment. The devices may also pave the way for portable *magnetic resonance imaging* (MRI) machines. (MRI is a medical imaging method closely related to NMR.) Such machines would no longer require patients to pass through a narrow tunnel surrounded by a huge magnet, as they must during many current MRI procedures.

Transforming with gold. In October 2005, scientists in the United Kingdom and the United States described a way to use tiny particles of gold to make manufacturing in the chemical industry more environmentally friendly, less expensive, and more productive. The researchers reported that gold particles can be used as catalysts to speed up a widely used class of reactions in which oxygen atoms are added to *hydrocarbons* (compounds of carbon and hydrogen). Chemical companies now use these reactions, known as *hydrocarbon oxidations,* to make dozens of products, from plastics and detergents to pesticides and drugs.

Currently, adding oxygen atoms to hydrocarbons requires the use of either chlorine or *oxidizing agents* (oxygen-releasing compounds) called *peroxides*. However, chlorine is a corrosive gas that produces toxic by-products that must be carefully disposed of, and peroxides are extremely expensive. Current oxidation processes also must take place in *organic* (carbon-containing) liquid *solvents* (substances that can dissolve other

MANY-COLORED BUBBLES

Zubbles, described as the world's first colored bubbles, float around inventor Tim Kehoe (right) and dye chemist Ram Sabnis. In 2005, Kehoe and Sabnis announced that they had perfected the chemistry of a colored bubble that does not leave a permanent stain. Sabnis created an unstable dye molecule that absorbs all colors of the visible spectrum except the one representing the desired bubble color. When water, air, or pressure is added to the molecule, its structure changes, allowing it to absorb all colors of the visible spectrum and so become transparent. Sabnis and Kehoe reported that bubbles made of these dye molecules disappear completely when popped.

CHEMISTRY continued

substances), which contribute to air and water pollution. According to a team of researchers led by chemist Graham J. Hutchings of Cardiff University in Wales, gold catalysts may offer a better way to oxidize hydrocarbons.

The gold the researchers used in their experiments was not the glittering metal found in earrings and bracelets. Instead, it consisted of tiny particles, about one *nanometer* (one-billionth of a meter) in size, that were sprinkled onto larger particles of graphite, a form of carbon used in pencils. When the investigators passed *unsaturated* hydrocarbons (those with two chemical bonds between a pair of linked carbon atoms) over the gold catalyst and added air bubbles, the oxygen in the air immediately reacted with the hydrocarbons. The reaction produced the same valuable oxygen compounds whose production currently requires chlorine and peroxides.

According to the researchers, it was possible to produce high yields of specific oxygen-containing compounds by changing the solvents used in the gold-catalyzed reactions. Another way to make high levels of a selected compound, the investigators found, was to add traces of the metal bismuth to the gold catalyst. As the bismuth levels increased, different oxidized compounds formed.

Hutchings and his team noted that some of the oxidation reactions promoted by the gold catalysts did not even require liquid solvents. Running reactions without solvents would be an environmental benefit, as would eliminating the use of chlorine. Making oxidized hydrocarbons without peroxides would also be far less expensive for manufacturers. Hutchings and his team announced that they would try to increase the yield and scale of their gold-catalyzed reactions to make the process more attractive to industrial companies.

Spongelike crystals. Scientists reported in 2005 and 2006 that a newly discovered group of solids that can absorb many times their weight in gases might offer an efficient and cost-effective way to reduce the build-up of carbon dioxide in the atmosphere. The materials were created by chemist Omar M. Yaghi of the University of California, Los Angeles, and colleagues at the University of Michigan in Ann Arbor. Carbon dioxide is a greenhouse gas that traps heat in Earth's atmosphere. Most *climatologists* (scientists who study climate) agree that a build-up of carbon dioxide in the atmosphere because of human activities has contributed to *global warming* (a gradual increase in Earth's surface temperature).

The new materials are crystalline solids called *metal-organic frameworks* (MOF's). They consist of clusters of metal atoms linked together by a three-dimensional network of connected *organic* (carbon-containing) molecules. Because the organic chains keep the metal clusters far apart, the MOF's are mostly empty inside. In fact, the MOF's have so many pores that 1 gram (0.03 ounce) of these materials has about the same surface area as a football field. MOF's are useful because huge quantities of gas molecules can easily fit inside their many pores. Once inside, the gas molecules bind chemically to the metal atoms and are not released unless the compounds are heated. The MOF's are essentially sponges that sop up gases instead of water.

MOF's are particularly good at absorbing carbon dioxide. In December 2005, Yaghi described one MOF compound that can absorb up to 140 percent of its weight in carbon dioxide, more than twice the carbon-dioxide storage capacity of other gas-absorbing solids. According to Yaghi, it might be possible to fit cylinders containing the newly discovered compound onto the flues of factory smokestacks or the outlet pipes of cars to absorb carbon dioxide and prevent its release into the atmosphere.

Another group of MOF's, described by Yaghi in March 2006, has the ability to store large amounts of hydrogen. Yaghi believes that their gas-absorption capacity is so great that they might offer a practical and safe way to store hydrogen, a flammable and explosive gas, in hydrogen-powered cars. (Unlike gasoline-powered vehicles, these cars would not release carbon dioxide.) However, the new MOF's can absorb huge amounts of hydrogen only when they have been cooled to 77 K (–321 °F). Yaghi, who has already made 500 varieties of MOF's, believes he can alter the chemistry of the hydrogen-absorbing compounds so that such extreme cooling will no longer be necessary.

Low cost is another reason the MOF's seem so promising. For example, typical raw materials for MOF's consist of zinc oxide, a common ingredient in sunscreen, and carbon compounds called *terephthalates,* which are used to make plastic bottles. ■ Gordon Graff

See also **ENGINEERING; PHYSICS.**

COMPUTERS AND ELECTRONICS

Dual-core computer chips dominated the personal computer industry in 2005 and 2006. Dual-core chips have two independent processing engines on a single chip—a major break from the decades-old practice of building a single engine on a single chip. In January 2006, Intel Corporation of Santa Clara, California, launched the Core Duo chip, the first dual-core processor for laptop computers.

Expensive and powerful computer servers have used dual-core chips since 2001. Chip makers introduced their first dual-core chips for desktop personal computers in 2005, but they were popular primarily with video gamers and other users of expensive, high-performance computers. The Intel Core Duo chip—best known by its code name, Yonah—was the centerpiece of a sweeping marketing campaign meant to highlight Intel's growing use of multicore chips. The company planned to have dual-core chips in more than 70 percent of its desktop and laptop computers by the end of 2006. Computer makers redesigned their products for the new chips, which consume more power and generate more heat than traditional chips. Manufacturers Acer Inc. of Taipei, Taiwan, and Sony Corporation of Tokyo, Japan, shipped laptop computers with the Intel Core Duo chip in January 2006.

Dual-core chips emerged from efforts by Intel and rival chip maker Advanced Micro Devices Inc. (AMD) of Sunnyvale, California, to increase the overall performance of their processors. The companies found that small increases in performance resulted in large increases in electric pow-

SEE ALSO CONSUMER SCIENCE, REMOVABLE STORAGE: MEMORY ON THE MOVE, PAGE 140.

er consumption, causing the machines to overheat. Dual-core designs, which typically run at slower clock speeds than their single-core counterparts, can, in principle, process twice as many instructions at once. For example, the single-core AMD Athlon FX chip, introduced in 2005, ran at a speed of 2.8 gigahertz (2.8 billion cycles per second) and consumed a maximum of 104 watts of electric power. Each core of the Athlon X2, a dual-core chip introduced in 2005, ran at 2.4 gigahertz. However, the entire chip consumed a maximum of only 110 watts of power.

Despite the popularity of dual-core chips, most personal computer software in use in 2005 and 2006 was not written to take advantage of the chips, and so many of their initial performance advantages were wasted. Dual-core chips are useful when a computer performs two tasks at the same time, such as playing digital music and surfing the Web. However, dual-core chips do not necessarily allow a computer to perform a single task, such as editing a photo, more quickly. As software compatibility increases, dual-core chips are likely to become standard equipment on even the least expensive personal computers.

Katrina slams technology industry. Hurricane Katrina swamped communication systems in three states when it struck the United

HUMAN-POWERED COMPUTING

A hand-cranked laptop computer designed for schoolchildren in developing countries made its first appearance at the United Nations World Summit on the Information Society in Tunis, Tunisia, in November 2005. The One Laptop per Child initiative was created by the Massachusetts Institute of Technology Media Lab in Cambridge. The computer, which the lab claimed could be built for about $100, featured wireless technology that enabled it to connect to the Internet through nearby laptops, forming a wireless mesh network.

COMPUTERS AND ELECTRONICS continued

States Gulf Coast on Aug. 29, 2005, according to testimony by Federal Communications Commission (FCC) Chairman Kevin Martin at a hearing of the Senate Commerce Committee. (The FCC is an independent government agency that regulates communication by radio, television, wire, satellite, and cable.) The storm highlighted the dangers in Americans' dependence on computers, cell phones, and other electronic communication devices.

The FCC estimated that Katrina disabled about 3 million telephone lines and knocked out at least 1,000 cellular sites and 38 emergency call centers in Louisiana, Alabama, and Mississippi. The hurricane's storm surge flooded electric power stations, and back-up generators and batteries proved to be insufficient replacements. Thousands of Gulf Coast residents could not contact friends and relatives for days. Katrina killed at least 1,800 people and caused tens of billions of dollars in damage, making it one of the most destructive natural disasters in U.S. history.

Hand-powered computer. An inexpensive, hand-cranked laptop computer, designed to be distributed free of charge to millions of schoolchildren in developing countries, was unveiled at the United Nations World Summit on the Information Society in Tunis, Tunisia, in November 2005. The computer was designed by the One Laptop per Child initiative, a project created by the Massachusetts Institute of Technology (MIT) Media Lab in Cambridge. The computer is about the size of a textbook and was designed to connect to the Internet through wireless networks. The prototype shown at the summit had a processor that operated at 500 megahertz (500 million cycles per second), about half the speed of an average commercial laptop. One minute of hand cranking was expected to supply about 10 minutes of power for the laptop.

The MIT Media Lab hoped to build the computer for about $100. In 2006, the lab was still designing the machine and working to build a low-cost, durable screen. The lab's goal was to provide the computers to children in developing countries who cannot afford to buy their own computers. The group wanted to sell the machines to governments—in primarily the Middle East, Asia, Africa, and Latin America—that would then distribute the laptops in schools.

Nano and video iPods. Apple Computer, Inc., of Cupertino, California, introduced the iPod nano, the fourth and smallest member of its iPod line of digital media players, in September 2005. The 42.5-gram (1.5-ounce) nano, which replaced the discontinued iPod mini, was about the size of a business card and thinner than a pencil. The nano's battery lasted up to 14 hours, and its 3.8-centimeter (1.5-inch) screen could display up to 65,536 colors. The device originally came in two sizes—a 2-gigabyte version for $199 that stored about 500 songs, and a 4-gigabyte version for $249 that held about 1,000 songs. In February 2006, Apple introduced a 1-gigabyte nano for $149 that stored about 240 songs. All nano models used *flash memory*—a form of electronic memory that has no moving parts—making the devices sturdy and skip resistant. After some users complained that the nano's screen was easily scratched and cracked, Apple vowed to replace any devices with defective screens.

In October 2005, Apple debuted its fifth iPod product, a video-capable iPod that was about the size of a deck of playing cards. Its 6.4-centimeter (2.5-inch) screen could display video with 262,144 colors. Although the player could also transfer video to a television via a cable, many consumers complained about the poor quality of the television images. The video iPod came in two models. The 60-gigabyte version, which had a 20-hour battery life and held 150 hours of video or 15,000 songs, sold for $399. A 30-gigabyte model had a 14-hour battery life and half as much storage as the larger version.

The video iPod was the latest hit from Apple, which sold 42 million iPods from the product's debut in October 2001 to January 2006. Video iPod users could buy video content from Apple's online music store. In 2005 and 2006, the store's offerings included music videos and television programs from such entertainment companies as the Walt Disney Company of Burbank, California. Apple had sold 1 billion songs and 15 million videos by February 2006.

Intel inside Macs. In June 2005, Apple announced that it would stop using PowerPC chips made by IBM Corporation of Armonk, New York, and Motorola, Inc., of Schaumburg, Illinois, in its Macintosh computers. Instead, Apple switched to chips from Intel, whose processors had long powered computers based on the Windows operating system made by the Microsoft Corporation of Redmond, Washington. Some Macintosh users were shocked and felt that Apple had betrayed IBM and Motorola after years of Apple advertisements praising their chips.

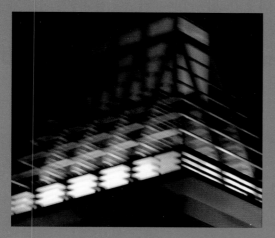 

HOLD IT STEADY

A night exposure of a lighted building taken by a handheld camera (above, right) appears crisp rather than blurry thanks to image-stabilization lenses developed for digital cameras. Many major camera makers released lenses or cameras with so-called antishake technology in 2005 and 2006. The technology includes electronic sensors that detect tiny hand movements made by the photographer. A computer compensates for these motions by altering the image on the camera's light detector. The new lenses allow camera users to take sharper pictures with longer exposure times.

The switch followed continuing problems with the PowerPC processors. In 2003, Steve Jobs, Apple's chief executive officer, promised that Macs based on the PowerPC G5 chip would run at speeds of up to 3 gigahertz (3 billion cycles per second) by 2004, but the chip makers failed to deliver on that promise. Apple also was disappointed because the PowerPC chips consumed too much energy and produced too much heat for Apple's laptop computers. Power-hungry chips are a problem for laptops because such processors quickly use up battery power. Intel's processors perform at similar speeds to the PowerPC chips but use less energy.

In January 2006, Apple debuted its first Intel-based computers—the MacBook Pro laptop and the iMac Core Duo desktop. Both computers used Intel Core Duo chips.

Podcasting. One of the biggest trends in computing in 2005 and 2006 was the sharp rise in the popularity of podcasting. In fact, the *New Oxford American Dictionary* named *podcast* as its "Word of the Year" for 2005. The word refers to the ability to "broadcast" digital audio and video files over the Internet so that others can listen to them on an iPod or other digital media player. Listeners can also play podcasts on a computer.

Anyone with a microphone, computer, software, and an Internet connection can record and distribute podcasts. Many podcasts consist of individual commentaries on topics ranging from

home improvement and cooking to world politics and religion. Internet users can download these programs and play them whenever they wish. In contrast, people who listen to programs on the radio must listen to whatever is being broadcast at the moment. Nor can they easily record programming for later listening.

Businesses and other organizations had also adopted the podcast format. In 2005 and 2006, National Public Radio and the British Broadcasting Corporation distributed daily radio programs as podcasts. The White House began podcasting U.S. President George W. Bush's weekly radio addresses in summer 2005.

Apple boosted the popularity of podcasting in June 2005 when it added podcasts to its iTunes software and music store. Most podcasts were available for free downloading on the Internet, but some of the most popular programs moved to paid-only formats. British comedian and writer Ricky Gervais, creator of the television hit "The Office," moved his weekly podcast from a free iTunes download to Audible.com, which charged $1.95 per episode or $6.95 for the season. By early 2006, just before he switched to a paid format, "The Ricky Gervais Show" was named the world's most popular podcast by London-based Guinness World Records Ltd.

Voice over Internet Protocol (VoIP) use grew dramatically in 2005. Also known as *Internet telephony* or *digital phoning*, VoIP routes

COMPUTERS AND ELECTRONICS continued

voice conversations over the Internet instead of over traditional telephone lines. According to a study published in January 2006 by Scottsdale, Arizona-based research firm In-Stat, worldwide VoIP subscriptions grew by 62 percent to 16 million in 2005. The survey predicted at least 55 million VoIP users by 2009.

Companies that provide VoIP service let users make and receive phone calls from anywhere in the world for only pennies per minute. VoIP costs from $15 to $40 per month for nationwide and local calling. International calls cost as little as 3 to 4 cents per minute—much less than traditional long-distance service.

The best-known provider of residential VoIP service is Vonage Holdings Corp. of Holmdel, New Jersey. Telecommunications giants Time Warner Cable and AT&T Corp. were also entering the VoIP market. Another VoIP provider, Luxembourg-based Skype Technologies S.A., which offers free calls between personal computers, had more than 75 million registered users by early 2006. Online auction company eBay Corp. of San Jose, California, announced in September 2005 that it would purchase Skype. EBay sought to incorporate the Skype service for voice communication during its auctions. Executives said they would preserve Skype's stand-alone services, such as videoconferencing.

Despite its low cost, VoIP had a number of technical flaws. VoIP service can be cut off during a power failure—unlike regular phone service, which keeps working in all but the most catastrophic situations. In addition, because VoIP uses the Internet, the service may slow down due to heavy online traffic, leading to poor-quality calls.

The biggest problem for VoIP in 2005 was that emergency call centers could not identify the location of a VoIP call. If a caller could not provide an address, the center could not dispatch an ambulance or other emergency responder. The FCC ruled in May 2005 that VoIP providers had to deliver "Enhanced 911" (E911) service to their customers. The service associates a physical address with the calling party's telephone number. E911 ensures that emergency responders receive the caller's location even if the caller hangs up or cannot provide the address. By December 2005, Vonage had complied with the FCC ruling.

Mashups. Hybrid Web sites known as *mashups* sprouted all over the World Wide Web in 2005. Mashups integrate content from at least two different sources but appear to users as a single, seamless site. The term comes from pop music artists who have long "mashed" vocal or musical tracks from one song with tracks from other songs to create new music.

Sites whose content is often used in mashups include Web-based calendars, photo- and video-sharing sites, and mapping services. Three computing companies that provide Internet mapping services—Google Inc. of Mountain View, California; Yahoo Inc. of Sunnyvale, California; and Microsoft Corporation—each released the computer code behind their services, allowing people to create mashups. Some of the more popular mashups included "Home Price Maps," an online real estate service that combines data on home prices with Google Maps; "Earthquakes in the Last Week," a site that provides a real-time earthquake map with data from the U.S. Geological Survey; and "ChicagoCrime.org," which integrates crime data in the city of Chicago with Google Maps.

Mashup creators typically do not ask permission from the companies whose content they use. In addition, most mashups created in 2005 and 2006 were not used to make money. Mashups were part of a broader phenomenon called Web 2.0, which referred to an array of computer programs and Web sites that enabled Internet users to interact and share information. Web 2.0 was different from Web sites made in the 1990's and the early 2000's—called Web 1.0—because the older sites were relatively static.

The Web browser wars accelerated in 2005, when a free program called Firefox challenged Microsoft Corporation's dominant Internet Explorer Web browser. Firefox was developed by volunteers and engineers at Mozilla Corp. of Mountain View, California. In October 2005, Firefox 1.0 was downloaded for the 100-millionth time. In November, Mozilla introduced Firefox 1.5, and within the first 36 hours, it was downloaded more than 2 million times. Users praised Firefox's simple interface and pop-up advertisement blocker. Another popular feature called *tabbed browsing* allowed Web surfers to open multiple Web pages in the same browser window.

Firefox was developed as an *open-source* project—one in which the project's computer code is published on the Internet so that anyone can copy, change, or distribute it without paying a licensing fee. In contrast, Microsoft and many other companies have fiercely guarded their source code and charged individuals or corporations large fees to use the programs. Programmers have made some of their code available to the public since the 1960's, but the phenomenon has grown dramatically since the 1990's. Some of the most popular open-source projects include the operating system Linux; the office software suite OpenOffice, made by Sun Microsystems of Mountain View, California; and the Apache Web server, which many corporations used to run their Web sites.

Microsoft's Internet Explorer was still the most popular Web browser on the market in 2005 and 2006. However, critics said Microsoft's browser had security problems that exposed users to harmful software, including *spyware* (programs that monitor a user's habits and transmit personal information to an outside source) and *computer viruses* (self-reproducing programs that can infect and damage other computer programs and files). In November 2005, 8.13 percent of people who used Web browsers used Firefox, up from 3.03 percent one year earlier, according to statistics compiled by research company WebSideStory Inc. of San Diego, California. During the same period, Internet Explorer's share of the market dropped by more than 4 percent—from 92.9 percent in November 2004 to 88.6 percent in November 2005. ■ Rachel Konrad

■■ CONSERVATION

In February 2006, Canadian officials formally designated a 6-million-hectare (16-million-acre) area on British Columbia's Pacific coast as a preserve known as Great Bear Rainforest. The move helped to ease long-standing tensions between conservationists, who wanted to protect the huge trees of the region, and the logging industry, which wanted to harvest the trees for their valuable timber. The trees of the rain forest included Douglas fir, Sitka spruce, western hemlock, and western red cedar.

The preserve includes more than twice as much land as Yellowstone National Park, which covers parts of Idaho, Montana, and Wyoming. Under the agreement, the trees, along with a variety of wildlife, will be fully protected on 1.8 million hectares (4.4 million acres). Government forestry officials will manage the remaining area as an *ecosystem* where *sustainable*

logging will be permitted. An ecosystem includes the living things in a particular place and the nonliving things important to their existence. Sustainable logging involves producing timber using environmentally sound practices that protect soil and water resources and wildlife. Area wildlife included grizzly bears, mountain goats, moose, wolves, spawning salmon, and many other species. The area is also a key habitat for "spirit bears," a rare form of black bear with white fur. Spirit bears, also called Kermode bears, are not *albinos* (animals unable to produce a coloring substance called pigment). Instead, a rare gene produces their white color.

In addition to addressing the interests of conservationists, government officials, and

FUNDING JUMP FOR ENDANGERED ATLANTIC SALMON

An Atlantic salmon, one of an endangered species in Maine, springs out of native waters. In 2005, only about 80 Atlantic salmon—half the number in the 1990's—returned from the ocean to eight Maine rivers for *spawning* (reproduction). In December 2005, United States officials pledged $34 million toward efforts to restore the Atlantic salmon population in Maine.

CONSERVATION continued

loggers, the agreement represented the concerns of Native people who have long inhabited the lush forest. The accord protects traditional sites where cemeteries, medicinal herbs, and trees suitable for canoes and totem poles may be found. Supporters of the agreement hope that the establishment of the Great Bear Rainforest will fulfill the appeal made, according to tradition, by a Native American raven deity that 1 of every 10 black bears shall be white—a request that future generations preserve the rain forest.

Funds for amphibians. About $404 million is needed to prevent additional declines in the world's amphibian population, according to a group of 500 scientists from 60 countries who met in Washington, D.C., in September 2005. The group included representatives from the International Union for the Conservation of Nature and Natural Resources (IUCN), which is based in Gland, Switzerland. The scientists called for more research and action in response to disease, habitat loss, and other causes that have put almost one-third of the world's frogs, toads, and other kinds of amphibians on the IUCN's list of globally threatened species.

Although scientists recognized 5,743 species as amphibians in 2005, that number may represent only about half of the actual number. Of the known species, conservationists claimed, at least 43 percent were declining in numbers and 122 species may have become extinct in the past 35 years.

Amphibians, which usually live part of their life in water, are unusually sensitive to environmental conditions because of their porous skins. In addition to habitat loss, a fungal disease called chytridiomycosis appeared to represent another reason for the amphibians' decline, which was particularly severe in Australia and North and South America. According to the scientists, the actions with the highest priority include finding and protecting key habitat, increasing the use of captive-breeding programs, and controlling chytridiomycosis.

Relocating Florida panthers. The United States Fish and Wildlife Service (FWS) in January 2006 announced a plan to move some Florida panthers from southern Florida to other states in the Southeast as the centerpiece of a recovery plan for the endangered animals. Florida panthers represent a rare subspecies of the more numerous mountain lion (also called the cougar or puma) that ranges over

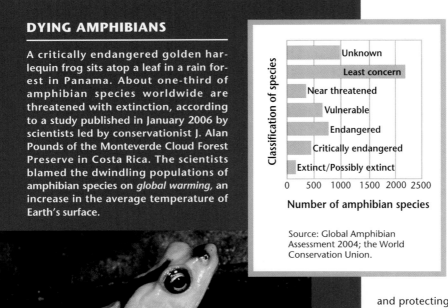

DYING AMPHIBIANS

A critically endangered golden harlequin frog sits atop a leaf in a rain forest in Panama. About one-third of amphibian species worldwide are threatened with extinction, according to a study published in January 2006 by scientists led by conservationist J. Alan Pounds of the Monteverde Cloud Forest Preserve in Costa Rica. The scientists blamed the dwindling populations of amphibian species on *global warming,* an increase in the average temperature of Earth's surface.

Classification of species (y-axis):
- Unknown
- Least concern
- Near threatened
- Vulnerable
- Endangered
- Critically endangered
- Extinct/Possibly extinct

Number of amphibian species (x-axis): 0 500 1000 1500 2000 2500

Source: Global Amphibian Assessment 2004; the World Conservation Union.

large areas of North and South America. Roughly 80 adult Florida panthers lived in the forests and swamps of southern Florida in 2006.

Because of these small numbers and their contained habitat, Florida panthers are protected by the Endangered Species Act of 1973, which prohibits hunting, collecting, and other activities harmful to the animals. However, human expansion has largely destroyed and continues to threaten the panthers' habitat. In addition, motorists traveling along "Alligator Alley," a major highway crossing Florida, killed at least 40 panthers from 1972 to 2006. The installation in 1993 of underpasses designed specifically for panthers and other wildlife has greatly reduced the number of panthers killed annually.

In 1995, in an effort to increase the dwindling Florida panther population, FWS employees imported mountain lions from Texas to breed with the panthers. Although this strategy helped to halt the sharp decline in panther numbers, FWS officials noted that the panther population might remain small unless the animals had additional territory in which to expand. Adult Florida panthers tend to live alone, except when mating, and require a range of about 712 square kilometers (275 square miles).

FWS officials reported that protected habitat in the Okefenokee National Wildlife Refuge along the Florida–Georgia border seemed especially promising as a relocation site. They also suggested that some areas in Alabama and Arkansas might be suitable, if the effort gained public support there.

Where the bison roam. The release of 16 bison, also known as American "buffalo," into northeastern Montana in November 2005 marked the initial step in the restoration of parts of the northern Great Plains to a self-sustaining prairie ecosystem. The Great Plains is a vast, dry grassland that extends from northern Canada into Texas and New Mexico in the United States. The release is part of a program sponsored by the American Prairie Foundation (APF) in Bozeman, Montana, and its partner, conservation group WWF, based in Geneva, Switzerland. The program's long-term goals include maintaining a fully functional prairie-based wildlife reserve.

The 16 bison represent a purebred and disease-free remnant of the millions of bison that once roamed North America's vast grasslands. The bison, which came from a herd at Wind Cave National Park in South Dakota, were released on more than 12,100 hectares (30,000 acres) of land owned or rented by the APF. The release site adjoins large areas of federal land, including a national wildlife refuge. The APF and the WWF hope to eventually create a 1.4-million-hectare (3.5-million-acre) preserve where bison as well as black-footed ferrets and other endangered species can flourish.

The agricultural economy of northeastern Montana is declining, and restoring a sustainable prairie ecosystem represents a new approach to land management. The prairie preserve also will feature cycling, hiking, and horseback-riding trails as well as camping sites and will be operated in ways that contribute to the local economy. As part of its efforts, the APF also is working to protect such historical sites as a sod-roofed, one-room school attended by the children of the area in the early 1900's.

Grizzly bears off the list? On Nov. 15, 2005, the FWS proposed removing grizzly bears living in Yellowstone National Park from the list of threatened U.S. animals. Threatened species, also known as vulnerable species, are abundant in some places but face serious challenges to their overall numbers. In 2005, the U.S. government classified grizzlies as threatened in every state except Alaska.

From the 1970's to 2005, the number of grizzly bears in Yellowstone rose from about 200 to at least 600. The recovery of the Yellowstone population provides ample testimony to the effectiveness of the Endangered Species Act, according to Tom France, director for conservation programs in the northern Rockies for the National Wildlife Federation in Reston, Virginia.

Approximately 50,000 grizzly bears wandered the western United States in the early 1800's. By the 1970's, relentless poisoning, shooting, and trapping, along with habitat loss, had reduced the western grizzly population to fewer than 800 animals.

In 1993, the FWS adopted a recovery plan to stabilize the Yellowstone grizzly population. One of the plan's goals was to record at least 15 adult females with cubs each year. In fact, each year since 1999, biologists have recorded at least 40 females with cubs. In 2005, the number of grizzly bear deaths attributed to human activities remained well within the limits that a stable population could withstand. The future of threatened grizzlies in other areas in the western United States was not as bright as in Yellowstone, except for one

CONSERVATION continued

location in and around Glacier National Park in Montana.

Protecting vultures. An international team of scientists in February 2006 identified a substitute for a widely used cattle medication that had caused catastrophic losses among vulture populations in southern Asia. The cattle medication, called diclofenac, was introduced into the Asian veterinary market in the 1990's. Conservationists estimated that since then, diclofenac, which causes kidney failure in vultures, had poisoned millions of the birds in Pakistan and Nepal and was responsible for a 97-percent reduction in the vulture population in India. The scientists claimed that the substitute drug, called meloxicam, is as effective as diclofenac in treating fever, inflammation, and pain in cattle but posed no threat to vultures.

In ecological communities, vultures have long played a role as removers of garbage, a role that was upset by diclofenac. In southeast Asia, livestock that die from injury or disease are typically left in the open for disposal by scavengers. The carcasses—increasingly tainted by diclofenac—have represented a major source of food for vultures. In many areas, rats and wild dogs have replaced vultures as the primary scav-engers, a change that may increase outbreaks of rabies. Vultures also play a cultural role in India, where a religious group known as Parsis traditionally relies on the birds to dispose of the bodies of dead members.

Diclofenac has imperiled three species of Asian vultures: the slender-billed vulture, the long-billed vulture, and the Oriental white-backed vulture. Scientists once considered white-backed vultures among the world's most common large birds of prey. In 2006, however, less than 1 percent of the birds still survived. Similarly, the world population of slender-billed vultures plummeted to no more than 200 birds in 2006. According to scientists, these losses represent some of the sharpest declines known for any birds, including species that eventually became extinct.

In March 2005, the Indian government promised to ban the sale of diclofenac, but the drug still was widely used by farmers in 2006. In a last-ditch effort to save the vultures from extinction, scientists placed 68 birds from the three threatened species in a captive-breeding program. ■ Eric G. Bolen

See also **BIOLOGY; ENVIRONMENTAL POLLUTION.**

DEATHS OF SCIENTISTS

Notable people of science who died between June 1, 2005, and May 31, 2006, are listed below. Those listed were Americans unless otherwise indicated.

Bahcall, John N. (1934–Aug. 17, 2005), astrophysicist, and **Davis, Raymond, Jr.** (1914–May 31, 2006), chemist, whose work led to a new way of studying the sun. In 1964, Bahcall suggested measuring the number of solar *neutrinos* that reach Earth to learn why the sun shines. (Neutrinos are a type of subatomic particle produced by nuclear reactions inside the sun.) However, Davis's observations of solar neutrinos, conducted from the 1960's to the 1970's, failed to match Bahcall's predictions. In 2002, Davis won the Nobel Prize in physics for capturing solar neutrinos. For decades, scientists struggled to resolve the discrepancy. Finally, in the 1990's, experiments proved that neutrinos change form, causing them to evade detection, a solution Bahcall had suggested. Bahcall also helped to create the Hubble Space Telescope, a powerful Earth-orbiting telescope that has provided images of celestial objects in unprecedented detail.

Cameron, Alastair G. W. (1925–Oct. 3, 2005), Canadian-born astrophysicist who, along with several collaborators in the 1970's, proposed a revolutionary idea of the moon's formation known as the Giant Impact or the "Big Whack." According to this theory, a planet-sized object struck Earth 4.6 billion years ago, sending a cloud of vaporized rock into orbit around Earth. Eventually, the space debris came together, forming the moon. The theory became accepted by most scientists.

Chamberlain, Owen (1920–Feb. 28, 2006), physicist who in 1955, along with

Alastair G. W. Cameron

Italian-born physicist Emilio Gino Segre, discovered the *antiproton*. An antiproton is a subatomic particle that carries a negative charge, the opposite of its corresponding particle—a proton—which carries a positive charge. Chamberlain and Segre shared the 1959 Nobel Prize in physics for the discovery. The team of scientists led by Chamberlain and Segre used a large particle accelerator (often called an "atom smasher") to study collisions between atomic nuclei. Chamberlain also worked on the Manhattan Project, a United States government-directed effort that in 1945 produced the first atomic bomb. In the 1980's, he became a founding member of the nuclear freeze movement, an effort to stop the production of nuclear arms by the world's nuclear powers.

Doll, Sir Richard (1912–July 24, 2005), British *epidemiologist* (scientist who studies epidemic diseases) whose findings conclusively connected smoking to lung cancer. Doll also proved smoking can cause emphysema as well as *cardiovascular disease* (a condition that affects the heart and blood vessels). In addition, his numerous studies provided evidence linking some forms of radiation to cancer and asbestos exposure to lung cancer.

Gould, Gordon (1920–Sept. 16, 2005), pioneer of laser technology who contributed to the invention of the most commonly used type of laser. Gould came up with the idea for his device in 1957 and is credited with coining the acronym *laser* from the words *light amplification by stimulated emission of radiation*. Lasers, which shoot out intense beams of light, are commonly used in such applications as communication, surgery, and distance measurement.

Howells, William W. (1908–Dec. 20, 2005), anthropologist credited with proving that modern human beings all belong to the same species. Howells based his conclusion on skull measurements. He also determined that

earlier people, including the Neandertals who lived as recently as 35,000 years ago, differed significantly from modern human beings.

Keeling, Charles D. (1928–June 20, 2005), *climatologist* (climate scientist) who provided the first direct link between an increase in the burning of *fossil fuels* (coal, oil, and natural gas) and a build-up of carbon dioxide in Earth's atmosphere. Carbon dioxide is a gas that helps regulate Earth's temperature by trapping heat in the atmosphere. Keeling measured the increasing amounts of carbon dioxide and convinced other scientists of the serious effects of global warming. Before his groundbreaking work began in 1956, most scientists believed plants could safely absorb all of Earth's excess carbon dioxide.

Kilby, Jack (1923–June 20, 2005), electrical engineer who in 1958 built the first microchip, or *integrated circuit.* His invention revolutionized electronics and launched the digital age. While at Dallas-based Texas Instruments, Kilby developed a method of putting all of the elements of an electronic circuit on a sliver of silicon about 13 millimeters (0.5 inch) long. The chips became commercially successful after Kilby invented the handheld calculator, and even more so as the chips became widely used in computers, appliances, and other types of electronic devices. Kilby won half of the Nobel Prize in physics in 2000 for his work.

Moog, Robert (1934–Aug. 23, 2005), physicist who in the 1960's invented one of the first commercially successful *synthesizers* (musical instruments that produce sounds electronically). Moog used controlled levels of *voltage* (strength of electrical force) to produce either purely electrical sounds or those that mimicked musical instruments. Such rock music bands as the Beatles, Pink Floyd, and Yes helped popularize Moog synthesizers.

Roberts, Anita (1942?– May 26, 2006), molecular biologist whose pioneering research helped determine the function of a protein known as T.G.F.-beta. In the 1980's, Roberts and a team of researchers discovered that the protein was critical in stimulating growth factors in the

Robert Moog

DEATHS OF SCIENTISTS continued

Joseph Rotblat

body to heal wounds, fractures, and many other kinds of injuries.

Rotblat, Joseph (1908–Aug. 31, 2005), Polish-born British physicist who shared the 1995 Nobel Peace Prize with the Pugwash Conferences on Science and World Affairs for their attempts to eliminate nuclear weapons. Rotblat, along with other scientists, founded the Pugwash Conferences in 1957. In 1944, he had been a member of the British team assigned to help the scientists involved in the Manhattan Project. He resigned from the project on moral grounds and devoted the rest of his life to opposing the use of nuclear weapons.

Shumway, Norman E. (1923–Feb. 10, 2006), surgeon who in 1968 performed the first human heart transplant in the United States. Shumway and his medical team improved surgical techniques and established ways of measuring and treating rejection in heart transplant operations. Shumway oversaw about 800 human heart transplants during his lifetime.

Smalley, Richard E. (1943–Oct. 28, 2005), physical chemist who shared the 1996 Nobel Prize in chemistry with two other scientists for their discovery of *fullerenes.* Fullerenes are forms of carbon in which large numbers of atoms link together in the form of closed cages. Smalley and his co-workers named the best known of these molecules *buckminster-fullerenes,* or buckyballs, after the American engineer R. Buckminster Fuller. Each buckyball consists of 60 carbon atoms bonded together in the shape of a soccer ball. The discovery of fullerenes helped bring about a new scientific

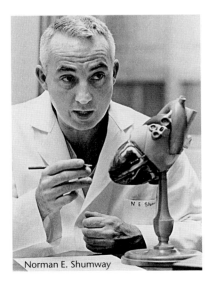

Norman E. Shumway

field called *nanotechnology* (the precise manipulation of individual atoms and molecules to create larger structures).

Sternbach, Leo (1908–Sept. 28, 2005), chemist born in what is now Croatia who helped develop a new class of anxiety-reducing drugs known as benzodiazepines. Beginning in the 1950's, Sternbach led in the development of such medications as chlordiazepoxide (sold under such brand names as Librium) and diazepam (sold under such brand names as Valium). During the 1970's, Valium became one of the best-selling drugs in the United States. However, concerns about the possibility of addiction after long-term use led some physicians to become more cautious about prescribing it.

Taube, Henry (1915–Nov. 16, 2005), Canadian-born chemist who won the 1983 Nobel Prize in chemistry for explaining how electrons are transferred among molecules during chemical reactions. In the late 1940's, Taube began to study reactions involving metals. Metals often form *complexes,* groupings in which other atoms cluster around a metal atom. The atoms bind together by sharing electrons. Taube discovered that during such a reaction, a temporary "bridge" of atoms often forms between metal atoms. Electrons can be transferred across this bridge, speeding up reactions that would otherwise happen only slowly or not at all. Taube's research was important not only to industrial processes but also to an understanding of biochemical processes.

Ullrich, Helen (1922– March 19, 2006), nutrition educator who effectively lobbied in 1969 for nutrition labels on food packaging in the United States and introduced an illustrated teaching tool called the food pyramid in 1988. Her suggestions led to the use of nutritional information labels on food beginning in 1974 and the creation of the first standardized U.S. food guide pyramid, published by the U.S. Department of Agriculture in 1992.

 Heather McShane

DRUGS

The United States Food and Drug Administration (FDA) approved 20 new drugs in 2005 and 5 more as of April 2006. Among these were new treatments for attention-deficit/hyperactivity disorder (ADHD), cancer, depression, diabetes, heart failure, rheumatoid arthritis, sleep disorders, and smoking cessation. In addition, the agency approved 6 new vaccines for the prevention of such infectious diseases as influenza and whooping cough. In 2005, the FDA also allowed the manufacture of 60 new versions of existing drugs, including different doses, different forms (extended release or inhaled rather than injected), and different combinations of drugs already on the market. In addition, the agency approved 452 *generic* medications (less-expensive copies of brand-name drugs whose patents had expired).

First race-related drug. In June 2005, the FDA approved the first drug for treating *heart failure* in black patients. (Heart failure is a condition in which the heart does not pump enough blood.) Sold under the brand name BiDil, the medication is a combination of two older drugs, hydralazine and isosorbide dinitrate. Neither of the older drugs was approved by the FDA for treating heart failure.

In the first two studies of BiDil among racially mixed groups of patients, the drug performed no better than a *placebo* (inactive substance). However, BiDil seemed to be more effective for patients who had identified themselves as black. Researchers then conducted a third trial of 1,050 self-identified black patients with severe heart failure who had already been treated unsuccessfully with the best available therapy. The group that received BiDil had significantly fewer deaths and hospitalizations than the group that received a placebo.

Gentler sleep aid. The first prescription sleep medication not designated as a controlled substance received FDA approval in July 2005. Ramelteon (sold under the brand name Rozerem) does not pose the risk of chemical abuse or dependency that other sleep aids do. The drug targets *melatonin* receptors in the brain, inducing more normal sleep. (Melatonin is a hormone that helps regulate the sleep-wake cycle.) Other sleep aids work by reducing the activity of the central nervous system in general.

Anticancer agents. Three new cancer drugs targeting specific molecules in the body were also approved by the FDA. Nelarabine (sold under the brand name Arranon) was approved in October 2005 for use in adults and children with specific types of *leukemia* and *lymphoma,* blood cancers caused by immune system cells called *T cells* that are multiplying out of control. Nelarabine blocks the cells' ability to reproduce.

Sorafenib (sold under the brand name Nexavar), which went on the market in December 2005, is intended for treating adults with advanced kidney cancer. It targets a number of *enzymes* (molecules that speed up chemical reactions in living things) in tumor cells to stop the cells from multiplying. It also blocks *angiogenesis,* the formation of new blood vessels that deliver oxygen and nutrients to the tumor.

In January 2006, for the first time, the FDA approved a single drug that can treat people with either of two types of cancer. Sunitinib (sold under the brand name Sutent) can be used for patients with a rare stomach cancer called gastrointestinal stromal tumors or for those with advanced kidney cancer. Sunitinib prevents certain enzymes in the cell from nourishing tumors while at the same time killing some of the cells that make up the tumors.

Inhaled insulin. Also in January, the FDA approved the first alternative to injections for treating diabetes since insulin was discovered in the 1920's. (Insulin is a hormone that regulates the body's use of sugars and other nutrients.) Recombinant human insulin (sold under the brand name Exubera) is a powder form of insulin that is inhaled through the mouth using a specially designed inhaler.

Exubera may be used by adults with both type 1 diabetes (usually called juvenile diabetes, in which the body produces virtually no insulin) and type 2 diabetes (the more common form of the disease, which usually develops in middle-aged and older individuals, in which the body does not produce enough insulin). Exubera is a rapid-acting form of insulin that is generally used before or after meals. As a result, many patients need to continue using injected, longer-acting insulin at other times to better control blood-sugar levels.

Treatment for chest pain. In January 2006, the FDA announced the approval of ranolazine (sold under the brand name Ranexa), the first of a new class of drugs for treating pa-

DRUGS continued

tients with chronic *angina* (chest pain). Angina occurs when the heart muscle does not get enough oxygen, usually during exercise. The most common cause of angina is *coronary artery disease,* in which the arteries that supply the heart with oxygen-rich blood are blocked.

Although researchers do not understand exactly how ranolazine works, the drug seems to switch the heart's fuel source from *fatty acids* (components of fat) to *glucose* (a type of sugar). Glucose uses less oxygen. Some older types of medications for angina, such as calcium-channel blockers and nitrates, widen blood vessels and allow more blood and oxygen to reach the heart. Angina drugs called beta-blockers slow nerve impulses that prepare the heart for activity by causing it to beat faster.

Depression patch. In February 2006, the FDA approved selegiline (sold under the brand name Emsam), the first skin patch for depression. The once-a-day patch delivers a type of antidepressant called a *monoamine oxidase inhibitor* (MAOI). MAOI's work by blocking the activity in the brain of a protein called monoamine oxidase. Monoamine oxidase breaks down chemical messengers in the brain after they have delivered their messages. Increasing the concentration of these chemicals can ease the symptoms of depression.

Monoamine oxidase also eliminates a molecule called tyramine, which helps to control blood pressure. In a patient taking an MAOI, levels of tyramine rise, causing potentially fatal increases in blood pressure. Certain foods—including aged cheeses, pepperoni, and red wine—contain tyramine, and so people taking MAOI's must follow a diet that avoids or limits those foods. At its lowest strength, Emsam does not raise tyramine levels significantly, so patients do not have to watch what they eat. However, people taking higher doses of the medication must closely observe dietary restrictions.

Rheumatoid arthritis drugs. A cancer drug called rituximab (sold under the brand name Rituxan) was approved in March 2006 for treating rheumatoid arthritis (RA). RA is an *autoimmune disease,* in which the body's immune system attacks its own tissues. RA causes pain and swelling in the joints and may erode bone and connective tissue.

Rituximab, first approved in 1997 to treat the blood cancer lymphoma, suppresses a spe-

cific type of immune system cell. It is one of a new wave of cancer drugs that target specific features of cancerous cells, making the medications more precise and less toxic than older chemotherapy drugs.

Another drug, approved in December 2005, offers an additional option to patients with RA. Called abatacept (sold under the brand name Orencia), the medication is administered intravenously, generally once every four weeks. Abatacept is intended for patients with moderate to severe RA who have not responded to other medications.

ADHD patch. In April 2006, the FDA approved the first skin patch for treating children with ADHD, a condition found primarily in children. People with ADHD often have unusual difficulty paying attention, sitting still, or controlling impulses.

The patch, named Daytrana, contains a full-day's dose of the medication methylphenidate (also sold under such brand names as Ritalin). It is designed to be worn for up to 9 hours a day, in which case its effects would last for 12 hours. Daytrana is intended for children between the ages of 6 and 12 who have difficulty swallowing pills. The patch is applied to alternate hips daily and is made to withstand normal activities, including swimming and bathing.

In several studies with a total of more than 300 children, those who wore the patch showed significant improvement in ADHD symptoms over children who had been given a placebo. For example, children who wore the patch completed—or at least attempted—more mathematics problems during a timed test than children without the patch.

New medication to stop smoking. A new type of medication to help cigarette smokers quit smoking was approved by the FDA in May 2006. Varenicline tartrate (sold under the brand name Chantix) works in two ways. The drug lessens withdrawal symptoms by partially activating reward centers in the brain that respond to nicotine with feelings of pleasure. In addition, if a patient continues to smoke while taking the medication, the drug blocks nicotine effects in the brain, reducing the pleasure the smoker would normally feel and lessening the strength of the addiction. The treatment is generally prescribed for 12 weeks, but patients may continue to take the medication for an additional 12 weeks if necessary. ■ Judy Peres

ECOLOGY

Debates raged in 2005 and 2006 over the reported rediscovery of a bird believed to have become extinct in the United States in 1944. The last known U.S. population of the ivory-billed woodpecker (*Campephilus principalis*) lived in a forest in northeastern Louisiana. Scientists believed that population became extinct while the area was being logged, despite pleas from conservationists to spare the woodland. Since then, there have been reported sightings of the bird, sometimes by reputable bird watchers. However, these sightings were called into question because a similar large woodpecker, the pileated woodpecker (*Dryocopus pileatus*), can easily be confused with the ivory-billed woodpecker. A population of ivory-billed woodpeckers survived in Cuba after 1944, but scientists believe that population became extinct in the 1980's.

In the June 3, 2005, issue of *Science,* scientists led by John W. Fitzpatrick, an *ornithologist* (scientist who studies birds) from the Cornell Laboratory of Ornithology in Ithaca, New York, offered evidence that a population of ivory-billed woodpeckers still exists in the Cache River National Wildlife Refuge in Arkansas. The authors presented photographic and videographic evidence of the birds. They also captured audio evidence of tree drumming consistent with the sounds made by ivory-billed woodpeckers. Altogether, ornithologists reported seven convincing sightings.

Addressing the question of why there had been no earlier sightings in the refuge, the scientists noted that the bird refuge is part of a remote area comprising about 220,000 hectares (540,000 acres). In addition, the birds have always lived somewhat solitary lives, even when not on the verge of extinction.

The news that the ivory-billed woodpecker still existed was greeted with great excitement by the scientific community. However, in 2006, other ornithologists who studied the evidence argued that Fitzpatrick and his colleagues had misinterpreted the data and that the bird the team recorded was, in fact, a pileated woodpecker.

Fitzpatrick's team stood by its original claim. Scientists hoped that future research and analysis would shed more light on what could be the most significant ornithological find of the 2000's.

SEE ALSO THE SPECIAL REPORT, **NEW GUINEA'S HIDDEN PARADISE,** PAGE 54.

Greenhouse gases in the canopy. New research in 2005 demonstrated that *greenhouse gases* have different effects on different parts of an ecosystem. (Greenhouse gases, including carbon dioxide, ozone, and methane, trap heat from the sun and contribute to a warming of the atmosphere.) In the past, scientists have studied the effects of additional quantities of atmospheric carbon dioxide on the growth of forest trees. However, because of technical limitations, scientists were able to study the effects of carbon dioxide only on small trees whose tops were readily accessible. The use of construction cranes that can lift ecologists into the forest *canopy* (the branches, leaves, and twigs that make up the top layer of a forest) has given scientists the ability to study tall trees.

Plant ecologist Christian Körner of the Institute of Botany at the University of Basel in Switzerland and colleagues studied the effects of increased atmospheric carbon dioxide on plant growth in a Swiss forest. Their findings appeared in the Aug. 26, 2005, issue of *Science*. The scientists used a construction crane to gain access to the canopies of various trees. They snaked tubes that released measured quantities of the gas into the tops of the trees. The researchers then measured plant growth over the next four years.

Previous studies with smaller trees had indicated that additional carbon dioxide promoted tree growth. The results with larger, mature trees were different. Although scientists found differences in the growth rates of different tree species, the additional carbon dioxide seemed to have no effect on overall plant growth.

Scientists found these results interesting for several reasons. First, they showed that in order to draw conclusions about an ecological phenomenon (added carbon dioxide in this case), it is necessary to investigate the entire ecosystem from below ground to the tops of the tallest trees. Second, they demonstrated the importance of studying different species in an ecosystem (temperate forest in this case) in

ECOLOGY continued

order to draw conclusions. Finally, they showed how a new technology (such as the use of a high crane in a forest) can open up new and exciting areas of study.

Construction cranes have become invaluable tools in modern ecological research. The first research crane was erected in 1992 at the Smithsonian Tropical Research Institute in Panama by ecologist Alan P. Smith of the Smithsonian Institution in Washington, D.C. By 2006, ecologists were using 10 cranes worldwide and the use of at least 5 more was planned.

One of the outcomes of the crane work has been international cooperation on canopy and forest research, with several databases currently available to scientists worldwide. Scientists working with research cranes also have found many new animal and plant species and have discovered more about the interrelationships of many species. An estimated 40 percent of terrestrial species of the forest live in the canopy.

"Devil's gardens" decoded. Scientists in 2005 announced finding an explanation for the forest phenomenon known as "Devil's gardens." These are areas of trees within Amazon River basin rain forests that consist almost exclusively of one species of tree, *Duroia hirsuta.* Devil's gardens are formed and maintained by ants, according to research reported in the Sept. 22, 2005, issue of *Nature* by biologists Megan E. Frederickson and Deborah M. Gordon of Stanford University in California and colleague Michael J. Greene of the University of Colorado at Denver. The research documented the first known instance in which an ant modified its environment using *herbicides* (chemical compounds that kill plants).

Ants of the species *Myrmelachista schumanni* live in small hollows in the stems of *D. hirsuta* leaves. The sci-

EROSION TRIUMPHS OVER FLOODING

Sandbars and beaches along a stretch of the Colorado River in the Grand Canyon remain eroded after a failed attempt to restore the areas to more natural sizes. In 2004, scientists had released a four-day controlled flood from Arizona's Glen Canyon Dam in an attempt to rebuild the sandbars and beaches, which serve as habitats for wildlife and plants. The 1964 construction of the dam had reduced by 98 percent the amount of sediment carried by the river. As a result, natural erosion outweighed the small gain from river-borne sediment. Researchers reported in November 2005 that while the flood had created some sandbars, the effects were short-lived.

entists observed the ants systematically visiting seedlings of other tree species in the area. The ants bit small holes in the leaves of these other species and inserted the tips of their abdomens into the holes. The ants then injected the leaves with a formic acid secretion, which act-

ed as an herbicide, killing the seedlings. As a result, *D. hirsuta* plants could grow without competition for water, food, and light and so provided more habitat for the ants.

■ Robert H. Tamarin
See also **BIOLOGY; CONSERVATION.**

ENERGY

A technique for using industrial carbon-dioxide emissions to enhance oil recovery from existing oil fields was first implemented in 2005. Engineers pumped 5 million metric tons (5.5 million tons) of industry-produced carbon dioxide into the Weyburn oil field in Saskatchewan, Canada, and doubled the rate at which oil was being recovered from the field, the United States Department of Energy (DOE) announced in November. The "Weyburn Project" used a technique called *enhanced oil recovery* (EOR) to inject carbon dioxide into the oil field to force more oil to the surface, while trapping the gas in the underlying geologic formations, a process known as *carbon sequestration*.

The carbon dioxide was piped from the Great Plains Synfuels Plant near Beulah, North Dakota, where it is formed as an undesired by-product of *coal gasification*. (Coal gasification is the process of combining coal with steam and oxygen to produce a combustible mixture of gases.) Instead of being vented to the atmosphere, where it could contribute to *global warming*, the carbon dioxide was piped to the Weyburn field. (Global warming is a gradual increase in the average temperature of Earth's surface that began in the 1800's.)

Traditional oil recovery uses natural ground pressure and pumps to bring oil to the surface, a technique that permits recovery of only about 10 percent of the reserve. Engineers have also flooded wells with water, increasing oil output by from 20 to 40 percent. The DOE estimated that EOR could increase an oil field's lifetime production by up to 60 percent, while extending the life of the field by decades.

Scientists project that the technique will enable the Weyburn oil field to produce an additional 130 million barrels of oil, operate for 20 additional years, and sequester as much as 30 million metric tons (33 million tons) of carbon dioxide. The DOE forecast that, if applied to all the oil fields of western Canada, the technique

SEE ALSO CONSUMER SCIENCE, LIGHTING: "LED"ING THE WAY, PAGE 148.

could produce billions more barrels of oil and reduce carbon-dioxide emissions by an amount equal to that produced by 200 million cars in one year.

The Weyburn Project is led by Canada's Petroleum Technology Research Centre in Regina, Saskatchewan, and cosponsored by EnCana of Calgary, Alberta, the oil field operator. The project is also funded by government and industry groups in Canada, Japan, the European Union, and the United States.

Hydrogen power. The construction of the first hydrogen-fueled power plant in the United States was announced in February 2006 by two major energy companies. The plant will be constructed by BP, a global energy company based in London, and Edison Mission Group, owned by Edison International of Rosemead, California. The $1-billion facility will be located near BP's Carson refinery, about 30 kilometers (20 miles) south of Los Angeles.

The plant will be able to generate 500 megawatts of electric power, enough to serve 325,000 homes in southern California. Engineers will use carbon sequestration to store most of the plant's carbon-dioxide emissions underground, preventing about 3.6 million metric tons (4 million tons) of the gas from entering the atmosphere each year. Instead, the carbon-dioxide will be used to increase production from California oil fields.

The plant will process *petroleum coke,* a solid residue produced during petroleum refining, converting it to gas streams of hydrogen and carbon dioxide. The hydrogen gas will fuel a *turbine* to generate electric power. (A turbine is a device with a rotor turned by a moving fluid

ENERGY continued

or other energy source.) About 90 percent of the carbon-dioxide gas will be captured and transported by pipeline to an oil field.

Storing sunlight. The first *flow battery* designed for use in homes and small businesses hit the market in September 2005. A flow battery is a device that converts electrical energy into chemical energy for storage purposes and then converts it back to electrical energy for use as a power supply. VRB Power Systems, a developer of energy storage systems in Vancouver, Canada, sold the first of a new line of small flow battery systems to the Phil Little Design Foundation, a sustainable housing developer based in Queensland, Australia.

Flow batteries can be used to store energy from the sun, wind, and other *renewable energy sources* (sources that cannot be used up). For example, electric power from a wind turbine generator fed into the flow battery may be stored as chemical energy. Flow batteries have tanks of liquid *electrolyte* (substance that conducts electric current) in which chemical energy is stored.

The most common type of flow battery is the vanadium redox battery (VRB), in which the metal vanadium is used to make liquid electrolytes. Two electrolytes made of different *ions* (charged atoms) of vanadium are pumped from storage tanks into *flow cells*, chambers where *electrons* (negatively charged particles) move from one electrolyte to another. This movement creates a current that can serve as an electric power source. The process is reversible, allowing the battery to be discharged and charged again.

The VRB energy storage system can be installed in a solar-powered home to provide a constant supply of electrical energy. The system will store the sun's energy during the daytime in a flow battery. The energy will then be available at night, reducing the need for electric power from commercial power suppliers.

Slush hydrogen. A high-performance, low-volume fuel being developed for rockets may someday provide a low-pollution fuel source for cars and trucks. Equipment used to produce a fuel from *slush hydrogen*, a portable

A FUSION FIRST

The first fusion reactor that is expected to release more energy than it uses was approved for construction in May 2006. The project—named ITER, a Latin word meaning *the way*—is an international collaboration estimated to cost as much as $12 billion. ITER, shown in an artist's rendition, consists of a 30-meter (100-foot) tall machine called a *tokamak* that confines and heats a *plasma* (form of matter made of charged particles) until the particles combine and release enormous amounts of energy. Scientists predict that ITER will generate about 500 megawatts of power, the output of a large electric power plant.

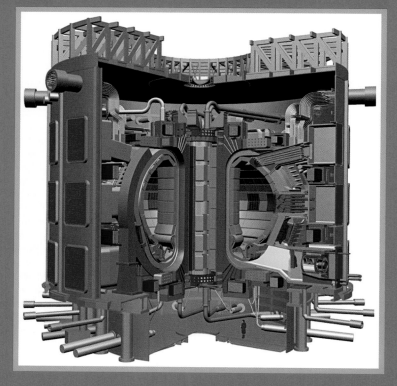

mixture of solid and liquid hydrogen, was successfully tested in March 2006. The slush hydrogen was developed as part of a study for the European Space Agency. The equipment was developed by ET EnergieTechnologie GmbH, located near Munich, Germany, in cooperation with automobile manufacturer Magna Steyr of Oberwaltersdorf, Austria.

As a fuel, slush hydrogen has advantages over liquid hydrogen. The slush has a density that is from 10 to 20 percent higher than the liquid, enabling it to hold more energy per unit volume, thereby reducing the space needed to store the fuel. In addition, slush hydrogen can be stored for longer periods of time with less loss from evaporation.

Magna Steyr developed a generator that produces slush hydrogen from liquid hydrogen in the same manner that a snow canon produces artificial snow from liquid water. The equipment uses compressed gas to force liquid hydrogen through a nozzle. Tiny droplets of the liquid are thrust into an ultracold environment, below hydrogen's freezing point of -259.14 °C (-434.45 °F), where they turn into slush hydrogen. ■ Pasquale M. Sforza

See also **ATMOSPHERIC SCIENCE; CHEMISTRY.**

ENGINEERING

The United States Corps of Engineers worked in 2005 and 2006 to repair levees and other flood control structures in New Orleans, Louisiana, damaged by Hurricane Katrina. The Corps won praise for finishing an $800-million restoration of the city's levees to their pre-Katrina condition in late May 2006, just before the official start of the hurricane season on June 1. The work included repairing or rebuilding damaged levees and fortifying concrete flood walls. The Corps was also constructing navigable floodgates across canals that drain into Lake Pontchartrain, which lies north of the city. However, preliminary reports released in 2006 by two independent panels as well as a report from a government task force agreed that the Corps was responsible for flaws in the design, construction, and maintenance of the system that led to most of the damage.

Hurricane Katrina, which struck on Aug. 29, 2005, ranked as one of the strongest storms on record. The hurricane drove a rush of seawater called a *storm surge* up the Gulf Intracoastal Waterway and into the city's Inner Harbor Navigation Canal. The force of the surge both overtopped and breached the city's levees. Breaches in other levees allowed water from Lake Pontchartrain to sweep into other neighborhoods. About 80 percent of New Orleans, much of which lies below sea level, was flooded.

The Corps of Engineers began constructing a hurricane protection system in New Orleans after Hurricane Betsy flooded the city in 1965. The Corps built the system to withstand a Category 2 storm—that is, a storm with winds of up to 177 kilometers (110 miles) per hour. Many government and public safety officials had unsuccessfully urged the Corps to build a stronger system.

A preliminary report on the 2005 disaster by the Corps blamed the failure of the hurricane protection system on unanticipated design flaws and the incredible power of the storm. In a letter to the commander of the Corps, released in March 2006, however, the members of one panel reported that the damage in New Orleans had resulted from inadequate engineering standards that were "too close to the margin" to protect the people of such a large city. That panel, which was charged with monitoring the Corps's restoration efforts, consisted of engineers chosen by the American Society of Civil Engineers in Reston, Virginia. Although Corps officials said they agreed with the letter's assessment, they noted that a lack of funding from the federal and Louisiana state governments had limited their ability to properly maintain the levees.

In May, a report from the second panel, which was financed by the National Science Foundation (NSF), a U.S. agency based in Arlington, Virginia, blamed the disaster on a "culture of inattention." The panel's lead investigator reported, "People didn't die because the storm was bigger than the system could handle, and people didn't die because the levees were overtopped. People died because mistakes were made and because safety was exchanged for efficiency and reduced cost."

A report from the federal Interagency Performance Evaluation Task Force, released on June 1, said that the New Orleans flood control

ENGINEERING continued

structures "did not perform as a system." The report noted that the system had been built in an unorganized way using flawed designs and engineering techniques as well as outdated information about the degree to which the city had sunk farther below sea level over the years.

After completing repairs to the levees, the Corps began upgrading the flood-protection system. Most of the levees around New Orleans consisted of low earthen mounds. Some of the mounds were topped with concrete walls called I-wall flood barriers. Anchoring the walls were interlocking *sheet piles* (thin sheets of steel driven into the ground). The walls stood about 4 meters (13 feet) high, with the piles extending about 3 meters (10 feet) into the mound. According to the NSF panel, these levees failed, in part, because the piles were not driven deeply enough into the ground and could not withstand either the shifting of the soils beneath them or the violence of the storm waters.

By 2010, the Corps planned to replace I-walls with new walls built in the shape of an inverted T or L, which have a stronger base. New walls will stand as tall as 4.6 meters (15 feet), with piles set as deep as 13.7 meters (45 feet) below sea level. In some places, the walls would have a bracing pile up to 28 meters (92 feet) long placed diagonally through the mound to provide extra stability and support. In addition, the levees would have concrete armoring at the base of the walls on the seaward side and other structures to prevent erosion. The Corps also planned to upgrade or build additional pumping stations.

Constructive computer programs. In 2006, building information modeling (BIM), a revolutionary method of visualizing construction projects, became increasingly popular among architects, contractors, and engineers. A BIM program allows various groups involved in designing and constructing a building to combine their plans and specifications into one highly detailed three-, four-, or five-dimensional model. BIM images may be detailed enough to show the bolts on steel beams. In 2006, BIM programs were being used for such U.S. projects as the Denver (Colorado) Art Museum and the California Academy of Sciences in San Francisco.

Building design and construction traditionally has involved a number of professionals who work independently on different aspects of the project. In contrast, BIM images allow all members of the construction team to see all elements of the project in one model. For example, the virtual model used by designers also can be used by the construction team for fabricating materials and scheduling subcontractors.

Supporters of BIM systems contend that the models improve productivity and quality control, streamline communication, and aid scheduling and organization. The systems can also help the project team reduce design errors, identify building materials that do not meet standards, and pinpoint problems before construction begins. BIM also allows customers to see alternative designs or materials quickly and at a lower cost than with paper-based models.

No task too small. The creation of the smallest robot that can be steered without wires or rails was announced in September 2005 by researchers led by computer scientist Bruce Donald of Dartmouth College in Hanover, New Hampshire. The microrobot is about as wide as a strand of human hair and about half the length of the dot above a printed "i." It is about 10 times as short and thousands of times as light as any other untethered, controllable robot, the researchers claimed.

The microrobot consists of a piece of silicon, two bits of memory, and two *actuators* (devices that convert energy into motion). One actuator controls the device's forward motion; the other, its turning motion. The device crawls forward like an inchworm, taking about 20,000 "steps" per second. The microrobot turns by extending a silicon "foot" and pivoting like a motorcyclist skidding around a tight turn, Donald elaborated.

The scientists control the microrobot's direction and velocity using the flat *electrostatic* grid (a surface with a nonmoving electric charge) on which it crawls. An increase in the charge makes the robot flex its "tail" and come in contact with the grid, which advances the robot's shorter front "leg" 10 nanometers. (A nanometer is 0.000000001 meter [$\frac{1}{25,400,000}$ inch].)

According to Donald, swarms of these microrobots could be used to repair other nanoscale devices as well as the circuitry in computer chips, explore hazardous environments, and manipulate individual cells.

Sizing things up. A new device that is attached to a camera enables engineers and scientists to determine the size of objects in a photograph lacking reference points, the National Aeronautics and Space Administration (NASA) announced in February 2006. The device, which is called the Laser Scaling and Measurement Device for Photographic Images (LSMDPI), was developed to help NASA engineers determine the size of dents from hail or space debris on the exterior of the space shuttle.

In order to determine the size of an object in a photograph, a viewer often must refer to another object of a known size near it. For example, a movie audience can recognize that a monster is huge because the people near it appear much smaller. NASA engineers used the same approach when developing the LSMDPI. When a user takes a picture, two laser beams shoot from the device and project a pattern of dots in the field of view. The dots, which appear in the photograph, then act as reference points for the object in question. The user downloads the photograph onto a computer equipped with scaling software, selects the reference points, and keys in the distance between the dots. The scaling software then calculates the size of the object.

The LSMDPI also has attracted the attention of law enforcement officials for analyzing and measuring objects in crime-scene photographs, including blood spatters on walls. The photographs also can be viewed from various angles for a more thorough analysis of the scene.

Fastest bullet train. A prototype of a high-speed passenger train that can reach speeds of 400 kilometers (250 miles) per hour was unveiled in April 2006 by the East Japan Railway Company in Tokyo. Company officials reported that they expected to put the new Fastech 360Z into operation between Tokyo and Aomori in 2011. The world's fastest wheeled train in operation in 2006, France's TGV, normally operated at 350 kilometers (220 miles) an hour but held the record for fastest wheeled train at 515 kilometers (320 miles) per hour.

The Fastech 360's engineering innovations include nose cones designed to minimize *drag* (an aerodynamic force that pushes in the direction opposite that of the object's motion) and peaked *spoilers* (flaps) that rotate out of the roof

THE SECRET OF THE BEAK

The secret to the lightweight strength of a toucan's beak is its foamy interior, researchers at the University of California in San Diego (UCSD) reported in January 2006. The interior consists of a network of strong, stiff bony fibers connected by drumlike membranes (shown in an image from a scanning electronic microscope, below). Because the huge beak is mainly air, it efficiently absorbs energy transmitted when the bird feeds or defends itself. The UCSD scientists reported that the foam could guide the development of lightweight, energy-absorbing materials for aircraft and automobiles.

500 μm

ENGINEERING continued

of each car to help slow the train in an emergency. The train also has more efficient, quieter connections to overhead electrical lines and cars that tilt inward on curves to ensure a smoother ride.

Catching an energy wave. The first of three wave-energy converters destined for the world's first commercial wave power farm arrived in Portugal for assembly in March 2006. The farm, which will use ocean waves to generate electric power, will be built about 5 kilometers (3 miles) off the northern coast of Portugal, near Póvoa de Varzim. The three devices will be able to produce 2.25 megawatts of electric power, enough for at least 1,500 nearby homes, beginning later in 2006. If successful, the farm will be expanded to 33 devices.

The sausage-shaped Pelamis converters are 140 meters (460 feet) long and 3.5 meters (11.5 feet) wide. They consist of cylindrical segments connected by hinged joints. The devices will be anchored at both ends to the ocean floor at depths from 40 to 100 meters (130 to 330 feet) to tap into deep, high-energy swells. As waves pass over the devices, the devices will move up and down and from side to side. The movement of the segments relative to one another will push oil through hydraulic motors at high pressures. The motors will power electric generators, which will produce electric power. A cable laid across the seafloor will carry the power to shore. In addition to producing *renewable* (replaceable) energy, the farm was expected to balance the 5,440 metric tons (6,000 tons) of carbon-dioxide emissions generated annually by a traditional power plant.   ■ Celeste Baine

See also **CHEMISTRY; COMPUTERS AND ELECTRONICS.**

ENVIRONMENTAL POLLUTION

Exposure to routine levels of two common forms of air pollution may cause a small increase in blood pressure in healthy adults, according to a report published in *Environmental Health Perspectives* in August 2005. While not hazardous to healthy adults, such an increase could cause serious problems in older individuals and people already suffering from high blood pressure or in individuals with weakened hearts.

Researchers led by Bruce Urch, an occupational and environmental health scientist at the University of Toronto in Ontario, exposed 23 healthy volunteers ranging in age from 19 to 50, to harmful *particulates* (small particles) and molecules of *ozone* (a form of oxygen). The volunteers inhaled the pollutants in densities similar to those found in heavily polluted cities or in certain traffic situations—for example, traffic stuck in tunnels. Scientists then measured the subjects' blood pressure.

Blood pressure is the force that blood exerts against the walls of the arteries. Measurements of blood pressure include two numbers, such as 120/80. The first number refers to *systolic pressure,* the blood pressure while the heart is contracting. The second

number refers to *diastolic pressure,* the blood pressure while the heart is relaxing.

Exposure to the pollutants raised the diastolic blood pressure in the subjects an average of 9 percent. Although the volunteers' blood pressure fell quickly after they stopped breathing the pollutants, the researchers hypothesized that serious health effects could occur among city dwellers constantly exposed to such polluting agents.

Reproductive abnormalities. Pregnant women with moderate-to-high blood levels of four common industrial chemicals significantly increase their risk of giving birth to boys with abnormalities in their reproductive system, according to a June 2005 report in *Environmental Health Perspectives.* These chemicals, commonly known as plasticizers, belong to a group of chemicals called *phthalates (THAL aytz)* and are found in adhesives, flooring, nail polish, paints, perfume, and plastic containers, among other products.

The study, directed by epidemiologist and biostatistician Shanna Swan of the University of Rochester School of Medicine in Rochester, New York, measured blood levels of phthalates in the mothers of 85 newborn boys. The boys born to mothers with the highest levels of phthalates

Residents of Harbin, China, fill buckets with water from a tanker on Nov. 25, 2005, after a chemical spill in the Songhua River contaminated the water supply for millions of people. On November 13, an explosion at a chemical plant in the city ruptured storage containers, sending 91 metric tons (100 tons) of cancer-causing chemicals into the river. The toxic spill later flowed into the Amur River in Russia.

were the most likely to have a reproductive-system abnormality, including a small penis or *testicles* (male sex glands) that had failed to enter the *scrotum* (testicle pouch) before birth.

Researchers believe that phthalates in mothers may reduce the production of the male sex hormone testosterone in developing babies, in turn, causing abnormal sex organs. The effects, they suggested, may also lead to sexual and health problems for the boys later in life, including lower sperm production and an increased risk of testicular cancer.

Naturally good. Children who switch to organic foods experience a rapid and dramatic decline in pesticide levels in their bodies, according to a study published in *Environmental Health Perspectives* in February 2006. Organic foods are grown or prepared with natural fertilizers and without the use of pesticides and other chemicals. The pesticides in the study included malathion and chlorpyrifos, which belong to a group known as *organophosphates*. According to the Washington, D.C.-based Environmental Protection Agency (EPA), pesticides can alter the nervous system of children.

Environmental and occupational health sci-entist Chensheng Lu of Emory University in Atlanta, Georgia, led the study of 23 children from families living in Seattle. The children, who ranged in age from 3 to 11, ate conventionally grown foods for three days. For the next five days, their parents prepared meals using only organically grown foods, among them cereals, fruits, juices, pastas, and vegetables. The scientists then asked parents to resume meal preparation using conventionally grown food for six days.

Researchers monitored the levels of the pesticides in the children by testing their urine twice daily. The researchers found that pesticides vanished from most urine samples during the five-day period when meals were prepared with organically grown food.

Amphibians and pesticide drift. Pesticides applied to crops in California's agri-culturally rich Central Valley may be responsible for the deaths of some species of frogs and toads in the nearby Sierra Nevada mountains, according to a study presented at a meeting of the Society of Environmental Toxicology in November 2005. The study provided additional support for the theory that pesticides may play

ENVIRONMENTAL POLLUTION continued

a significant role in the massive die-off of amphibians that is occurring worldwide.

The migration of pesticides from the site where they are applied is known as *pesticide drift*. Prevailing winds carry some pesticides from the application site immediately. In addition, pesticides that land on crops may later become airborne through evaporation and be carried to sites downwind. Pesticide drift has raised concerns among environmental scientists because human and wildlife populations may become exposed to potentially dangerous levels of the toxic chemicals.

To test the effect of pesticide drift in the Sierra Nevada range, Donald W. Sparling, a wildlife toxicologist at Southern Illinois University in Carbondale, exposed amphibian eggs collected from ponds in the mountains to endosulfan, a pesticide commonly used in the Central Valley. Sparling found that one-half of the eggs from the foothills yellow-legged frogs died when exposed to endosulfan levels commonly found in the range's ponds and lakes. Sparling found that even lower concentrations of the pesticide caused higher-than-normal death rates among this species as well as one species of toad.

Researchers have long expressed concerns about the alarming death rates among amphibians throughout the world and the impact of this development on aquatic ecosystems. Scientists have proposed a number of explanations for the die-off, including ozone depletion and environmental pollution. Sparling's research suggests that pesticides may play an important role in this phenomenon.

Feminized fish. Male cod that live in the open ocean far from sources of human-made pollution produce a female protein that also has been found in male cod and other fish that inhabit waters contaminated by pollution from sewage-treatment plants and paper mills. That finding, reported in November 2005, suggests that certain pollutants may be working their way through the ocean's *food chain* (the system by which energy is transferred from one living thing to another in the form of food). In female cod, the protein, known as vitellogenin, is produced by the liver and deposited in eggs.

When ingested by male cod, the pollutants mimic the female hormone estrogen, interrupting male sexual development and fertility. Researcher Alexander P. Scott of the Centre for Environment, Fisheries, and Aquaculture in Weymouth, England, found the same hormone imbalance in larger, older male cod that had spent their entire life at sea.

According to Scott, the chemicals are carried by ocean currents to the deep sea, where they become deposited in sea-floor sediments. There, bottom-dwelling eels and other fish upon which the cod feed ingest the chemicals. As the male cod grow older, concentrations of the estrogenlike pollutants build up in their tissues, including the liver, where they stimulate the production of vitellogenin. Scott contended that such exposure could reduce the ability of this commercially important food source to reproduce. The findings also underscore how preventing and controlling pollution at the local level may avert serious global impacts.

New Orleans floodwaters. Samples of floodwater collected in New Orleans in the wake of Hurricane Katrina were less toxic than many health experts had feared, according to a study published in *Environmental Science and Technology* in October 2005. Environmental engineer John Pardue at Louisiana State University in Baton Rouge and colleagues found high levels of bacteria and certain chemicals, including lead and arsenic, in the floodwaters. They concluded, however, that these levels were similar to those in typical storm-water runoff and posed no real danger to people. Slightly elevated levels of gasoline raised some concern among the researchers, however.

Nonstick coating concerns. The EPA asked eight U.S. companies in January 2006 to significantly reduce their use of a suspected cancer-causing chemical called perfluorooctanoic acid (PFOA) in popular consumer goods. Numerous scientific studies found that 95 percent of people living in the United States have PFOA in their bloodstream. Research has linked PFOA to cancer and birth defects in animals.

The eight manufacturers—including DuPont, based in Wilmington, Delaware, and 3M, based in St. Paul, Minnesota—used PFOA in the nonstick coating Teflon, in stain- and water-resistant fabrics, and in fast-food packaging, among other items. By March 2006, the companies had reduced their use of PFOA by 95 percent. The EPA requested that they completely eliminate its use by 2015. ■ Daniel D. Chiras

See also **CONSERVATION; ECOLOGY.**

Rachel Carson

In her classic book *Silent Spring* (1962), ecologist and writer Rachel Carson imagined a place where a deadly cloud creeps across the land, stilling all birdsong. That powerful image—and Carson's accompanying research that carefully documented the destructive effects of pesticides on birds and other wildlife—helped launch the modern environmental movement. Her work led to the establishment of the Environmental Protection Agency and restrictions on the use of certain pesticides as well as the passage of the Clean Air Act of 1963 and other laws that form the backbone of government conservation efforts. Recollections of this legacy as well as of Carson's passionate belief in the interrelation of all living things and the importance of the natural world to human welfare will certainly mark events in 2007 celebrating the centennial of her birth.

Born in Springdale, Pennsylvania, Carson grew up on her family's 26-hectare (65-acre) rural homestead overlooking the Allegheny River. Her mother's intense appreciation of the natural world and the many hours Carson spent in the gardens and fields around her home inspired her love of nature. The fossilized shells she found— the remnants of an ancient ocean—helped kindle her lifelong interest in the sea. Carson thought of being a writer from an early age and often wrote poems and stories about nature. She published her first article when she was only 10 years old.

Carson entered the Pennsylvania College for Women (now Chatham College) in Pittsburgh as an English major. A class in biology, however, drew her toward science. She graduated in 1929 with a bachelor's degree in zoology. In the summer of 1929, before starting graduate school at Johns Hopkins University in Baltimore, Carson received a scholarship to study at the Marine Biological Laboratory in Woods Hole, Massachusetts. There she saw the ocean for the first time. Her fascination with tide pools as well as the creatures and plants that inhabited the boundaries between water and land formed the core of her professional work. She completed her master's degree in science in 1932. Carson considered pursuing a doctorate, but the need to care for her mother led her to take part-time teaching positions at Johns Hopkins and the University of Maryland in Baltimore. After her sister died, Carson took in her two nieces.

A SENSE OF WONDER

Rachel Carson uses binoculars to examine the beauty of the woods near her Maryland home in 1962. She inspired many others to admire nature's offerings and to question human acts that might harm other living things.

CONTINUED

In 1936, Carson, now the head of a household, took a part-time job writing radio scripts for the Bureau of Fisheries (later the U.S. Fish and Wildlife Service [FWS]). The following year, she was hired as a full-time junior aquatic biologist, only the second woman to receive a professional position at the bureau. Appreciative of her writing skills, the bureau gave her numerous opportunities to produce a variety of materials, from pamphlets to technical journals. From 1949 to 1952, Carson served as the FWS's editor-in-chief. Her work earned her a Distinguished Service Award from the Department of the Interior.

Carson also wrote for sources outside the bureau. In 1937, the *Atlantic Monthly* published her article "Undersea," which won high praise and became the basis of her first book, *Under the Sea-Wind* (1941). In 1950, Carson won a writing fellowship, which supported the research for her next book, *The Sea Around Us* (1951). In addition to becoming a best seller, the book brought her a National Book Award and an award for Woman of the Year in Literature in 1951. Now financially secure, Carson resigned from her government job. She bought a home in Silver Springs, Maryland, where nearby salt marshes and the ocean enriched her work. Her third book, *The Edge of the Sea* (1955), explored the terrain between sea and land.

In the mid-1950's, Carson's attention turned with increasing urgency to the effects of human beings on the environment. In particular, she became concerned about the indiscriminate use of such pesticides as dichloro-diphenyl-trichloroethane, more commonly known as DDT. *Silent Spring*, which she worked on from 1958 to 1962, gave voice to those concerns. In the book, she argued that DDT killed both harmful and beneficial insects as well as birds. She also worried about the pesticide's effect on human food supplies.

In addition to collecting critical acclaim, the book aroused a firestorm of criticism, particularly from the chemical industry. Articles attacked Carson personally, and industry scientists contradicted her work. In fact, Carson's recommendations in *Silent Spring* were very modest. Her recommendations included labeling pesticides, regulating their use, undertaking more research on the health effects of chemicals on people, and using biological means to control insect pests. In testimony before Congress in 1963, Carson stated: "If we are ever to solve the basic problem of environmental contamination, we must count the many hidden costs of what we are doing and weigh them against the gains or advantages." In 1966, Congress responded to public concern about the widespread use of pesticides by severely limiting the use of DDT and then banning it from the United States altogether in 1972.

Rachel Carson died in 1964. President Jimmy Carter posthumously awarded her the Presidential Medal of Freedom in 1980 for creating "a tide of environmental consciousness that has not ebbed."

■ Patricia M. DeMarco

FOSSIL STUDIES

A study of marks left on fossilized leaves revealed that South America, which is more biologically *diverse* (has a larger variety of species) than modern North America, was also more diverse than North America 52 million years ago. This finding was reported by paleontologists Peter Wilf of Pennsylvania State University in University Park; Conrad C. LaBandeira of the Smithsonian Institution in Washington, D.C.; Kirk R. Johnson of the Denver Museum of Nature and Science in Colorado; and N. Rubén Cúneo of the Museo Paleontológico Egidio Feruglio in Trelew, Argentina.

The scientists described fossil leaves and other plant remains that show marks made by a wide array of insects millions of years ago. The paleontologists' samples came from 52-million-year-old deposits from the Eocene Epoch at Laguno del Hunco, Patagonia, in southern Argentina; Washington state; Wyoming; and Utah. The early Eocene was a time of extreme global warmth that, scientists had theorized, promoted the diversity of plant and insect species. The scientists confirmed this theory using the traces that ancient insects left in fossil plants.

The researchers identified 52 types of insect-produced leaf damage, including feeding scars, piercing and sucking marks, and *galls* (abnormal growths on plants caused by organisms that feed on the plant). The scientists noted that the diversity and abundance of plant damage inflicted by insects were similar to those seen in modern environments. Notably, the sample from Argentina, containing nearly 3,600 leaves, showed the highest diversity of plants and the greatest number of feeding marks. The traces of feeding marks and other damage by insects have provided a rich source of data demonstrating that current animal-plant relationships apparently had developed to a large extent at least 52 million years ago.

SEE ALSO THE SPECIAL REPORT, *TIKTAALIK:* A LANDMARK DISCOVERY, PAGE 62.

Spider blood in amber. The discovery in 2005 of ancient spider "blood" in amber could help researchers further their understanding of the evolution of arachnids. Amber, derived from the hardened resins of plants, is noted for its ability to preserve the delicate structures of ancient organisms. Insects and

ULTRAVIOLET *ARCHAEOPTERYX*

The remains of a fossilized *Archaeopteryx*, seen in ultraviolet light, have enhanced scientists' understanding of the relationship between birds and dinosaurs. *Archaeopteryx* was a creature that lived about 130 million years ago and exhibited features of birds as well as those of dinosaurs. The ultraviolet light revealed minerals in the fossilized bones, allowing scientists a look at the creature in greater detail than ever before.

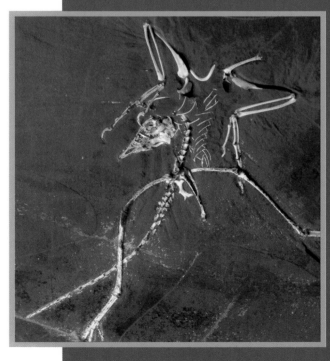

FOSSIL STUDIES continued

spiders often become trapped in blobs of the sticky resin and are gradually entombed within them. Generally, the process occurs slowly, allowing the early stages of decay to take place.

However, in September, paleontologist David Penney of the University of Manchester in the United Kingdom, described a remarkably preserved spider that included drops of *hemolymph, a* bluish substance that functions as a spider's blood. The specimen is from 20 million to 15 million years ago and comes from the famous amber deposits in the Dominican Republic. From the position of the spider's body in the amber, Penney surmised that it must have been entombed almost instantly in flowing resin as it climbed up a tree. Penney noted that it might be possible to extract DNA (deoxyribonucleic acid, the molecule that directs growth in all living organisms) from the sample.

Crested dinosaur from China. An odd dinosaur with a crest on its head appears to be the oldest member of the family that includes *Tyrannosaurus rex.* Scientists reported in February 2006 on their discovery of the fossils, which are approximately 155 million to 160 million years old, in the Xinjiang region of western China.

In February 2006, paleontologist Xu Xing of the Institute of Vertebrate Paleontology and Paleoanthropology in Beijing, China, and several colleagues described *Guanlong wucaii* ("crowned dragon of the five-colored rocks") based on two well-preserved skeletons. The fossils measured about 3 meters (10 feet) long and are among the most complete tyrannosaurlike skeletons ever discovered. The skulls of *G. wucaii* showed a distinctive crest—a tall projection with hollow depressions. The scien-

CRESTED TYRANNOUSAUR

A crest filled with air sacs adorns the head of the oldest known tyrannosaur, shown in an artist's re-creation based on 160-million-year-old fossils found in western China. The discovery of the fossils was reported in February 2006 by an international team of researchers. The adult tyrannosaur, named *Guanlong wucaii,* measured only about 3 meters (10 feet) long and 2 meters (6 ½ feet) tall. It lived more than 90 million years before *Tyrannosaurus rex.*

tists suggested that the crests functioned for such displays as mate attraction. *G. wucaii* was only about one-fourth the size of *Tyrannosaurus rex* and had well-developed arms with three fingers, in contrast to the short, two-fingered arms of its famous relative. The new dinosaur helps confirm that the tyrannosaurids were closely related to slender raptors called coelurosaurs and their descendants, birds.

Jurassic beaverlike mammal. The discovery of a fossilized aquatic mammal in China has demonstrated that specialized mammals existed early in their evolution. Mammals lived alongside dinosaurs for at least 100 million years, but scientists have long believed that these small furry animals were unspecialized insect-feeders. However, in February 2006, paleontologists led by Qiang Ji of the Chinese Academy of Geological Sciences in Beijing reported the discovery of a mammal only about 50 centimeters (20 inches) long, in 164-million-year-old (Jurassic-Period) lake sediments in northern China. The well-preserved fossil shows impressions of skin with fur, webbed feet, and a scaly, broadened tail somewhat like a beaver's.

Although the mammal was not closely related to modern otters or beavers, it may have had a somewhat similar lifestyle. The scientists named the new fossil *Castrocauda lutrasimilis,* meaning "beaver-tailed animal that looks like an otter." The animal probably weighed from 500 to 800 grams (1 to 1.8 pounds), making it the largest known Jurassic mammal. The scientists speculated that it may have resembled and been loosely related to the platypus.

Several of *Castrocauda*'s features led the paleontologists to conclude the animal may have lived along riverbanks and swam to capture fish for food. Soft tissue around the animal's feet suggests that they were webbed. This feature, along with its tail, would have greatly aided its ability to swim. In addition, *Castrocauda* had sharp, thin teeth, somewhat similar to those used by modern-day seals to eat fish.

Dinosaurs ate grass. Grasses existed at the time of the dinosaurs, according to research reported in November 2005. Until recently, scientists believed that grasses did not evolve until about 55 million years ago, some 10 million years after dinosaurs became extinct. As a result, the discovery of distinctive *phytoliths* (tiny pieces of silica in plant tissues) representing five types of grasses in dinosaur *coprolites* (fossilized dung) came as a surprise

to paleobotanist Vandana Prasad of the Birbal Sahni Institute of Paleobotany in Lucknow, India, and several colleagues.

The scientists reported finding phytoliths in dinosaur coprolites from the Late Cretaceous Period in India. Not only did this discovery greatly extend the range of modern grasses back perhaps 30 million years but also offered clues to dinosaur diets. The coprolites contained grass remains as well as the remnants of conifers and palms. The dinosaurs, probably long-necked sauropods, apparently fed freely on these different types of vegetation.

Predation in ancient shellfish. Paleontologists in 2005 discovered evidence that some types of ancient shellfish may have preyed on other types of shellfish by accident or in confusion. Carnivorous *gastropods* (snails) drill through shellfish, including other gastropods and clams, with a rasplike tongue called a *radula,* leaving a record of their predatory act in the form of a circular hole in the shell. Brachiopods are a group of shellfish unrelated to clams that have a *bivalved* (two-hinged) shell. Unlike clams, brachiopods are not particularly nutritious prey. However, the fossil record of brachiopods, which is extensive, shows drill holes made by gastropods going back nearly 500 million years.

In June 2005, paleontologists Michal Kowalewski and Alan P. Hoffmeister of Virginia Polytechnic Institute and State University in Blacksburg; Tomasz K. Baumiller of the University of Michigan in Ann Arbor; and Richard K. Bambach of the Harvard University Botanical Museum in Cambridge, Massachusetts, reported on the abundance of drill holes in fossil brachiopods over the past 450 million years. Although the number of brachiopod shells with drillings increased slightly through time, the frequency remained low, generally less than 5 percent of shells in any given population of brachiopods.

In contrast, the frequencies of drillings in clams greatly increased over this time. In modern populations, as many as half of the individuals may be drilled. The paleontologists concluded that the low frequency of drillings in brachiopods was largely the result of accidental predation.

According to the researchers, gastropods do not prefer brachiopods, but they do not actively reject them. Instead, the researchers noted, gastropods occasionally drill brachiopods by mistake when stressed, confused, or inexperienced, or when other food is not available. This study

FOSSIL STUDIES continued

showed how some evolutionary trends may be an incidental effect of others. The great increase in gastropod drilling in clams fostered a "ripple effect" on other nonpreferred prey species.

Plesiosaur diet. An examination of two newly discovered specimens of Cretaceous elasmosaurid (plesiosaur), a long-necked, carnivorous marine reptile, revealed *bromalites* (preserved masses of fecal material) in their abdominal cavities that offered clues about the foods they ate. In October 2005, Australian paleontologists Colin R. McHenry of the University of Newcastle in New South Wales, Alex G. Cook of the Queensland Museum, and Stephen Wroe of the University of Sydney reported that the bromalites contained large amounts of *gastroliths* (stomach stones). The paleontologists traced the stones to a site about 300 kilometers (190 miles) from the place where the remains of the plesiosaurs were found.

The plesiosaurs gulped the gastroliths from near-shore gravels and used them to help grind their food. Gastroliths are an unusual find in elasmosaurs because the stones are generally found in *herbivores* (plant-eaters), which use them to grind tough plant matter. Scientists had thought that elasmosaurs fed exclusively on soft prey, including fish and squidlike animals. However, the newly discovered bromalites also contained the remains of *crinoids* (sea lilies), bivalves, and gastropods that lived on the sea bottom. Their presence indicates that the plesiosaurs could use their long necks to grasp prey from the sea bottom, which they digested using gastroliths.

Short-necked sauropod. Sauropods, which include *Apatosaurus* (also known as *Brontosaurus*) and *Brachiosaurus,* are well known as huge, long-necked, herbivorous dinosaurs. However, a newly discovered fossil, *Brachytrachelopan mesai,* described in June 2005 by paleontologist Oliver W. M. Rauhut of the Bavarian State Collection for Paleontology and Geology in Munich, Germany, had a stubby, short neck. The new species comes from 150-million-year-old Jurassic beds in Patagonia.

Paleontologists have suggested that the long necks in most sauropod species evolved as a response to limited food supplies. These necks allowed sauropods to reach vegetation in high treetops. *Brachytrachelopan* may have had a short neck to give it easy access to low-lying food sources.

Prosauropod paleobiology. Scientists from Germany made a startling discovery while applying a technique similar to that used to determine the age of trees to efforts to calculate the chronological age of several specimens of prosauropod. Prosauropods were moderately long-necked dinosaurs that flourished near the beginning of the "age of dinosaurs" in the late Triassic and early Jurassic periods.

In a report published in December 2005, paleontologists P. Martin Sander and Nicole Klein, both of the Institute of Paleontology at the University of Bonn, Germany, used growth rings in the bones of the late Triassic (220-million-year-old) prosauropod *Plateosaurus* to determine the age of individual dinosaurs. They discovered that, unlike most dinosaurs, adult prosauropods varied in size. They based this conclusion on their finding that prosauropods of different sizes were the same chronological age.

Modern *warm-blooded* animals, such as birds and mammals, reach their full adult size in a fixed period of time. (A warm-blooded animal almost always has about the same body temperature, regardless of the temperature of its surroundings.) Apparently, prosauropods could change their growth rates if temperatures cooled or conditions became stressful. This pattern is typical of *cold-blooded* animals, especially modern lizards, snakes, and other reptiles. (A cold-blooded animal has little physiological control over its body temperature.) The evidence provided by the prosauropod study shows that these dinosaurs may represent an intermediate form of dinosaur, exhibiting some traits of warm-blooded animals and some traits of cold-blooded animals.

Fossil penguin. The discovery of fossils of a giant penguin that stood at least 2.5 meters (8 feet) tall and weighed at least 100 kilograms (220 pounds) was reported in February 2006, by naturalist and paleontologist Chris Templer of the Kawhia Museum at Te Kauri Lodge, New Zealand. The bones of this animal, the largest penguin that ever lived, were discovered protruding from a sandstone platform exposed at low tide in Kawhia Harbor, located on the North Island of New Zealand. Templer and his colleagues used rock saws to extract the partial skeleton of the huge bird. The researchers estimated that the huge penguin lived about 40 million years ago and was the largest of at least 12 extinct species of New Zealand penguins.

Dinosaur eggs and embryos. A discovery reported in July 2005 gave scientists important insight into the early development of dinosaurs. Vertebrate paleontologist Robert Reisz of the University of Toronto at Mississauga, Canada, and several colleagues described a dinosaur nest with six fossil eggs the size of hens' eggs from 190-million-year-old (Jurassic-Period) sediments in South Africa. Five of the eggs contained bones representing the remains of the oldest dinosaur embryos ever recovered. The scientists believed that the sixth egg had already hatched at the time the nest was covered by the sediments.

Scientists identified the bones as those of the prosauropod *Massospondylus carinatus.* The embryos were toothless or had weakly developed teeth, demonstrating that the young animals could not have nipped tough vegetation. This indicates that hatchling dinosaurs, at least in this species, required parental care.

Long front legs, a large head, and a horizontally held neck indicate that when these juveniles hatched, they were *quadrupedal* (walked on all four legs), even though scientists believe the adults were *bipedal* (walked upright on two legs). The proportions of the limbs more closely resemble those of later sauropods, suggesting that the sauropods may have retained their juvenilelike proportions as their species grew increasingly larger.

Ancient venomous mammals. Recently described canine teeth from an early mammal suggest that these animals used venom in much the same manner that rattlesnakes do. Few modern mammals use venom to help subdue prey or ward off predators. However, this may not have been so for some ancient mammals.

In June 2005, paleontologists Richard Fox and Craig Scott, both from the University of Alberta in Edmonton, Canada, described the 60-million-year-old fossilized canine teeth of *Bisonalveus browni,* a mouse-sized mammal found in what is now Alberta. The teeth bear a groove along one side that widens to form a circular area at the base. Fox and Scott surmised that the circular depression represents the site of a venom gland and that the venom could run out along the groove to the sharp tip of the tooth, the way it does in rattlesnake teeth. These teeth also resemble those of certain living shrews that use *hypodermic* (skin-piercing) injections of venom to immobilize insect prey.

Early mammals, which may not have been as quick and agile as later mammals, may have greatly benefited from using this type of prey injection, the scientists theorized. The evolution of more sophisticated mammals may have reduced the need for hypodermic teeth.

■ Carlton E. Brett

See also **GEOLOGY.**

A GIANT, TOOTHY FLYING REPTILE

A new species of flying reptile with large, fanglike teeth is depicted in an artist's re-creation based on fossil remains. In 2006, paleobiologists from the University of Portsmouth in the United Kingdom described the 65-million-year-old fossils, which they named *Caulkicephalus trimicrodon.* The fossils were discovered on the Isle of Wight in 2003.

GENETICS

The first version of the chimpanzee *genome* was published in September 2005 by an international group of researchers known as the Chimpanzee Sequencing and Analysis Consortium. According to the researchers, chimpanzees and human beings share about 98.5 percent of the genetic material found in each groups.

A genome is the complete genetic code of an individual. It is made up of chemical subunits called *nucleotides* that are abbreviated using the letters A, C, G, and T. In a genome, these nucleotides are strung together in a specific order called a *sequence* to "spell out" individual genes. The published chimpanzee genome sequence consists of nearly 3 billion individual nucleotides.

Most scientists had long considered chimpanzees to be the closest living relatives of human beings. For this reason, one of the goals of the consortium was to compare chimpanzee and human genes using the chimp genome sequence and information obtained from the Human Genome Project, which was completed in 2003. The researchers found differences in just over 1 percent of the individual nucleotides in the two genomes. These include individual differences called *single nucleotide polymorphisms,* or SNPs (pronounced *snips*). In genomes as large as those of human beings and chimpanzees, such a small percentage nevertheless represents about 35 million changes.

In addition to the SNPs, the team identified roughly 5 million changes that represent blocks of genetic material present in one genome but missing in the other. Scientists have often found such changes when comparing genomes, but their significance is not fully understood.

Most genes make proteins. The researchers also compared the makeup of the proteins produced by the genes in each genome. They found that about 29 percent of the proteins examined were identical in chimpanzees and human beings. Most of the other proteins were so similar that their counterparts in each genome could be identified easily.

Despite these findings, scientists have not discovered why human beings are different from chimpanzees. In the 1970's, geneticist Mary Claire King and biochemist Allan Wilson (both then at the University of California at Berkeley) proposed that most of the differences between people and chimps could be explained by a relatively small number of changes in key genes that

SEE ALSO THE SPECIAL REPORT, **CHRONIC INFLAMMATION: PATHWAY TO DISEASE,** PAGE 68.

control the actions of other genes in the genome. With the completion of the human and chimp genome projects, scientists now have the tools to examine this and other ideas to better understand what makes human beings unique.

Hapmap project. The first version of a catalog of human genetic differences was published in October 2005 in the scientific journal *Nature.* The catalog, known as a *haplotype map*—or *hapmap* for short—was assembled by an international group of researchers called the International Hapmap Consortium. The scientists created the hapmap because even though all modern human beings belong to the same species, individuals display genetic differences.

The human genome, like the chimpanzee genome, contains nearly 3 billion SNPs. One person's genetic makeup may differ from another's by hundreds of thousands—or even millions—of SNPs. In part, these differences define each person as a unique human being. In creating the hapmap, the researchers sought to establish the relevance of blocks of SNPs, rather than trying to learn the purpose of each SNP. In particular, the scientists hoped to use the map to look for blocks of SNPs that form the same pattern in different individuals. Because many human disorders— such as cardiovascular disease, cancer, and dementia—likely involve many genes, the blocks of information from the hapmap may help reveal how certain combinations of genes contribute to these problems.

To develop the hapmap, the researchers analyzed the genetic makeup of 269 people from four different human populations. The groups included Yoruba people in Ibadan, Nigeria; North Americans in Utah; Han Chinese in Beijing; and Japanese in Tokyo. The groups were chosen for practical reasons and because of established knowledge about their genetic makeup. They were not chosen to represent all human groups. As the project continues, the scientists plan to add new individuals and groups. The researchers stressed that the hapmap was intended only as a beginning and that, as it grows, they expect to find new uses for the information.

Cloning fraud comes to light. In a stunning series of disclosures reported in December 2005 and January 2006, the groundbreaking research on human stem cells described by Hwang Woo Suk and his research team at Seoul National University in Korea was exposed as scientific fraud. In January 2006, the journal *Science,* which had previously published the team's work, formally retracted its papers.

The papers had described two major accomplishments in *embryonic human stem cell* research. Embryonic stem cells are cells that, in theory, can develop into any of the cell types that make up the tissues and organs of the body. Many researchers believe such cells hold great promise for creating new treatments for diseases. The first claim made by Hwang's group was that they had successfully used a *cloned* (genetically duplicated) human embryo to produce a new line of stem cells that could

be grown and maintained in the laboratory. The group claimed that, using voluntarily donated eggs, they had made dramatic progress in developing more efficient cloning techniques.

The second claim was that the team had used the same method to produce genetically matched lines of stem cells for nine patients with problems that included spinal cord injuries, diabetes, and immune-system disorders. The researchers hoped that the stem cells produced from these patients could replace their damaged cells and cure the disorders.

The fraud was uncovered primarily by other researchers at Seoul National University. In June 2005, one of the researchers working with Hwang sent an anonymous message to a Korean television news program called "PD Notebook" about his concerns with Hwang's work. The tipster told the reporters that some of Hwang's claims about how he had obtained

LIVING SELF-PORTRAIT

An image of an enlarged *E. coli* bacterium appears on "biological film" held by a researcher at the University of Texas at Austin. Students at the University of Texas and at the University of California at San Francisco created the film by growing billions of genetically engineered *E. coli* bacteria in a dish of *agar* (a gel-like material used as a growth medium). The students, whose work was reported in November 2005, manipulated the bacterium's genes to develop *E. coli* that would respond to light. They also manipulated the bacteria's digestive system so that when the microbes are in darkness, they digest sugar, producing black pigment. When they are in light, they do not digest sugar and produce no pigment. To create the image, the students developed a camera in which light is projected through a microscope slide containing an *E. coli* bacterium onto a dish of bacteria maintained at the temperature of a human body. As the bacteria in the dish grew, those that were in the dark digested sugar in the agar and produced black pigment, forming an outline around the image on the microscope slide. Bacteria on which the light had been projected did not digest sugar and thus did not produce black pigment. The students' work is part of a new scientific field called *synthetic biology,* a discipline in which principles of engineering are applied to biology.

GENETICS continued

egg donations for the cloning work were untrue and that the group had not actually created patient-specific stem cell lines.

Through interviews with researchers involved in the work and independent testing by other laboratories in Korea, the journalists established that everything the tipster had said was true. Contrary to Hwang's frequent assertions, some of the eggs used in the procedures had come from paid donors; others had come from junior researchers in the lab who had been pressured to

donate them. Hwang later admitted that these facts were true. The investigation also brought to light the fact that Hwang's group had used many more eggs than initially reported, an indication that their methods were not as efficient as the group had claimed.

In addition, independent testing showed that most of the new stem cell lines that Hwang said came from specific patients had, in fact, come from a stem cell line not even made in Hwang's laboratory. A committee of scientists set up to investigate the work reported that they were unable to determine where the reported stem cell lines had come from. Neither could they confirm that the stem cell lines had been created in Hwang's laboratory in the manner he described.

The committee did confirm that Hwang's group had made some progress toward creating cloned embryos and that Hwang's earlier work involving the first successful cloning of a dog was valid. However, the extent of the fraud surrounding the stem cell claims destroyed Hwang's credibility in the scientific community. In March 2006, Hwang was fired from his position at Seoul National University. In May, he was charged by the Seoul Central District prosecutor's office with embezzling—for using fabricated data to obtain research funding—and with violating a bioethics law by purchasing eggs for research.

The origin of HIV. HIV-1, the virus that causes most cases of human AIDS, came from living wild chimpanzees in southern Cameroon, Africa. An international team of researchers led by Beatrice Hahn, a *virologist* (scientist who studies viral diseases) from the University of Alabama at Birmingham, reported the finding in May 2006.

Researchers had long suspected that HIV-1 developed from a closely related virus in chimpanzees called SIVcpz (simian immunodeficiency virus from chimpanzees). However, they had been able to detect the virus only in a few captive chimps of subspecies that live in the wilds of Cameroon, Congo (Brazzaville), and Gabon. Collecting blood samples from chimps in the wild is difficult because the animals are few in number, live in remote jungle areas, and generally avoid people.

Hahn's team employed trackers who collected nearly 600 samples of chimp droppings from the forest floor in Cameroon. The researchers found evidence of SIVcpz in droppings from 5 of the 10 sites tested. A genetic analysis of the virus revealed that one community of chimps, located

TRACING THE ANT'S FAMILY TREE

Ants evolved more than 40 million years earlier than previously thought, researchers reported in April 2006. The scientists compared DNA (deoxyribonucleic acid—the molecule genes are made of) from 139 different ant *genera* (a unit of the scientific classification system just above the species level) and found that ants evolved sometime between 168 million and 140 million years ago. DNA *mutates* (changes) at a calculable rate. By using these calculations, as well as examining the fossil evidence, the scientists were able to reconstruct the ant's family tree. They learned that ants diversified into hundreds of species about 100 million years ago, when flowering plants appeared. The plants provided new habitats—as well as new food sources—for the ants.

near the Sangha River, carried a form that most closely matched HIV-1. The team theorized that a person who became infected with HIV by eating a chimp with SIVcpz could have spread the disease by traveling down the Sangha River to Congo (Kinshasa). The first known case of HIV-1 has been traced to a man in Kinshasa whose blood was stored in 1959 as part of a medical study.

Genetic defect in Lincoln's family. The discovery of a genetic defect involved in a disorder that may have afflicted Abraham Lincoln was reported in January 2006 by researchers led by geneticist Laura P. W. Ranum at the University of Minnesota in Minneapolis. In this disorder, called spinocerebellar ataxia type 5, nerve cells in the part of the brain that controls movement die, causing loss of coordination in limbs and eye movements, slurred speech, and swallowing difficulties.

To identify the defect, the researchers analyzed genetic material and historical records from members of a family descended from President Lincoln's paternal grandparents. Although it was not possible to analyze material from a direct descendant of President Lincoln (none are living), descendants of one of his father's brothers and one of his sisters were included in the study. From an analysis of 299 family members, the team identified 90 individuals affected by the disorder.

In all the affected individuals—as well as in 35 other family members who did not show any obvious symptoms—the researchers found a mutation in a gene called beta-III spectrin. This gene normally produces a protein that helps provide an internal structural framework in cells. The mutated copy of this gene found in the Lincoln family was missing a segment of material, the loss of which could affect the ability of the protein to function properly.

The researchers also analyzed genetic material from two other families (one French, one German) with the disorder. In all cases, the affected individuals also had mutations in the same spectrin gene, though the mutation was different in each of the families. None of the mutations was found in genetic material obtained from a large number of people not affected with the disorder.

The researchers could not say for certain whether Lincoln himself was afflicted with this form of ataxia, though his chance of inheriting the mutation was 25 percent. However, they cited historical accounts referring to the president's "loose, irregular, almost unsteady gait." The re-

searchers expressed hope that their investigations would increase public awareness of ataxias and other neurodegenerative disorders.

Pregnancy disorder gene. A defective gene that can cause a rare pregnancy disorder called a *molar pregnancy* was described in February 2006. The work was conducted by researchers led by geneticist Rima Slim of the McGill University Health Center in Toronto, Canada. In a molar pregnancy, a mass known as a *hydatidiform mole* forms in the uterus in place of an embryo. Serious complications become apparent around the 10th week of pregnancy, and often the molar tissue must be surgically removed from the uterus. About 1 in 1,000 women experiences a molar pregnancy. Such pregnancies occur more often in women who have had two or more miscarriages and may increase the risk of developing certain kinds of cancer.

The most common type of molar pregnancy is known as a *complete hydatidiform mole,* which occurs in an unpredictable manner. The researchers led by Slim studied a less common form of the disorder known as *recurrent hydatidiform moles.* The development of such moles tends to run in families, and researchers believed they may occur because of defects in specific genes. By analyzing the genetic makeup of a family in which several women had experienced such moles, the researchers had previously learned that the problem seemed to lie with a gene somewhere within a relatively small region of chromosome 19. By examining all the genes in this region, the McGill researchers discovered that the affected women in one family had *mutations* (changes) in a gene known as NALP7. In addition, they found that women from two other families with a history of molar pregnancy had mutations in the same gene. These mutations—five in all—were not present in the NALP7 gene from several hundred women with no history of molar pregnancy.

The NALP7 gene normally produces a protein that is closely related to other proteins involved in *inflammation* (the body's reaction to injury or infection). The exact relationship between the mutations in NALP7 and molar pregnancy is not clear. However, the researchers hoped that further work on NALP7 and other genes that may be involved in molar pregnancies will lead to a better understanding of events that may cause a broad range of pregnancy-related problems—and possibly even cancer.

■ David Haymer
See also **SCIENCE AND SOCIETY.**

GEOLOGY

Scientists studying ancient sea temperatures discovered in 2006 that during the Cretaceous Period, which lasted from 145 to 65 million years ago, the ocean may have been hotter than a modern hot tub. The data come from three *cores* (cylindrical samples) drilled in 2003 by the Ocean Drilling Program, an international expedition studying Earth beneath the ocean floor. The cores came from sections of seafloor in the Atlantic Ocean off the South American country of Suriname.

At the 2006 annual meeting of the American Association for the Advancement of Science, paleoclimatologist Karen Bice of the Woods Hole Oceanographic Institution in Woods Hole, Massachusetts, reported on analyses of oxygen *isotopes* in microscopic shells found in the cores. (Isotopes are atoms of the same element that differ in the number of neutrons in the nucleus.)

The fossilized shells belonged to organisms called foraminifera. The ratio of two isotopes of oxygen found in foraminifera shells directly correlates to the temperature of the water in which the organisms lived. The fossilized shells can, therefore, be analyzed to determine the temperature of ancient seawater. Bice and her team concluded that 90 million years ago, sea surface temperatures may have reached 42 °C (108 °F). This is 14 Celsius degrees (25 Fahrenheit degrees) warmer than the modern tropical Atlantic and hotter than water in a hot tub.

An analysis of organic matter in the cores suggests that the carbon-dioxide content of the atmosphere during the Cretaceous was between three and six times as great as it is today. Carbon dioxide is a *greenhouse gas* (gas that traps heat in Earth's atmosphere). Bice believes the high carbon-dioxide levels and the high water temperatures are connected and suggested that scientists studying climate change are underestimating the link between carbon dioxide and *global warming,* the gradual increase in the average temperature of Earth's surface that began in the 1800's.

Analysis of Chilean earthquake. Scientists studying the cause of a 9.5-magnitude earthquake that struck Chile in 1960 discovered that the quake did not fit with generally accepted ideas about how earthquakes occur. That 1960 quake was the largest ever recorded and produced a tsunami that crested at 15 meters (49 feet) and reached Japan.

According to the theory of plate tectonics,

SEE ALSO THE SPECIAL REPORT, **THE EARLY EARTH,** PAGE 84.

Earth's outer shell consists of 30 moving plates of varying sizes. When one plate slips under another in a *subduction zone,* earthquakes occur. Earthquakes occur in Chile because the Nazca Plate, located between the East Pacific *Rise* (an underwater ridge) and South America, moves steadily eastward and is forced below the western edge of the South American continent at a rate of 8 meters (26 feet) per 100 years.

Plate tectonic theory states that the longer a subduction zone goes without an earthquake, the greater the build-up of energy within the zone and the more powerful the next earthquake in that zone will be. However, *agronomist* (soil and crop scientist) Marco Cisternas of the Pontificia Universidad Católica de Valparaíso, Chile, and an international team of geologists, geophysicists, and historians, reported in September 2005 that the 1960 Chilean quake did not meet these criteria.

Cisternas and his team examined written records and investigated sedimentary deposits left by tsunamis from earthquakes that occurred over the last 2,000 years. Although written records exist for smaller earthquakes in 1575, 1737, and 1837, the sedimentary record extends back much further. The researchers dug trenches near the mouth of Chile's Río Maullín in the center of the earthquake region to examine the *stratigraphic layers* (layers of sand and rock deposited over time) for evidence of earlier tsunamis. The researchers discovered layers of sand deposited by tsunamis generated by eight earthquakes that occurred before 1960. They then used *radiocarbon dating* to determine the approximate years of the tsunamis. (Radiocarbon dating is a method of determining the age of organic material by measuring its level of radioactive carbon.) The scientists discovered that the interval between the four historical earthquakes averaged 128 years, while the average interval between major quakes over the past 2,000 years has been 285 years, more than twice as long.

The researchers calculated that the 1960 quake released an amount of energy that should have required from 250 to 350 years to build up.

They concluded that when earthquakes occur at shorter time intervals, they may not release all the pent-up energy in the subduction zone. This energy remains in the zone and is released in a major earthquake later.

The researchers suggested that this finding may explain why the 2004 Sumatra–Andaman earthquake that generated the most disastrous tsunami of modern times was so large. Its magnitude is now estimated to have been between 9.0 and 9.3. Its *fault* (fracture in Earth's crust) moved about 10 meters (33 feet). The quake occurred in an area that had last ruptured in 1881, producing an earthquake with an estimated magnitude of 7.9. The plate movement between 1881 and 2004 was less than 4 meters (13 feet). The scientists believe that the 2004 quake released accumulated plate motion that the previous earthquake had left unspent. The research could help researchers more accurately predict the scale of future earthquakes.

New dates for the Sahara. Desert sands have blown in the Sahara region of Africa for millions of years longer than previously thought, according to research reported in 2006. The Sahara is now the largest desert on Earth, yet only 6,000 years ago it was covered with plants.

Scientists have long known that during the past 2 million years, as Earth alternated between glacial periods (when the planet is partly covered in ice) and interglacial periods (when the ice melts and retreats), the Sahara changed from grassy plain to sand desert and back again. A group of geologists led by Mathieu Schuster of the Université Bretagne Occidentale in Plouzané, France, reported in February on the discovery of an extensive area of petrified dune sands in the Chad Basin, which is located in the Djurab region of central Chad. The scientists unearthed fossils of mammals from deposits overlying the dunes. They deduced that because the fossils are from animals that are known to have lived 7 million years ago, the petrified dunes must be older than that. This discovery greatly pushes back the earliest known date of desert formations in the Sahara region, which scientists were previously able to date only to 86,000 years ago.

Yellowstone magma. Scientists studying the volcano that lies beneath Yellowstone National Park in northwestern Wyoming discov-

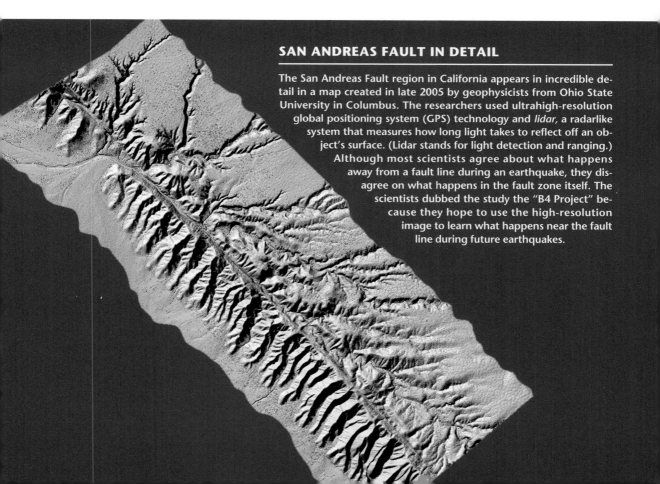

SAN ANDREAS FAULT IN DETAIL

The San Andreas Fault region in California appears in incredible detail in a map created in late 2005 by geophysicists from Ohio State University in Columbus. The researchers used ultrahigh-resolution global positioning system (GPS) technology and *lidar,* a radarlike system that measures how long light takes to reflect off an object's surface. (Lidar stands for light detection and ranging.) Although most scientists agree about what happens away from a fault line during an earthquake, they disagree on what happens in the fault zone itself. The scientists dubbed the study the "B4 Project" because they hope to use the high-resolution image to learn what happens near the fault line during future earthquakes.

GEOLOGY continued

ered in 2006 that the cause of the geological changes currently affecting the region might be far deeper than previously believed. The largest known volcanic eruption in Earth's history occurred about 640,000 years ago in what is now Yellowstone. The gigantic explosion blew about 1,000 cubic kilometers (240 cubic miles) of rock into the air. It left a huge depression, called a *caldera,* measuring 43 by 77 kilometers (27 by 48 miles). By comparison, the eruption of Mt. St. Helens in Washington state in 1980 involved about 1 cubic kilometer (0.25 cubic mile) of material. The greatest eruption in historical times, the eruption of the Indonesian volcano Mount Tambora in 1815, spewed about 40 cubic kilometers (9.6 cubic miles) of rock into the atmosphere. Yellowstone has a history of cataclysmic explosions, with one occurring 1.2 million years ago and another at 2.1 million years ago. To better understand Yellowstone's volcanic activity, the United States Geological Survey (USGS) in Menlo Park, California, has established a systematic method of watching the Yellowstone volcano.

In March 2006, USGS geologists Charles W. Wicks, Wayne Thatcher, Daniel Dzurisin, and Jerry Svarc reported on their examination of the movement of *magma* (molten rock) beneath the surface in Yellowstone. The scientists measured the movement of the land surface using a technique known as *interferometric synthetic aperture radar* (InSAR). InSAR collects the echo returns from many radar pulses emitted by a satellite and processes them into a single radar image. The resulting *interferogram* provides a precise measurement of changes in elevation over time.

Interferograms created using information collected by the European Space Agency's ERS-2 satellite from 1992 to 1997 showed that by 1995, some areas of the caldera were rising, reversing the calderawide sinking that began in 1985. By 1997, the whole caldera floor had begun to rise. The scientists found that surface movements became even more dynamic from 1996 to 2002, when an area along the northern rim of the caldera rose by about 130 millimeters (5 inches). The upward motion stopped abruptly in 2003.

The researchers attributed the uplift to the intrusion of a *sill,* a flat sheet of molten volcanic rock. Using data from many small earthquakes that occur regularly in the Yellowstone region, the scientists calculated that the sill lies about 18 kilometers (11 miles) below the surface. The movement of the surface may have affected geysers and hot springs at the park's Norris Geyser basin. ■ William W. Hay

See also **ATMOSPHERIC SCIENCE; FOSSIL STUDIES; PHYSICS.**

SHIFTING GRAVITY

Areas of changing gravity during a December 2004 earthquake in Southeast Asia are shown in a map compiled from satellite data and released in December 2005 by the United States National Aeronautics and Space Administration. The earthquake caused a tsunami that killed more than 216,000 people in coastal areas of Southeast Asia, South Asia, and East Africa. In the areas colored blue and purple on the map, Earth's gravity decreased slightly. In the red and yellow areas, Earth's gravity increased slightly. These areas lie northeast of the quake's *epicenter* (the spot on Earth's surface directly above the earthquake's point of origin).

MEDICAL RESEARCH

Alzheimer's disease (AD) appears to be mainly an inherited condition, according to a study reported in February 2006 by an international team of researchers. AD is a progressive, *degenerative* (gradually worsening) brain disorder that causes *dementia* (a loss of memory, reasoning, and thinking functions severe enough to interfere with daily life). Dementia affects about 10 percent of people over age 65. About two-thirds of those with dementia have AD.

Researchers led by psychologist Margaret Gatz at the University of Southern California in Los Angeles sought to identify links between AD and genetic and environmental influences. The researchers evaluated 11,884 pairs of twins from the Swedish Twin Registry. The registry, which was established in the late 1950's, tracks more than 140,000 individuals who are *fraternal* or *identical* twins. (Fraternal twins are those that develop from separate embryos that form when sperm fertilize two different eggs. Identical twins grow from a single fertilized egg that splits into two embryos. Identical twins share the same genes; fraternal twins do not.)

Gatz and her team identified 392 sets of

SEE ALSO

THE SPECIAL REPORT, **CHRONIC INFLAMMATION: PATHWAY TO DISEASE,** PAGE 68.

THE SPECIAL REPORT, **TURNING UP THE VOLUME ON HEARING LOSS,** PAGE 100.

PUBLIC HEALTH CLOSE-UP, **LESSONS FROM THE SPANISH FLU,** PAGE 248.

twins in which one or both members had AD, as well as a control group in which neither twin had dementia. Investigators then compared AD rates in the two groups that had people with dementia. Based on these numbers, the researchers estimated that Alzheimer's disease was inherited in from 58 to 79 percent of cases.

The researchers then examined the age at which each of the twins with AD had been diagnosed with the condition. Those ages were significantly closer in identical twins—an average of

WIRING A BRAIN ANEURYSM

Filling a bleeding aneurysm in the brain with wire coils may be a safer long-term treatment for some patients than a traditional procedure that involves drilling into the skull and sealing off the aneurysm with a metal clip, according to a study reported in September 2005. An aneurysm is an abnormal bulge in a blood vessel, the rupture of which can lead to disability or death. In the coil procedure, surgeons insert a springy platinum wire into a blood vessel and guide the wire through the vessel into the brain aneurysm (top). The wire fills the aneurysm with tangles (middle), which stop the bleeding. After the wire is cut (bottom), cells in the vessel wall seal off the aneurysm. British researchers found that seven years after treatment, 31 percent of patients with the clips had died or become disabled, compared with only 24 percent of the patients with the wire. Patients with certain types of brain aneurysms cannot be treated using the coils, however.

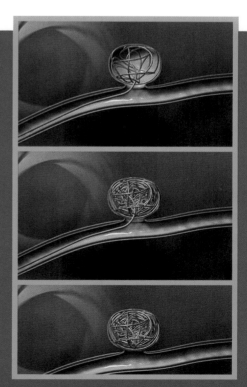

MEDICAL RESEARCH continued

3.66 years—than in fraternal twins—an average of 8.12 years. The shorter gap between the diagnoses dates for identical twins indicated that genes also play an important role in the timing of AD, the researchers concluded.

The researchers noted that though AD involves a strong inherited component, environmental factors also play a role in whether and when the condition develops. Eating a healthful diet, exercising, and controlling inflammation may reduce the risk or delay the onset of AD.

The first successful lab-grown organs. For the first time, researchers successfully implanted laboratory-grown organs—in this case, bladders—in human patients. Researchers led by Anthony Atala, director of the Institute for Regenerative Medicine at Wake Forest University School of Medicine in Winston-Salem, North Carolina, reported the details of the procedures in April 2006.

Atala's study involved seven patients ages 4 to 19 with a *congenital* (present at birth) birth defect. The defect prevents the spine from closing completely, leading to a loss of nerve function, including in those nerves that control the bladder. The condition causes high pressure in the bladder that can put extra pressure on the kidneys and lead to kidney damage. In addition, patients with the condition experience frequent urinary leakage.

The patients could have received a standard treatment in which nonfunctioning bladder tissue is repaired with intestinal tissue. But the tissues in the two organs differ greatly. The bladder excretes waste products, while the intestines absorb nutrients. People who undergo such a procedure often experience problems later in life, including higher-than-normal rates of cancer, *osteoporosis,* and kidney-stone formation. (Osteoporosis is a condition in which bones become abnormally fragile, increasing the risk of fractures, pain, and disability.)

Atala's team sought a better solution. First, they performed biopsies in which they took samples of each patient's muscle cells and of cells that line the bladder wall. Then, they grew the cells in a culture in the laboratory. When they had a sufficient number of cells, the researchers placed them onto a bladder-shaped mold called a *scaffold* on which the cells continued to grow. About seven or eight

THE FIRST BIONIC ARM

Jesse Sullivan, who lost both arms in a work-related accident, controls a new bionic arm using his own thought-generated nerve impulses in a demonstration in June 2005. Physicians at the Rehabilitation Institute of Chicago attached four major nerve endings from Sullivan's amputated arm to his chest muscles, where the nerves grew into the muscles. Electrodes positioned above the site pick up impulses sent by Sullivan's brain to the arm nerves and transmit them to a computer that controls the bionic arm. The new arm allows Sullivan to eat, shave, open jars, and even throw and catch a ball. Physicians hope that the technology, developed by Rehabilitation Institute physician and biomedical engineer Todd Kuiken, can be adapted for use by other types of amputees and for stroke victims.

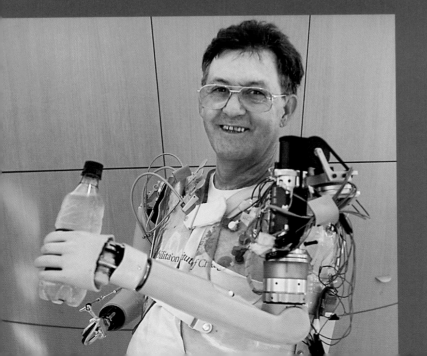

weeks after the biopsy, the lab-grown bladders were sewn to the patients' natural bladders. As the lab-grown bladders fused with the original bladders, the scaffold decomposed.

The researchers implanted the first of the bladders in 1999. Over the next seven years, the laboratory-grown bladders functioned as well as bladders repaired with intestinal tissues. However, because the bladders were grown from the patients' own cells, the patients' immune system did not reject them, nor did the patients experience the side effects common in the standard procedure. The researchers hoped to eventually perfect techniques for growing other organs and tissues in the laboratory, including blood vessels and hearts.

Calcium/vitamin D and health. Taking daily calcium and vitamin D supplements may be less beneficial to the health of older women than previously thought. In February 2006, researchers led by osteoporosis expert Rebecca D. Jackson at Ohio State University in Columbus found that the two supplements seemed less effective at keeping bones healthy than physicians and many consumers had believed. Moreover, taking the supplements increased the risk of kidney stones while having almost no effect on the risk of colorectal cancer. Previous studies had suggested that taking calcium and vitamin D supplements lowered the risk of developing colorectal cancer.

Researchers recruited 36,282 *postmeno-pausal* women ranging in age from 50 to 79 from the Women's Health Initiative (WHI) study. (Postmenopausal women are those who have passed the age of menopause, a time during which women stop menstruating.) The WHI was established in 1991 by the United States National Institutes of Health to study the three most common causes of death, disability, and impaired quality of life in postmenopausal women—cancer, cardiovascular disease, and osteoporosis.

The researchers divided the women into two groups. One group received a *placebo* (inactive substance). The women in the other group received 1,000 milligrams of calcium in the form of calcium carbonate and 400 IU (international units) of vitamin D_3 daily. Calcium is a major component of bone. Vitamin D helps the body absorb calcium through the intestines. Previous studies had suggested that taking calcium supplements reduced the risk of developing osteoporosis and increased bone density. Researchers had found little evidence that the supplements reduced the risk of bone fractures, however.

The researchers monitored the women's bone health for seven years. Among women older than 50 in the supplements group, bone density increased by a significant 1.06 percent. However, compared with the control group, the women taking supplements reduced their risk of suffering a hip fracture by only 12 percent. In addition, their risk of experiencing other types of fractures fell only slightly. In contrast, as the women aged, the beneficial effect of the supplements on the risk of hip fracture increased for those who continued to take the full amount of the supplements. Women in the supplement group as a whole suffered 21 percent fewer fractures than the women not taking supplements.

The study also determined that participants who took the daily calcium plus vitamin D had virtually the same risk of developing colorectal cancer as did those taking a placebo. The risk of developing kidney stones for women taking the supplements increased by 17 percent, however.

The researchers noted that the study may not have lasted long enough for the effects of the calcium and vitamin D supplements to become apparent. Both osteoporosis and colorectal cancer can take from 10 to 20 years to develop. The researchers suggested that women should continue to take calcium and vitamin D supplement because even a slight reduction in the risk of osteoporosis can produce a positive effect on overall health.

Estrogen therapy and breast cancer. Estrogen therapy, a type of hormone therapy, does not increase the risk of breast cancer in postmenopausal women, according to a study published in April 2006. The female hormones estrogen and progesterone have long been used to treat the hot flashes and other symptoms that some women experience during menopause. However, studies reported in the early 2000's that seemed to show a link between hormone therapy and a higher risk of stroke and breast cancer led many women to stop taking the hormones.

The results of the 2006 study were based on data from the WHI Estrogen-Alone trial. During that trial, researchers had followed 10,739 postmenopausal women ages 50 to 79 from 1993 to 1998. All the women had had a *hysterectomy* (surgery to remove the uterus). Women with an intact uterus have an increased risk of developing *endometrial cancer*—cancer of the uterine lining—when taking estrogen. The women were randomly selected to receive hormone therapy or a placebo. The estrogen used was conjugated equine estrogens (CEE), the most common

MEDICAL RESEARCH continued

type. Researchers halted the WHI Estrogen-Alone trial in February 2004—earlier than planned—because of a higher-than-normal number of strokes among the women in the hormone group. In addition, the scientists found the rate of coronary artery disease—another supposed benefit of the therapy—was about the same in both groups.

For the 2006 study, researchers reexamined the data from the WHI Estrogen-Alone trial. After seven years, 237 of the women had developed breast cancer. The rate of all types of breast cancer among women taking CEE was 18 percent lower than for the women in the placebo group. The rate of *invasive breast cancer* (cancer that has spread outside the milk ducts to other breast tissue) was 20 percent lower. However, the women in the CEE group were 50 percent more likely to have a mammogram that indicated an apparent abnormality that required follow-up tests. The abnormalities did not turn out to be indicators of cancer.

The researchers concluded that taking CEE for seven years does not increase the risk of breast cancer for postmenopausal women with a hysterectomy. CEE may even decrease the risk of early-stage breast disease and *ductal carcinoma* (cancer of the milk ducts), the most common form of breast cancer. The researchers cautioned that estrogen therapy may not be safe for all older women with a hysterectomy. Women and their physicians, they recommended, should consider the potential risks and benefits of the therapy on an individual basis.

Omega-3 fatty acids—no magic bullet. Omega-3 fatty acids, consumed by many people to reduce their risk of cancer and *cardiovascular disease* (diseases of the heart and blood vessels), apparently provide no protection against these diseases and have no effect on an individual's life span. That conclusion was reported in 2006 by two groups of researchers who reviewed numerous studies investigating the health effects of omega-3 fatty acids. Animal studies had found that eating a diet rich in omega-3 fatty acids could lower the levels of cholesterol and other heart-damaging fats in the body. The fatty acids also seemed to thin the blood and so help to reduce the risk of clots that can cause strokes or heart disease. However, some researchers had cautioned that not all people might get the same benefits by consuming omega-3's.

Omega-3 fatty acids, which are essential to human health, are carbon-rich molecules that help form *lipids* (organic compounds that include fats and oils). The human body cannot manufacture these fats—which can be found in fish, flax, walnuts, and other foods.

The first study, reported in January 2006, focused on omega-3 fats and cancer. It was conducted by investigators led by Catherine MacLean of Rand Health, an independent health policy research program in Santa Monica, California. The researchers reviewed 38 articles published from 1966 to 2005 that analyzed the effects of omega-3 consumption (either from foods or dietary supplements) on various types of cancer. The articles described studies of 20 different groups of people from 7 countries with 11 types of cancer. The study populations ranged in size from 6,000 to 121,000 people.

The studies offered conflicting results regarding cancer in general. For example, one study found evidence that greater consumption of omega-3 increased the risk of breast cancer. Three other studies, however, suggested that increasing omega-3 intake lowered the risk of breast cancer. Seven studies showed no significant link between breast cancer and omega-3 intake. Similarly, one study showed that consuming more omega-3 increased the risk of lung cancer, while another showed a decreased risk, and four others showed no significant effect.

Based on the review, the authors concluded that increased omega-3 consumption "is unlikely to prevent cancer." They attributed this difference in results to differences between animal and laboratory studies and those involving people. Variations in the forms of omega-3 found in foods and dietary supplements may also have played a role.

The second review involved 89 studies investigating the link between omega-3 fats and heart attacks, cancer, and life span. All the studies reviewed by the investigators, who were led by nutritionist Lee Hooper of the University of East Anglia in Norwich, the United Kingdom, involved a treatment group and a control group. Again, the researchers found no evidence that omega-3 fats benefit health.

Government health guidelines in the United States and the United Kingdom advise people to consume more omega-3 fats, especially after a heart attack. Although the researchers did not recommend changing this advice, they suggested that more omega-3 studies were needed.

Vaccine for cervical cancer. The first vaccines designed to prevent cancer reached critical milestones in 2006. Both vaccines—Gardasil, manufactured by Merck & Co., Inc., of Whitehouse Station, New Jersey, and Cervarix, produced by GlaxoSmithKline of Brentford, the United Kingdom—target two sexually transmitted strains of the human papilloma virus (HPV) that cause most cases of cervical cancer. About 500,000 women worldwide develop cervical cancer annually. The disease kills about 300,000 of them, most of whom live in developing nations without screening programs.

There are more than 100 different types of HPV. Both Gardasil and Cervarix protect against the two deadliest strains, types 16 and 18. Together, HPV-16 and HPV-18 account for 70 percent of all cases of cervical cancer. The strains can also cause other forms of genital cancers.

Gardasil was approved by the United States Food and Drug Administration (FDA) in June 2006. The vaccine is administered as a series of three injections. Merck researchers tested Gardasil over a period of 10 years in nearly 21,000 women ages 16 to 26 throughout the world. The vaccine proved to be 100-percent effective in preventing cervical, vaginal, and vulvar cancers caused by HPV-16 and HPV-18. Gardasil is also 99-percent effective in protecting against two other HPV strains— 6 and 11—which cause about 90 percent of all cases of genital warts, a painful—but noncancerous—condition.

Previous studies involving at least 1,000 women in the United States, Canada, and Brazil had shown that Cervarix—which also is administered in three doses—was 100-percent effective against HPV types 16 and 18. A follow-up study of Cervarix, reported in April, revealed that the vaccine was still effective 4 ½ years later. Cervarix also protects against HPV-45 and HPV-31, the third and fourth most common forms of HPV, respectively. Both of these forms have been linked to cervical cancer.

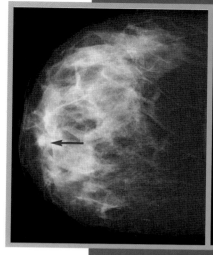 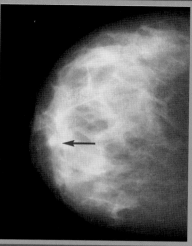

A CLEARER PICTURE

A cancer *lesion* (abnormal area of tissue) is more visible in a digital mammogram (arrow, above, left) than in a mammogram performed with standard X-ray technology (arrow, above, right). In a digital mammogram, the breast is scanned using an electronic X-ray detector. The information is then sent to a computer that creates an image and improves the contrast between any abnormal tissue and the background tissue. Researchers led by radiologist Etta D. Pisano of the University of North Carolina at Chapel Hill reported in October 2005 that, in a study of 42,760 women's breasts, digital mammograms detected 28 percent more cancers than standard mammograms did. Digital mammography is particularly beneficial for women under age 50 and for those with dense breast tissue. Nevertheless, digital technology is more expensive and, as of early 2006, only about 11 percent of mammography centers in the United States had access to it.

Despite the vaccines' effectiveness, several concerns remained. Gardasil and Cervarix provide protection only in females who are inoculated before they become sexually active and, thus, have not been exposed to HPV. As a result, most researchers recommend that the vaccines be given to teens or preteens. However, many parents and various religious and political groups are concerned that the vaccines may promote sexual promiscuity.

In addition, the vaccine does not protect against all forms of HPV, including about a dozen strains that cause from 25 to 30 percent of all cases of cervical cancer. For that reason, women who receive the vaccine would still need regular Pap smears, the primary screening test for cervical cancer. ■ Renée Despres

See also **DRUGS; NUTRITION; PSYCHOLOGY; PUBLIC HEALTH.**

NOBEL PRIZES

The 2005 Nobel Prizes in science were awarded in October for the discovery of the way carbon atoms change bonds like dance partners; the improvement of lasers by applying quantum physics to optics; and the discovery that bacteria, rather than stress, cause most ulcers. Each prize was worth about $1.3 million.

The prize in chemistry was shared by French chemist Yves Chauvin of the French Petroleum Institute in Rueil-Malmaison and American chemists Richard R. Schrock of the Massachusetts Institute of Technology in Cambridge and Robert H. Grubbs of the California Institute of Technology in Pasadena. The researchers were recognized for developing a chemical reaction called *metathesis.* Metathesis involves *organic* (carbon-based) compounds, most of which consist of complex frameworks of carbon atoms to which other atoms can be attached.

Metathesis allows chemists to break double bonds, which are much stronger than single bonds, by introducing a metal *catalyst* (substance that starts or speeds up a chemical reaction). New bonds are then formed in a way that has been compared with dancers switching partners. Metathesis is a simplified version of or-ganic synthesis, a method of rearranging groups of atoms within molecules. Metathesis provides a more efficient method of creating drugs and plastic products while reducing cost and using fewer harmful chemicals.

The prize in physics was awarded to Americans Roy J. Glauber of Harvard University in Cambridge, Massachusetts, and John L. Hall of the University of Colorado in Boulder and to German physicist Theodor W. Hänsch of the Max Planck Institute for Quantum Optics in Garching, Germany.

Glauber won half the prize for applying *quantum physics,* the study of atoms and sub-atomic particles, to *optics,* the branch of physics that deals with the properties of light. In the 1960's, Glauber demonstrated how light's particle nature affects its behavior under circumstances important in working sophisti-cated optical instruments.

2005 PHYSIOLOGY OR MEDICINE PRIZE WINNERS

The 2005 Nobel Prize in physiology or medicine was award-ed to gastroenterologist Barry J. Marshall at the University of Western Australia in Nedlands (left) and retired patholo-gist J. Robin Warren from the Royal Perth [Australia] Hospital. The scientists discovered that stomach ulcers and gastritis are caused by a bacterium known as *Helicobacter pylorus* (below) and not by stress or eating spicy foods.

Hall and Hänsch shared the other half of the prize for helping to develop *optical frequency comb technology,* a technique for making highly precise measurements of the frequency of light. The optical frequency comb technique combines pulses from a laser to produce light made up of regularly spaced peaks in frequency. The spectrum of this light is known as the *frequency comb* because the narrow peaks in the spectrum resemble the teeth of a haircomb.

Physicists can measure the frequency of light by comparing it with the evenly spaced peaks on a frequency comb. The applications of the three scientists' work include improvements in laser technology, Global Positioning System (GPS) technology, and long-distance communications.

The prize in physiology or medicine was awarded to Australians Barry J. Marshall, a *gastroenterologist* (physician who studies the stomach and intestines) at the University of Western Australia in Nedlands, and J. Robin Warren, a retired *pathologist* (physician who studies disease) from the Royal Perth Hospital in Perth. The pair received the prize for their surprise discovery in the 1980's that a bacterium, *Helicobacter pylorus,* causes stomach ulcers and *gastritis* (inflammation of the stomach lining). Their finding disproved a long-standing medical belief that ulcers resulted from an overproduction of stomach acid triggered by stress and spicy foods. Their research also has led to new discoveries about the role inflammation plays in many diseases.

In 1982, Warren discovered an unknown bacterium, later named *H. pylorus,* in samples of stomach tissue from patients with *peptic* (digestive system) ulcers. He observed that *H. pylorus* was associated with inflammation that occurs as peptic ulcers develop. Marshall and Warren conducted a study of 100 patients who suffered from peptic ulcers and found *H. pylorus* in most of them. Marshall grew the bacterium in a laboratory for further study and, in a risky experiment, drank a solution containing *H. pylorus.* He developed severe gastritis, proving his theory.

■ S. Thomas Richardson

See also **CHEMISTRY; MEDICAL RESEARCH; PHYSICS.**

Barry J. Marshall   J. Robin Warren

NUTRITION

The amount of heart-risky trans fats in processed foods appeared for the first time on food labels in the United States on Jan. 1, 2006. The Food and Drug Administration (FDA), a public health agency based in Rockville, Maryland, mandated the change to assist consumers in making healthy food choices. Foods containing trans fats include stick margarine and vegetable shortening as well as certain processed foods including cookies, crackers, and potato chips.

Numerous research studies have reported a link between eating a diet high in trans fats and increased blood levels of low-density lipoprotein cholesterol (LDL; commonly referred to as "bad" cholesterol). High levels of LDL cholesterol increase the risk of heart disease, the leading cause of death for people in the United States.

All fats contain fatty acids, which consist of chains of carbon atoms with hydrogen atoms attached. Trans fats, also called trans fatty acids, are produced by partially *hydrogenating* (adding hydrogen to) vegetable oils. This process causes the fatty acid chains to straighten out and stack closer together, forming fats that are more solid.

Hydrogenated fats in vegetable margarine and shortening are less likely to break down under high temperatures. Foods made with trans fats decay more slowly than foods made with vegetable oil and so have a longer *shelf life* (the amount of time during which food is at its best quality). In addition, some food manufacturers use trans fats to improve the flavor, freshness, and texture of their products.

According to the new labeling requirements, manufacturers must list trans fats under the Total Fat heading for almost all foods. The FDA made some exceptions, most notably for foods with less than 0.5 gram (0.02 ounce) of trans fats.

Toddlers on the fatty way. Toddlers who eat away from home or day care at least once each weekday consume higher amounts of trans fats than those who eat all their meals at home or consume at least one snack or meal at day care, according to a report published in the *Journal of the American Dietetic Association* in January 2006. Nutritionists led by Paula Ziegler, then at the Gerber Products Company in Parsippany, New Jersey, conducted a telephone survey of mothers and primary caretakers of 632 U.S. toddlers ages 15 to 24 months. The study found that 43 percent of the toddlers consumed all food and beverages at home and 8 percent ate at least one snack or meal at day care. The remaining 49 percent ate at least one snack or meal away from home or day care, in some cases inside an automobile. The researchers calculated the amount of trans fats each group ate. They found that the third group consumed the highest amount, an average of 1.5 times as much as the groups eating at home. All three groups consumed similar amounts of total fat, however.

FRUIT TATTOOS

Experimental "tattoos" etched onto the skins of fruits and vegetables using a new kind of laser are gaining popularity with growers and could some day completely replace stickers. The tattoos can provide consumers with the name and variety of the fruit or vegetable, its country of origin, and when it will be ripe and ready to eat. The government-approved process causes no damage to the food, and the tattooed areas are safe to eat.

SOMETHING FISHY

Mercury, a *toxic* (poisonous) metal, may accumulate in swordfish and some other fish to levels that are potentially harmful for people, according to a December 2005 report by the *Chicago Tribune.* Fish from local food stores was sent for testing to scientists at Rutgers University in New Brunswick, New Jersey. The researchers found that salmon had the lowest mercury level, while swordfish had the highest (results below, in parts per million). Mercury increases in the body with repeated exposure to levels that may damages nerve cells, including those in the brain. For this reason, the Food and Drug Administration recommends that young children and women of child-bearing age do not eat shark, swordfish, king mackerel, or tilefish and eat no more than one meal per week of albacore tuna.

MERCURY LEVEL	SWORDFISH	ORANGE ROUGHY	WALLEYE	YELLOWFIN TUNA	ALBACORE TUNA	GROUPER	SKIPJACK TUNA	SALMON
HIGH	3.07	1.38	1.74	0.94	0.51	0.76	0.31	0.16
AVERAGE	1.41	0.57	0.51	0.35	0.30	0.26	0.11	0.03
LOW	0.47	0.18	0.11	0.01	0.01	0.01	0.02	0.01

Some restaurants, particularly fast-food chain restaurants, rely on hydrogenated oils for frying foods. French fries were a major source of trans fats eaten by the toddlers. Away from home or day care, 6 percent of toddlers ate fries as a morning snack, while 35 percent of toddlers ate them for lunch. Other foods contributing trans fats to the toddlers' diets included deep-fried boneless chicken and cookies. The researchers encouraged caregivers, government programs, and restaurants to offer choices of fruits and vegetables in forms that appeal to children.

Teen-agers in sports eat healthier. Teen-agers who participated regularly in sports in 31 middle and high schools in Minnesota ate a healthier diet than those who were not physically active, according to a report published in the May 2006 *Journal of the American Dietetic Association.* Eating disorders specialist Jillian Croll and others at the University of Minnesota in Minneapolis analyzed data from the Project EAT (Eating Among Teens) study, in which students completed food and activity questionnaires. The students ranged in age from 11 to 18 years.

Researchers divided the 2,553 students into three groups based on the students' participation in sports for at least 3 months over a 12-month period. The power-related sports group included teen-agers who played baseball, basketball, football, hockey, soccer, or volleyball. Students in the position-related sports group were active in cheerleading, dance, gymnastics, ice skating, wrestling, and yoga. The study excluded students who participated in both power- and position-related sports. Students who were not engaged in any athletics on a regular basis made up the third group.

The students filled out questionnaires about the numbers and types of meals and snacks they ate each day. Researchers used that information to determine the students' consumption of five key nutrients: calcium, fat, iron, protein, and zinc.

The researchers found that teen-agers in all groups typically ate two snacks each day and ate dinner six nights each week. Those students in the two sport groups generally ate foods with more calcium, iron, protein, and zinc. They also ate breakfast more often than students who did not play organized sports.

NUTRITION continued

Students who did not participate in sports tended to come from lower socioeconomic groups, which may have limited their access to sports as well as to healthier foods and meals at home. The researchers categorized students' socioeconomic status using the parents' education level and employment status.

Although girls who played sports had better overall diets than the nonathletic girls, fewer than 30 percent in all three groups met the recommended daily intake for calcium or iron, even those girls taking vitamin-mineral supplements. The researchers recommended that students make up the shortfall in calcium by adding about one daily serving of such food as low-fat yogurt. The recommended calcium intake for girls and boys ages 9 to 18 years is 1,300 milligrams daily, and the recommended dairy intake is 3 cups daily. According to Surgeon General of the United States Richard Carmona, each year about 1.5 million people in the United States suffer a bone fracture related to *osteoporosis,* a disease in which bones become weak and brittle. Growing children should consume sufficient amounts of calcium in order to increase their bone mass as much as possible to help reduce their risk of osteoporosis when they are older.

For children ages 14 to 18, the recommended daily iron intake is 11 milligrams for boys and 15 milligrams for girls. Girls require more iron when they go through puberty because of the monthly loss of red blood cells during menstruation. Teens can increase their iron intake by eating more beans, dark-green leafy vegetables, dried fruit, iron-fortified cereals, and lean red meat.

The report by Croll and her colleagues showed a positive connection between following a healthy diet and playing sports. However, the study did not investigate dieting or such eating disorders as anorexia or bulimia that some athletes develop. ■ Catherine Klein

See also **MEDICAL RESEARCH.**

OCEANOGRAPHY

Illegal mining of coral reefs off the coast of Sri Lanka intensified damage to some areas of that island caused by a catastrophic *tsunami* that struck the Indian Ocean in December 2004, an international team of scientists reported in August 2005. (A tsunami is a series of powerful ocean waves produced by an underwater earthquake, landslide, or volcanic eruption.) The 2004 tsunami caused more than 216,000 deaths in southern Asia and left millions of people homeless. Coral reefs in Sri Lanka and other parts of the world have been destroyed by poachers who dynamite the reefs to collect fish and coral for sale.

Scientists from the United States and Sri Lanka, led by engineer Harindra Joseph S. Fernando of Arizona State University in Tempe, studied several towns damaged by the tsunami. They analyzed whether such coastal features as bays, cliffs, and river channels affected the amount of damage caused. The team found that these elements had little impact on the tsunami. They concluded that the presence of rock and coral reefs offshore was the most important factor in reducing flooding and loss of life. The sci-

SEE ALSO THE SPECIAL REPORT, **THE SECRET LIVES OF JELLYFISH,** PAGE 40.

entists found that healthy coral reefs off Sri Lanka absorbed some of the tsunami's energy before it struck land and deflected some of the waves in a direction parallel to the shore.

One town, Peraliya, suffered heavy damage because its reefs had been mined. A wave that was 10 meters (30 feet) tall slammed into the town, causing flooding that swept about 1.5 kilometers (1 mile) inland and carried a passenger train 50 meters (160 feet) from its tracks. About 1,700 people died in Peraliya. In contrast, just 3 kilometers (2 miles) away in the town of Hikkaduwa, wave heights measured 2 to 3 meters (7 to 10 feet) tall, with flooding reaching only 50 meters (160 feet) inland. That town had no casualties. A protective ring of intact reefs around Sri Lanka, the scientists concluded, could have prevented many deaths and significantly reduced property damage.

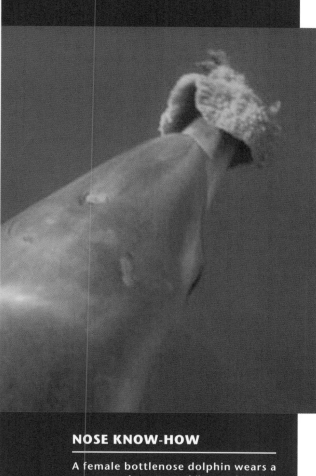

NOSE KNOW-HOW

A female bottlenose dolphin wears a sponge on her snout while searching for food along the seafloor of Shark Bay in Western Australia. In June 2005, biologists led by Michael Krützen of the University of Zurich in Switzerland reported that the dolphins had learned this behavior, making it the first instance of culture in marine mammals. The scientists believed that the sponges protect the dolphins from stinging fish or jagged coral.

Ocean acidity rising. Carbon-dioxide emissions from human activities are increasing the acidity of the oceans to levels that could threaten marine life within decades, according to a report published in September 2005. The main source of carbon-dioxide emissions from human activities is the burning of fossil fuels.

A team of 27 marine chemists and biologists from Australia, Europe, Japan, and the United States used global ocean data and computer simulations to study the relationship between atmospheric carbon dioxide and ocean acidity. Some regions of the oceans could become too

acidic by 2050 for small marine creatures to form shells, reported James Orr, a team member and oceanographer from the Laboratory for Science of the Climate and Environment in Gif-sur-Yvette, France. Previous studies had indicated that it would take hundreds of years for the oceans to become that acidic.

Orr reported that one of the first animals to be affected would be tiny winged snails called *pteropods* that live in the northern Pacific Ocean. These organisms build their shells using a delicate form of calcium carbonate called *aragonite* that is more sensitive to ocean acidity than other shell materials. Pteropods are a major source of food for many other marine animals. A major decline in their numbers would mean less food for larger species, including salmon, cod, and herring.

Giant squid seen in the wild. The first images of a giant squid in its natural habitat were published in September 2005. A team of Japanese scientists filmed an 8-meter- (26-foot-) long squid hunting in the ocean.

Zoologists Tsunemi Kubodera of the National Science Museum in Tokyo and Kyoichi Mori of the Ogasawara Whale Watching Association discovered the squid in waters off the Bonin Islands of Japan. They searched in this location because they had previously found the remains of giant squid near the islands. The scientists lured the animal to their submerged camera with bags of mashed shrimp. As the squid lunged at the bait, it caught one of its tentacles on a hidden hook. Eventually, the creature tore off the tentacle and swam away.

The researchers retrieved the severed limb, which measured 5.5 meters (18 feet) long and was lined with powerful suckers. They tested the DNA (deoxyribonucleic acid, the material that controls heredity) of the tissue and were able to confirm that the animal was a member of *Architeuthis*, the same group as the giant squid carcasses previously caught in fishing nets or washed ashore.

Ocean plagues. A worrisome increase in the number of marine animals dying from toxins in the sea was reported by a group of marine veterinarians and disease experts in February 2006. Frances Gulland, director of Veterinary Science at the Marine Mammal Center in Sausalito, California, reported that about 1,000 sea lions in California have died since 2001 from poisoning by domoic acid, a potent *neurotoxin* (toxin that can damage or destroy nerve tissue) produced by certain types of marine algae. Researchers first observed domoic acid poison-

OCEANOGRAPHY continued

ing in sea lions in 1998, when more than 70 of the animals died over one weekend.

The toxins are produced at dangerous levels during *algal blooms*—rapid population explosions—which have become more intense and more frequent since 2001. Many scientists linked this increase to several factors, including agricultural run-off of nutrients and rising temperatures from *global warming.* (Global warming is a gradual increase in the average temperature of Earth's surface.)

Gulland and her colleagues said that stranded sea lions are also dying of *urogenital cancer* (cancer of the urinary and genital organs), similar to cancers found in human beings. The scientists reported a correlation between the presence in sea lion tissue of cancer and high levels of persistent organic pollutants (POP's), which are slow-decaying chemicals produced by such industrial activities as waste burning and paper manufacturing. The researchers argued that POP's may weaken the animals' ability to fight disease.

Sea lions are top predators that eat anchovies, squid, salmon, and mussels. Because human beings eat some of the same animals, Gulland said, the sea lions' health problems could also signal human health risks.

Algal toxins that are killing endangered manatees in the Gulf of Mexico may also pose health threats to people, scientists reported in February 2006. Manatees become sick when they inhale too much of a poisonous gas produced by marine algae. The gas, called *brevetoxin,* can also make people ill.

Gregory Bossart, a marine biologist at the Harbor Branch Oceanographic Institution in Fort Pierce, Florida, and his colleagues reported that chronic exposure to brevetoxin can impair mammals' respiratory and immune systems, making them more susceptible to diseases. The scientists found that during recent toxic algal blooms off the west coast of Florida, emergency room admissions for pneumonia, asthma attacks, and other respiratory illnesses rose by 54 percent. Bossart called for more studies to explain the sharp rise in algal blooms.

Sewage killing coral. Rapid growth of bacteria caused by sewage is destroying coral reefs in the Caribbean, according to a February 2006 report. Normally, these bacteria are beneficial to

DOTING IN THE DEEP

A female squid of the species *Gonatus onyx* cradles a mass of thousands of eggs in the depths of Monterey Canyon off the coast of California. A team of biologists led by Brad A. Seibel of the University of Rhode Island in Kingston and the Monterey Bay Aquarium Research Institute in Moss Landing, California, announced in December 2005 that they had discovered five female squid holding and protecting their eggs. It was the first report to show that at least one squid species cares for its unborn offspring. Scientists had long believed that squid lay their eggs on the seafloor and then leave them to develop on their own.

A FURRY FIND

An eyeless shellfish with blond-haired claws, shown in an image released in March 2006, is so unique that scientists categorized it as the first member of a new family of *crustaceans* (ocean animals with a hard, external shell). The animal, named *Kiwa hirsuta*, is about 15 centimeters (6 inches) long. It was discovered about 2,300 meters (7,540 feet) below the surface of California's Monterey Bay during a diving expedition led by biologist Robert C. Vrijenhoek of the Monterey Bay Aquarium Research Institute.

coral animals, but an overgrowth of bacteria, like an infected wound, lowers the corals' resistance to disease-causing organisms. Caribbean reefs have lost about 80 percent of their coral cover since the 1970's.

David Kline, a postdoctoral fellow at the Smithsonian Tropical Research Institute in Ancón, Panama, said that as human populations near coastal areas have grown, sewage discharge has increased the amount of phosphorous and nitrogen in nearby waters. These compounds, he said, cause sugar-producing algae to grow. These sugars then promote the growth of coral-damaging bacteria.

Vanishing ice. Images taken by National Aeronautics and Space Administration (NASA) satellites suggest that glaciers on Greenland are melting twice as fast as previously thought, U.S. scientists reported in February 2006. The melting ice is contributing to a faster-than-expected rise in Earth's ocean level.

Greenland is now contributing 0.5 millimeter (0.02 inch) of the total 3.0 millimeters (0.12 inch) of global sea level rise each year, according to geoscientist Eric Rignot of NASA's Jet Propul-

sion Laboratory at the California Institute of Technology in Pasadena and colleagues. The scientists also found that the melting of Greenland's ice sheet increased from 50 cubic kilometers (12 cubic miles) per year in 1996 to 150 cubic kilometers (36 cubic miles) per year in 2005.

The Arctic Ocean could become ice-free during the summer within 100 years, according to an August 2005 report by a team of scientists led by geologist Jonathan T. Overpeck of the University of Arizona in Tucson. The team made its prediction by studying the interactions between sea ice, land ice, human population, and other elements of the Arctic system.

Overpeck and his colleagues found that some components within the Arctic system interact in ways that accelerate ice melting. For example, sea ice reflects some of the sun's radiation, partially preventing melting. However, as the ice melts, it exposes more of the dark ocean, which absorbs more sunlight and leads to more melting. The scientists could not identify any natural processes that would stop the accelerating loss of ice. ■ Christina Johnson

See also **BIOLOGY**.

PHYSICS

A fundamental theory in molecular biology explaining how organisms reproduce on the cellular level was confirmed by a team of physicists in a report published in November 2005. For their work, the scientists, led by Steven Block of Stanford University in California, developed an instrument that measures the motions of proteins on the molecular level, thus giving the scientists the ability to study biological processes that are fundamental to life.

The central concept of molecular biology is that DNA (deoxyribonucleic acid, the material that controls heredity) serves as a blueprint for constructing the protein molecules that are the working subunits of life. However, DNA does not participate directly in this manufacturing process. Instead, the information carried by the DNA molecule is copied to a related but simpler molecule called RNA (ribonucleic acid) that serves as the working template.

Proteins, DNA, and RNA are long, chainlike molecules made up of shorter subunits called *nucleotides*. A set of three RNA nucleotides carries the instructions for one *amino acid*, a number of which serve as the building blocks of proteins.

Scientists have assumed since the 1960's that a special protein called RNA polymerase (RNAP) assembles the RNA molecule one nucleotide at a time. This process had never been observed directly, however, because each nucleotide is separated by only 0.34 *nanometers* (billionths of a meter). Until Block's work, scientists could not observe processes occurring on such a small scale.

To solve this problem, Block and his colleagues used a technique called *optical trapping*, in which strong electric forces from sharply focused laser beams are used to immobilize biological molecules. They tied down the DNA molecule by stretching it between tiny plastic beads that served as anchors. One bead was attached to the DNA itself, while the other held the attached RNAP molecule. The physicists manipulated the laser beams to produce a constant pull on the DNA molecule, and the position of the bead attached to the RNAP molecule was carefully recorded.

The scientists discovered that the RNAP molecule advanced along the DNA strand in steps of approximately 0.37 nanometer, very close to the distance separating nucleotides in DNA. Based on this measurement, Block and his

SEE ALSO   SCIENCE STUDIES, NANOTECHNOLOGY: A REVOLUTION IN MINIATURE, PAGE 112.

colleagues concluded that RNAP moves along the DNA molecule one nucleotide at a time.

The researchers also demonstrated that the random motion of molecules in the environment propels the RNAP along the DNA. Chemical forces prevent the RNAP molecule from sliding backward, so that it moves in only one direction, an action known as *ratcheting*.

New president and precedent. John Hopfield of Princeton University in New Jersey took office as the new president of the American Physical Society, the premier organization for the promotion of physics research and education, based in College Park, Maryland, in January 2006. Many physicists saw Hopfield's presidency as a formal acknowledgment of the growing importance of biological systems in physics research. Hopfield was trained as a physicist and spent 16 years working in the field of *condensed matter physics* (physics of solid materials). He then switched his research focus to chemistry and biology. He is a member of Princeton's Department of Molecular Biology.

Many of the top physics departments in the world offer doctoral-level training in the physics of biological systems. However, most recipients of these degrees will not work in university physics departments. Instead, they, like Hopfield, will likely work in biology departments or at biomedical research institutions. Hopfield noted that the introductory physics courses offered at most universities are tailored to the needs of traditional physicists and engineers. He argued that these courses, which have changed little since the 1950's, are badly in need of revision.

Evidence for inflation. The first direct evidence supporting the theory of inflation was announced in March 2006 by scientists from the United States National Aeronautics and Space Administration (NASA) and several U.S. and Canadian universities. The theory of inflation proposes that the early universe went through a rapid period of expansion. The physicists based their conclusions on data collected by a satellite called the Wilkinson Microwave Anisotropy Probe (WMAP). WMAP was launched in 2001 to study the Cosmic Microwave Background (CMB) radiation, energy left over from the early universe. The instrument is a joint effort by Princeton University and the NASA Goddard Space Flight Center in Greenbelt, Maryland.

The CMB radiation was emitted by a hot soup of atoms when the universe was only 400,000 years old. Previous studies of the CMB radiation revealed that it is nearly uniform across the entire sky. However, the studies also identified tiny fluctuations in brightness that scientists believe represent areas where matter is slightly denser. Scientists think that these dense spots were the centers of attraction around which galaxies formed.

The inflation theory was proposed in the late 1970's by the American physicist Alan Guth. According to this theory, bright spots in the CMB radiation developed because of fluctuations in mass and energy that occurred in the early stages of inflation. According to *quantum mechanics*, the theory that describes the behavior of atoms and subatomic particles, the universe at that time was small enough to be strongly affected by random fluctuations. Inflation theory proposes that these fluctuations were blown up trillions of times and separated by millions to billions of light-years during inflation. (A light-year is the distance light travels in one year, about 9.5 trillion kilometers [5.9 trillion miles]). The theory predicts that, as a result, the CMB radiation should have greater variation in brightness over large regions of the sky than it does within small areas of sky.

The WMAP satellite produced a map of the CMB radiation with unprecedented resolution that the scientists used to examine variations in brightness over the entire sky. The researchers found that the brightness varies more over large areas of the heavens—corresponding to regions from about 1 billion to 10 billion light-years across—than it does within small regions only hundreds of millions of light-years across. The physicists concluded that the WMAP data supported inflation theory.

Modeling Earth's magnetic field. In 2005, scientists took an important step toward understanding the origin of Earth's magnetic field. A team of physicists headed by Cary Forest at the University of Wisconsin in Madison announced in December that they had found evidence supporting a long-held theory about how Earth generates its magnetic field.

Most scientists believe that Earth's magnetic field is created by electric currents circulating within the molten metal of the planet's liquid outer core. This effect is known as the *geodynamo*. The circulation is driven by the process of *convection*, in which lighter hot metal in the inner region of the core flows upward and then sinks as it cools and grows heavier. The flow is also influenced by Earth's rotation.

According to the mathematical equations that describe magnetism and fluid flow, Earth's magnetic field cannot be generated by a smooth and steady flow. Instead, the flow must be turbulent, containing whirls and eddies, like a swift mountain stream. The theory of turbulent flow is one of the most mathematically challenging areas of physics.

Physicists usually solve complex mathematical problems by using computer simulations. However, the fluid flows within Earth's interior occur on scales of about 10 meters (30 feet) while the magnetic phenomena become apparent on scales of hundreds of kilometers. The combination of these effects is so complex that even the most powerful supercomputers cannot model the system in sufficient detail. Instead, researchers must build laboratory devices that are crude models of Earth's interior.

Forest and his colleagues modeled Earth's liquid core with a sphere of liquid sodium measuring 1 meter (3 feet) in diameter. The researchers chose sodium because it is a liquid *conductor* (substance that transmits electric current) with a relatively low melting point, 98 °C (208 °F). Sodium also flows with about the same resistance as the iron in the Earth's core—nearly as freely as water. The physicists attempted to duplicate the motion of molten iron in Earth's core by rotating the sodium with propellers immersed in the liquid.

The scientists attempted to generate a magnetic field by creating a small "starter" field with electromagnetic coils placed outside the apparatus. A collection of devices called *Hall probes* then measured the magnetic field inside the sodium and on the surface of the sphere. The researchers found that the liquid sodium generated a magnetic field. The existence of this field proved that the liquid had turbulent motion. This result confirmed that liquid conductors flowing within Earth's core could generate the planet's magnetic field.

Radioactive Earth. The first measurement of Earth's internal radioactivity was announced in July 2005 by an international team of physicists conducting experiments in a Japanese mine. The scientists reported that they had detected *antineutrinos* (tiny, electrically neutral particles) produced by the *decay* (breakdown) of radioactive

PHYSICS continued

elements in Earth's interior. Many scientists believe that most of the planet's internal heat comes from the decay of such elements as uranium and thorium. However, because researchers cannot access matter deep in Earth, they have been unable to determine the amount of radioactive material present there.

The team turned to antineutrinos to obtain an indirect measure of the amount of radioactive matter within Earth. These tiny particles are produced when the *nuclei* (cores) of radioactive elements decay. Antineutrinos interact so feebly with matter that they can pass through Earth with only a small chance of being absorbed by the planet. As a result, physicists must use large detectors containing many tons of sensitive material to detect the absorption of an antineutrino.

The researchers worked with a detector known as KamLand, located in the Kamioka mine in central Japan. The detector must be located deep underground to avoid being bombarded by *cosmic rays*—high-energy particles and radiation that rain down from space or Earth's atmosphere. The detector contains 1,000 metric tons (1,100 tons) of *liquid scintillator*, a substance that *emits* (gives off) a brief flash of light when it absorbs radiation. When antineutrinos enter KamLand, some may interact with a *proton* (positively charged particle) in the liquid scintillator, producing a *positron* (positively charged electron) and *neutron* (neutral particle). When the liquid absorbs the positron, it emits a flash of light that signals the presence of an antineutrino.

The scientists found that 16.2 million antineutrinos were passing through each square centimeter (0.15 square inch) of the detector every second. Based on this finding, the researchers calculated that the radioactive processes producing the antineutrinos are most likely generating about 24 *terawatts* (trillions of watts) of power, more than half of the heat that geologists estimate is produced inside Earth. Future studies at KamLand may allow researchers to determine if radioactivity generates all of the planet's heat or if other such processes as the solidification of liquid metal in the outer core make significant contributions.

Another Earth? The discovery of the lightest planet ever detected around a sunlike star was announced in January 2006 by the Probing Lensing Anomalies NETwork (PLANET), a collaboration of 18 institutions from 10 countries. The object is only about five times as massive as Earth and orbits a star 20,000 light-years from the sun, near the center of the Milky Way Galaxy.

Nearly 200 *extrasolar* (beyond the solar system) planets have been discovered since 1995. Scientists cannot see these planets directly because the reflected starlight from the planets is lost in the bright glow of their stars. As a result, most of these planets have been observed using indirect methods.

The most common of these methods is the *Doppler shift method.* In this technique, scientists infer the presence of a planet by observing changes in the motions of its parent star caused by the gravitational pull of the planet. As the planet orbits the star, the planet's gravity causes the star to wobble. As the star moves closer to or farther from Earth, the *Doppler effect* causes a slight shift in the *frequency* (number of cycles per second) of the light waves it emits. (The Doppler effect is the change in the frequency of sound, light, or radio waves caused by the relative motion of the source of the waves and the observer.)

The bigger the planet's mass, the bigger the wobble and resulting frequency shift. However, because larger shifts are easier to detect, researchers were able to find only giant planets, about the size of Jupiter, using this technique.

Scientists discovered the Earth-sized planet using a different technique called *gravitational lensing.* This effect was first described by German-born physicist Albert Einstein in 1915 as a consequence of his general theory of relativity. The theory predicts that the path of a light beam will be affected by nearby massive objects. For example, light from distant galaxies passing close to a nearby galaxy cluster on its way to Earth will be distorted by the gravity of the cluster.

Lensing may also occur if the orbit of an extrasolar planet takes it through the line of sight between a background star—located far behind the planet's parent star—and Earth. Light rays that pass at the right distance from the planet will emerge in nearly parallel lines, making the image of the remote star appear brighter. However, the distorted image of the background star is too small to be observed, and so this phenomenon is called *gravitational microlensing.*

The timing and magnitude of the brightening allow an observer to calculate the size of the orbit and the *mass* (amount of matter) of the planet. The brightening lasts for a few days in the case of a giant planet, but for only a few hours with a smaller planet. For this reason, an observer using this method to search for a planet must be fortunate. The orbital plane of the planet must be aligned with Earth's line of sight, and the telescope must observe the star at the right time.

The brightening event caused by the Earth-sized planet was spotted on July 11 by the Optical Gravitational Lensing Experiment (OGLE), a joint project of Princeton University and Warsaw University in Poland. OGLE scientists scan the sky each night using a telescope at Las Campanas Observatory in Chile. Although the scientists were not able to identify the planet, they named it OGLE-2005-BLG-390Lb in recognition of the observation.

PLANET observed the same brightening phenomenon on August 10 and 11. The organization watches the sky in the Southern Hemisphere nearly 24 hours a day using five telescopes located throughout the world. PLANET scientists focus on the center of the Milky Way Galaxy, which has the highest density of stars and so offers the greatest likelihood of observing gravitational microlensing events. Two PLANET telescopes were looking in the right place at the right time to observe the lensing from the Earth-sized planet—the Perth telescope in Australia and the Danish telescope in La Silla, Chile.

The observations enabled the scientists to determine that the planet's mass is about 5.5 times that of Earth. The planet circles its parent star every 10 years in an orbit about 2.6 times the diameter of Earth's.

Although the planet is Earth-sized, it is probably not Earthlike. The planet is so far from its parent star, which is much smaller than the sun, that its temperature must be about −220 °C (−430 °F). Some scientists think that the planet, like Pluto, is covered with frozen water.

Scientists have observed only two other planets using the gravitational microlensing technique, both of them giant planets. However, because lensings are rare events and because one in three observed events has revealed an Earth-sized planet, some scientists believe that Earth-sized planets actually may be common in the cosmos. The discovery brought hope to researchers looking for evidence of life elsewhere in the universe. ■ Robert H. March

See also **ASTRONOMY; GEOLOGY; NOBEL PRIZES.**

SPAGHETTI-SHATTERING SCIENCE

A stick of dry spaghetti hit with a metal slug (below, left) undulates and then snaps at regular intervals in a series of images taken with a high-speed video camera. In 2005, a team led by mathematician Andrew L. Belmonte of Pennsylvania State University at University Park investigated the unique breaking patterns of rods of spaghetti. Scientists had long puzzled over the fact that uncooked spaghetti does not snap into two equal pieces when bent past its breaking point. Based on these images, the researchers concluded that the spaghetti buckled in a regular, wavelike pattern. The stick then broke at all of its highly curved points— the wave's peaks and troughs—at nearly the same time.

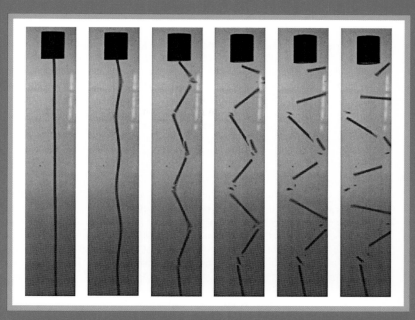

PSYCHOLOGY

About half of all Americans will develop recognized symptoms of mental illness at some time in their life, according to a large survey reported in June 2005. The survey, conducted by researchers at Harvard University in Cambridge, Massachusetts; the University of Michigan at Ann Arbor; and the United States National Institute of Mental Health (NIMH) in Rockville, Maryland, sought to determine how common mental disorders are in the general U.S. population. The survey also found that at least half the people with a mental disorder developed symptoms during childhood or adolescence, and that most people with mental disorders are either untreated or inadequately treated. Finally, only a small group of people experience the most serious mental disorders, according to the researchers, but people in this group tend to have several disorders.

The survey, called the U.S. National Comorbidity Survey Replication, involved interviews with 9,282 people over age 18. Follow-up interviews were conducted with 5,692 survey subjects. The survey replicated similar studies conducted from 1990 to 1992 and in 1994. The surveys are the only investigations of the symptoms of mental illness in the general population. Most other studies of mental illness have focused on people who describe themselves as mental health patients.

The researchers measured how common mental disorders are in two ways: according to *lifetime prevalence* (the likelihood of developing a disorder at some point in life) and *12-month prevalence* (the likelihood of developing a disorder in the next 12 months). They found that of the nearly 50 percent of people who will become mentally ill in their lifetime, 28.8 percent will develop an anxiety disorder (such as panic disorder); 20.8 percent, a mood disorder (such as depression); and 14.6 percent, a substance abuse disorder (such as drug abuse).

Among the people studied for lifetime prevalence, the researchers looked at *age of onset* (the age at which symptoms appeared). Symptoms appear earlier for anxiety disorders (11 years) than for substance abuse disorders (20 years) and mood disorders (30 years). They also found that, overall, half of lifetime cases start by age 14 and three-fourths start by age 24.

Over a 12-month period, 26.2 percent of Americans will meet the criteria for some sort of psychiatric disorder; 18.1 percent will develop anxiety disorder; 9.5 percent, a mood disorder;

and 3.8 percent, a substance abuse disorder. Of the 12-month cases, 22.3 percent were classified as serious; 37.3 percent, as moderate; and 40.4 percent, as mild.

The researchers were particularly interested in *comorbidity* (a diagnosis of more than one illness at the same time). During the 12-month period, 55 percent of survey respondents had a single diagnosis; 22 percent had two diagnoses; and 23 percent had three or more diagnoses. The researchers concluded that serious mental illness is concentrated among a relatively small proportion of people who often have more than one condition.

Finally, the researchers studied whether—and how quickly—people received treatment for mental disorders. They found that though over their lifetime people usually did get treatment, they delayed doing so. This period of delay ranged from 6 to 8 years for mood disorders to from 9 to 23 years for anxiety disorders. Among the 12-month cases, fewer than half the people with symptoms had received treatment in the previous 12 months.

Antidepressants. Several studies published in the first half of 2006 addressed the controversy over whether antidepressant drugs cause suicidal feelings and behavior, especially in adolescents. The controversy began in 2004, when researchers at the U.S. Food and Drug Administration (FDA) warned that both adults and children taking any of 10 newer antidepressants could experience worsening depression or increased suicidal feelings and behavior.

The 2006 studies varied in their findings. Researchers at the FDA, who published their results in March, pooled data from pediatric patients in 24 antidepressant drug trials, all but 1 of which were sponsored by drug companies. In the studies, some patients received a real drug while others were given a *placebo* (inactive substance). In all, 4,582 people participated in the trials, which studied whether antidepressants worked in treating children and adolescents for depression, obsessive-compulsive disorder, generalized anxiety disorder, attention deficit/hyperactivity disorder, and social anxiety disorder. No suicides occurred during the studies. However, pediatric patients taking antidepressants experienced a modest increase in suicidal thinking or behavior.

The researchers pointed out that FDA studies in the 1990's of adults taking antidepressants

TRAUMA AND MENTAL HEALTH

Researchers studying the psychological effects of the Sept. 11, 2001, terrorist attack on the World Trade Center in New York City reported varying results. According to a report published in March 2006, a majority of survivors have shown remarkable psychological resilience. (Resilience was defined as having no more than one stress-related symptom—such as sleep difficulty—during the six months following the attack.) Researchers at Columbia University in New York City conducted telephone interviews with more than 2,700 people living in or near New York City in mid-2002. They found that 65 percent of those interviewed had either no or one symptom of *post-traumatic stress disorder*, a condition caused by experiencing or witnessing a traumatic event. However, researchers at the U.S. Centers for Disease Control and Prevention in Atlanta, Georgia, reported in April 2006 that 64 percent of survivors who had witnessed the collapse of the towers, the deaths of victims, or injuries to other people experienced depression, anxiety, or emotional problems for up to three years. The researchers interviewed more than 8,400 people in 2003 and 2004 who were near the collapsing buildings on Sept. 11, 2001.

found no evidence of an increased risk of suicide. They also noted that the rate of adolescent suicide in the United States has declined. The rate for males ages 15 to 19, for example, fell from 17.6 per 100,000 in 1992 to 12.2 per 100,000 in 2002. These two findings led the researchers to conclude that, because there are many more adolescents being treated with antidepressants in the 2000's than there were in the 1990's, the drugs may have an overall benefit of decreasing the number of suicides.

Another study, published in January 2006, examined suicide risk during antidepressant treatment by looking at computerized health plan records. The researchers, at Harvard Medical School and at Brigham and Women's Hospital, both in Boston, studied 65,103 patients who had been treated with antidepressants at some time from 1992 to 2003. In the six-month period after the patients were prescribed antidepressants for the first time, 31 committed suicide. The risk of a suicide attempt was 314 per 100,000 in children and adolescents and 78 per 100,000 in adults.

The researchers determined that the risk of death by suicide was highest in the month before starting antidepressant treatment and declined after starting medication. In addition, their data showed that the 10 newer antidepressants that the FDA had warned about in 2004 actually posed less risk of suicide than older drugs in the first months of treatment. According to the data, the risk of suicide while taking an antidepressant is about 1 in 3,000, and the risk of a serious suicide attempt is 1 in 1,000. The team concluded that people beginning treatment with newer antidepressant drugs have no significant increased risk of suicide or of a serious suicide attempt.

In May 2006, however, the British-based drug company GlaxoSmithKline sent letters warning doctors that its antidepressant drug paroxetine (sold under the brand name Paxil) may increase the risk of suicide attempts in young adults ages 18 to 30. The company's researchers had studied 8,958 people who took paroxetine and 5,953 who took a placebo for either depression or such disorders as panic attacks and obsessive-compulsive disorder. The researchers found that of 3,455 people who took paroxetine for depression, 11 reported attempting suicide. Only 1 person in 1,978 of those taking a placebo reported a suicide attempt. The researchers found no reports of suicide attempts in individuals over age 30.

PSYCHOLOGY continued

Hormones and infant bonding. Children who are neglected during infancy produce abnormally low levels of certain hormones that seem to affect social behavior. The low levels may make it difficult for such children to form emotional attachments later in life, according to a study reported in November 2005 by researchers led by psychologist Seth D. Pollak of the University of Wisconsin in Madison.

Pollak and his team studied 18 children about 4 ½ years of age who had been adopted from Russian and Romanian orphanages and who had lived in the United States for up to three years. First, the researchers measured levels of the hormones oxytocin and vasopressin in the children's urine. Both hormones are proteins produced by the pituitary gland in the center of the brain. Laboratory tests in rodents and monkeys have indicated that the proteins may affect the ability to interact with others. Researchers believe, for example, that vasopressin plays a role in the ability of infants to recognize certain people as familiar to them and is critical for forming bonds with parents and other caregivers. The orphans who experienced neglect early in life had lower levels of vasopressin than did children who had been raised by their family.

Then, both the children who had lived in an orphanage and children who were raised in families sat on either their mother's lap or the lap of an unfamiliar adult while they played an interactive computer game. After 30 minutes, oxytocin levels had risen in children raised by their family who had sat on their mother's lap. But oxytocin levels in the children initially raised in orphanages who sat on their mother's lap did not rise. This finding may explain why previously neglected children often turn to any nearby adult—rather than to adoptive parents—when frightened, while children raised by their parents normally turn to their parents. The researchers theorized that there may be a critical time early in life for developing the ability to form a strong bond with another person and that later relationships depend on the levels of certain hormones having been set correctly during this time. The previously neglected children may have missed this window of development, or it may take them longer to develop a normal hormonal response.

■ Thomas A. M. Kramer

PUBLIC HEALTH

The worst outbreak of mumps in the United States since the late 1980's hit the Midwest in early 2006. By late April, public health officials had reported at least 1,200 confirmed or suspected cases in Iowa and another 500 cases in eight other states.

Mumps is caused by a virus and is transmitted in saliva spread by coughing, sneezing, or sharing eating utensils or food. The disease may cause such symptoms as fever, headache, difficulty swallowing, and swelling of the *salivary* (saliva-producing) glands in the cheeks and jawline. Although considered a common childhood disease, mumps occasionally can cause more severe conditions, including *encephalitis* (swelling of the brain) or deafness.

Most children in the United States receive two vaccinations against the mumps virus, one at the age of 12 to 15 months and another at 4 to 6 years. The majority of those affected by mumps in 2006 were young adults from 18 to 25 years old, ages by which, health officials speculated, the effects of the vaccines may have worn off.

Although the source of the U.S. mumps outbreak remained unknown, health officials noted that the strain was the same as one responsible for an epidemic in the United Kingdom from 2004 to 2006. By April 2006, more than 70,000 cases of mumps had been confirmed in the United Kingdom. Those cases chiefly involved people ages 15 to 24, most of whom were enrolled in colleges or universities. Most of these young people did not receive routine mumps vaccinations when they were children because such vaccinations were first required in the United Kingdom in 1998.

Hepatitis A vaccine. In October 2005, a national vaccine panel at the U.S. Centers for Disease Control and Prevention (CDC) in Atlanta, Georgia, recommended that all children ages 1 to 2 be vaccinated against the hepatitis A virus. Hepatitis A is a highly contagious disease that causes fever, chills, vomiting, and *jaundice* (a yellowish discoloration of the skin and of the whites of the eyes that occurs when the liver is not functioning properly). The virus is spread through contact with the feces of an infected person, usually through contaminated food or water.

Researchers led by Beth P. Bell, an *epidemiologist* (infectious disease specialist) at the CDC, reported in July 2005 that a hepatitis A vaccine used since 1999 had brought cases of the illness to historic lows in the 17 states with the highest incidence of the disease. Hepatitis A rates dropped by 88 percent in those states. By 2003, 2.5 people per 100,000 had contracted the disease, compared with 21.1 people per 100,000 in 1992, the lowest annual rate in the years prior to the introduction of the vaccine.

Unvaccinated rabies victim survives. The first documented case of a person who survived infection with the rabies virus without undergoing the standard treatment of injections was reported in June 2005. Doctors led by pediatric epidemiologist Rodney E. Willoughby, Jr., at the Medical College of Wisconsin and Children's Hospital, both in Milwaukee, treated the 15-year-old girl using a novel combination of medications.

Rabies is an infectious disease caused by a virus. It is transmitted in saliva, usually through the bite of an infected animal. The virus attacks the central nervous system, causing swelling of the brain and, ultimately, death. Although few cases of rabies occur in the United States, worldwide about 60,000 people are infected annually. The standard treatment for rabies—a series of injections that must be given before symptoms develop—is nearly 100-percent effective.

The 15-year-old girl had been bitten by a bat in September 2004 and did not seek treatment immediately. One month later, she experienced such symptoms as fever, double vision, vomiting, and drooling. The physicians devised a treatment in which they administered a combination of drugs to induce a week-long coma, giving the girl's immune system time to kill the virus. The girl eventually recovered fully. Willoughby and his team said that they were unsure whether the coma itself, one of the medications the patient received, or the combination of drugs was responsible for the cure. They hoped that other doctors would attempt this same treatment in a similar situation to help determine which factors made the treatment a success.

Cancer death decline. Deaths from cancer in the United States declined for the first time since record keeping began in 1930, according

to a February 2006 report by the American Cancer Society (ACS), headquartered in Atlanta. The decline represented a drop of 369 cases, from 557,271 cancer deaths in 2002 to 556,902 deaths in 2003 (the latest date for which figures were available). The ACS attributed the downturn to a decrease in the number of men who smoked and to an increase in the number of people who benefitted from early detection and treatment for prostate, breast, and colorectal cancer. Although the number of men who died of cancer declined by 778, the number of women increased by 409. According to the report, the increase among women occurred primarily because of smoking.

In 2006, the report stated, lung cancer was expected to remain the leading cause of cancer deaths in both men and women. The second leading cause of cancer death was breast cancer for women and prostate cancer for men. Colon and rectal cancer together ranked as the third most common cause of cancer death among both men and women.

Avian flu vaccine. Researchers in March 2006 reported disappointing results for a new vaccine developed to prevent a *pandemic* (widespread epidemic) of *avian* (bird) flu among human beings. Bird flu is a deadly respiratory illness caused by the H5N1 virus. By the end of April 2006, it had appeared in birds in 51 countries. From December 2003 to April 2006, 205 people had become ill with bird flu and 113 of them had died. They had contracted the disease by handling or eating infected poultry. Health care workers feared that eventually the H5N1 virus would *mutate* (change genetically), causing human beings to pass the virus to one another.

The researchers, led by immunologist John Treanor of the University of Rochester in New York, reported that only 54 percent of people who received the highest dose of the new vaccine—180 micrograms—showed an immune response. That is, their bodies developed *antibodies* (immune system cells that fight foreign invaders) in sufficient quantities to fight the disease. In contrast, the flu shots many people receive annually, which contain 15 micrograms of vaccine, provide protection in 75 to 90 percent of recipients.

Researchers also expressed concern that the high dosage required would limit treatment to only 4 million people, given the current supply of vaccine. In addition, this vaccine was based on a 2004 version of the H5N1 virus. Scientists feared that the tested vaccine could become even less effective as the virus mutated. ■ Deborah Kowal

Lessons from the Spanish Flu

In September 1918, World War I was raging in Europe. At Camp Jackson in South Carolina, 21-year-old United States Army Private Roscoe Vaughan was battling a different enemy. In less than a week, he was dead.

Private Vaughan died of *influenza* (flu). In 1918, a particularly vicious *strain* (variety) of the influenza virus caused a *pandemic* (widespread outbreak of disease) that killed from 20 million to 50 million people worldwide. Because Spain was one of the first countries to publicly acknowledge the spread of the illness, the disease became known as the Spanish flu. Where did this virus come from? Why did it kill so many people? What can we do to keep so many people from dying in the future? A little piece of Private Vaughan's lung is helping scientists trying to answer those questions.

People who die from influenza are usually very young (with immature immune systems) or old (with weakened immune systems). Things were different in 1918. The Spanish flu struck hardest at young adults between the ages of 20 and 40. Overall, one-third of the world's population became infected. Those who caught the virus were 20 times more likely to die than are flu victims during an average year.

In 1918, health officials knew that influenza is chiefly a respiratory disease that spreads easily from person to person, and so they did what they could to contain the virus. In many cities, it became illegal to appear in public without a cloth mask covering the nose and mouth. Public officials closed schools, churches, theaters, and other places where large groups of people might gather.

Eventually, the pandemic ran its course. Although new, milder strains of the virus that caused the Spanish flu continued to sweep across the world every year, the 1918 virus was gone. That disappearance frustrated efforts by modern scientists to learn why the 1918 influenza virus was so much more deadly than ordinary flu viruses, including even the viruses that caused the flu pandemics of 1957 and 1968.

In 2005, however, the virus that killed Private Vaughan and so many others was brought back to life. Terrence M. Tumpey, a microbiologist at the Centers for Disease Control and Prevention (CDC) in Atlanta, Georgia, re-created the virus. For his work, Tumpey used a "blueprint" of the virus discovered by molecular pathologist Jeffery K. Taubenberger and colleagues at the Armed Forces Institute of Pathology in Rockville, Maryland. In 1995, Taubenberger's team had obtained a tiny slice of

Nurses in Lawrence, Massachusetts, treat patients infected with the Spanish flu outside hospital tents. The nurses hoped that fresh air and sunshine would help to cure their patients.

Private Vaughan's lung that had been preserved in chemicals. They knew that the sample would not contain a living influenza virus, but they hoped to find some of its genetic components. Taubenberger's group also found genetic components of the virus in the lungs of two other victims of the 1918 pandemic: the first, another soldier, and the second, an Inuit woman who had been buried in permanently frozen ground in what is now Brevig, Alaska.

Influenza viruses are simple, made up of only eight genes surrounded by a membrane and some protein. By examining the virus's genes and comparing them to the genes of other influenza viruses, Taubenberger hoped to learn where the 1918 virus came from and why it was so deadly.

The researchers pieced together the genetic material in their samples until, in October 2005, they announced that they had found the *genetic code* for making all eight genes of the 1918 influenza virus. (The genetic code is the set of instructions that directs the production of proteins by genes.) When they compared the genetic code of the 1918 virus to those of less virulent influenza viruses, they found that the 1918 virus looked more like a virus that typically infects birds than one that infects human beings. In fact, they identified 10 small *mutations* (genetic changes) between the code of the 1918 virus and those of other bird flu viruses that may have allowed the 1918 strain to jump from its bird host directly to people and survive. (In contrast, the viruses responsible for the 1957 and 1968 pandemics were mixtures of bird and human influenza virus genes produced when a host was infected by both viruses.) The avian influenza (bird flu) that appeared in 1997 has some of the same genetic changes that enabled the 1918 strain to infect people.

The re-creation of such a deadly virus and the publication of its complete blueprint troubled some people. They feared that the virus could accidentally escape from the laboratory or be stolen. Some feared that terrorists could use the blueprint to produce a deadly epidemic. Scientists working with the virus recognized that concern and worked under strict security with the approval of the U.S. National Science Advisory Board for Biosecurity. Anthony S. Fauci, director of the National Institute of Allergy and Infectious Diseases of the National Institutes of Health, and Julie L. Gerberding, director of the CDC, issued a joint statement supporting the work. They concluded that knowledge about the 1918 virus is the best defense against natural influenza pandemics or bioterrorism. Such information could help scientists identify a flu virus with pandemic potential soon after mutation and so take steps to contain its spread. Moreover, modern antiviral medications work against viruses like the 1918 strain.

The modern world is so mobile that a pandemic form of influenza could spread much faster than it did in 1918. However, modern physicians have a much wider arsenal of vaccines and drugs to fight not only flu viruses but also the pneumonia and other secondary infections they may produce. Private Vaughan lost his battle against influenza in 1918, but the information he left behind may help scientists win the war against the next killer flu.

◼ Jacqueline Houtman

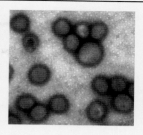

The influenza virus is a relatively simple organism that consists of only eight genes surrounded by a membrane and some protein.

SCIENCE AND SOCIETY

The *intelligent design* movement in the United States, which offers itself as an alternative to the theory of evolution, was dealt a major blow in 2005. A federal judge in Pennsylvania ruled that intelligent design theory is not science and has no place in a public school science classroom.

According to intelligent design theory, human beings are so complex that they could have developed only under the direction of an "intelligent designer," who may be the Biblical God or some other superior being. Supporters of intelligent design want it taught along with or instead of the theory of evolution in U.S. schools. This latter theory, popularized by British naturalist Charles Darwin in the mid-1800's, states that all living things evolved from a few common ancestors by means of natural selection.

Intelligent design is closely associated with *creationism*. Creationists reject the idea that human beings evolved from other forms of life and take literally the Biblical notion that God made people in his own image. The U.S. Supreme Court ruled in 1987 that creationism is a religious belief and that teaching it in public schools violates the constitutional principle of separation of church and state.

In October 2004, the school board of Dover, Pennsylvania, adopted a policy requiring teachers in ninth-grade biology classes to read a statement to their students casting doubt on evolution and proposing intelligent design as an alternative. When the teachers refused, school administrators read the statement. A group of 11 local parents filed a lawsuit against school officials, claiming the policy violated their religious liberty.

The trial began on Sept. 26, 2005. In his decision, rendered on December 20, U.S. District Judge John E. Jones III ruled that intelligent design may not be taught in public school science classrooms as an alternative to evolution. Meanwhile, eight of the nine Dover school board members were defeated in their bids for reelection in November 2005 by a slate of candidates critical of their actions. The new board said it would not appeal the court decision.

The Dover decision, described as a landmark ruling, did not end the controversy over teaching evolution. Also in November, over the objections of scientists, the Kansas State Board of Education voted to endorse revisions to statewide science teaching standards. The new

SEE ALSO SCIENCE STUDIES, NANOTECHNOLOGY: A REVOLUTION IN MINIATURE, PAGE 112.

standards redefined science to allow for studies beyond natural explanations of the physical world. Three months later, in February 2006, a local school board near Kansas State University in Manhattan became the first school board in the state to reject those standards.

In early 2006, legislatures in a number of states—including Alabama, Michigan, Missouri, Oklahoma, and South Carolina—were considering bills regarding various aspects of teaching evolution or intelligent design. On February 27, the Utah House of Representatives narrowly defeated an anti-evolution bill. Also in February, the Ohio State Board of Education voted to eliminate a model biology lesson plan promoted by intelligent design advocates that had singled out evolution for "critical analysis."

Research misconduct. Two cases of misconduct by research scientists received worldwide attention in late 2005 and early 2006. In South Korea, geneticist Hwang Woo Suk's status as a national hero began to erode in November 2005, after his major U.S. collaborator, biomedical researcher Gerald Schatten of the University of Pittsburgh in Pennsylvania, announced that he was ending his association with Hwang. Hwang had claimed to have achieved success in cultivating a line of *stem cells* from a *cloned* (genetically duplicated) human embryo. (Embryonic stem cells have the capability to develop into any of the cell types that make up the tissues and organs of the body.) Schatten said that Hwang, a professor at Seoul National University (SNU), had misled him. Hwang had claimed that the eggs used in the experiments had been donated by volunteers, when, in fact, the donors had been paid, a violation of ethical standards.

More serious charges soon followed. In December 2005, Hwang notified the editors of the journal *Science* that a number of images supposedly depicting 11 different stem cell lines in a widely heralded paper published in that journal in June 2005 represented a single cell line. Although Hwang claimed the photos were the result of a mix-up, the controversy refused to die. In December, the president of SNU appointed a committee to investigate Hwang's research. The committee reported in January 2006 that its members had found no evidence that Hwang had actually done the research he described in his *Science* paper and concluded that the data had been fabricated. In March, Hwang was dismissed from his faculty position. Six of his collaborators were punished with suspensions or pay cuts.

In January, officials at the Norwegian Radium Hospital in Oslo reported that cancer researcher Jon Sudbø had fabricated data for an article published in the prestigious British medical journal *The Lancet* in October 2005. The study had found that anti-inflammatory drugs including ibuprofen (sold under such brand names as Motrin and Advil) and naproxen (sold under such brand names as Naprosyn and Aleve) reduced the risk of cancer of the mouth. The results led some doctors to shift away from surgery in favor of drug therapy as a treatment for oral cancer. Sudbø claimed that his findings were based on data from 908 patients. However, an internal investigation by the hospital revealed that most—if not all—of the data had been faked.

In light of the Hwang and Sudbø cases, scientific journals began examining their article review procedures. Journal reviewers operate on the assumption that other scientists are honest. Most observers agreed that in 99 percent of cases, this is a valid assumption. Spotting fabrications in the remainder is a problem that the science community was now having to face.

Visa issues anger foreign scientists. Difficulties faced by a renowned Indian scientist in getting a visa to enter the United States highlighted the problems remaining in visa processing systems following the restrictions imposed after the Sept. 11, 2001, terrorist attack on the United States. On Feb. 9, 2006, organic chemist Goverdhan Mehta, former director of the Indian Institute of Science in Bangalore and president of the International Council of Science (ICSU), applied for a visa at the U.S. Consulate in Chennai, India. (The Paris-based ICSU is an international organization that fosters exchanges of ideas and information among scientists.) Mehta had been invited to speak at an international conference at the University of Florida in Gainesville, where he had been a distinguished visiting professor in 2001.

During what he expected to be a routine interview, Mehta was questioned about whether his research might be used for chemical warfare. He replied that it could not. Mehta reported feeling "humiliated" after being asked to supply additional information on his research before his

RULING AGAINST INTELLIGENT DESIGN

A court ruling involving Dover Area High School in Pennsylvania in December 2005 determined that *intelligent design* theory is not science and has no place in a public school science classroom. According to intelligent design theory, human beings are so complex that they could have developed only under the direction of an "intelligent designer." The Dover school board had adopted a policy that required ninth-grade teachers to read a statement to their students casting doubt on evolution and proposing intelligent design as an alternative.

SCIENCE AND SOCIETY continued

application would be considered. The scientist, who said he had visited the United States about 20 times, angrily dropped his travel plans. Two other prominent Indian scientists subsequently reported similar experiences while applying for visas at the U.S. Consulate in Chennai. A U.S. State Department representative emphasized that consular officials are required by law to conduct more intense screenings on visa requests for individuals involved in certain "technology-alert" specialties.

The incidents, which were widely reported in the Indian press, occurred shortly before a visit to India by U.S. President George W. Bush. On Feb. 24, 2006, the United States issued Mehta a visa. The U.S. ambassador to India also telephoned Mehta personally to apologize. Nevertheless, the incidents embarrassed the U.S. government as it was working to strengthen relations with India.

In March 2006, the Council of Graduate Schools reported that foreign applications to U.S. graduate schools had increased after two years of decline. Applications from India, in fact, rose by 23 percent in 2005, though they remained below their 2003 levels.

Animal rights extremists convicted. On March 2, 2006, a federal jury in New Jersey convicted a militant animal rights group and six of its members of several charges, including interstate stalking and conspiracy to violate the U.S. Animal Enterprise Protection Act. The act had been amended in 2002 to equate certain offenses with terrorism. Those convicted faced fines of up to $250,000 and jail terms of up to seven years.

The group, known as SHAC (Stop Huntingdon Animal Cruelty), began in the United Kingdom in 1999 but also has chapters in the United States. The jury agreed that SHAC had harassed employees of Huntingdon Life Sciences (HLS)—the largest animal testing facility in the United Kingdom—as well as employees of insurance companies and other firms that did business with HLS. (HLS maintains a branch in New Jersey.) SHAC's Web site posted such personal information as the names and home addresses of HLS employees and their families, as well as information about employees of HLS's business partners. The site also praised such acts of violence as vandalizing and smoke bombing the offices of partner companies. ■ Albert H. Teich

SPACE TECHNOLOGY

Space exploration moved in important new directions in 2005 and 2006. The United States space shuttle returned to flight for the first time since the Columbia accident in 2003. China launched a crew of two in a Shenzhou capsule for a five-day mission in Earth orbit. Russia launched six people into space in two Soyuz capsules. At a historic meeting in Florida, the heads of space agencies from around the world made final plans for completing the International Space Station (ISS) in 2010.

Robots also made impressive contributions to space exploration. One robotic spacecraft sent an impactor smashing into a distant comet and analyzed the dust kicked up by the impact. Another brought samples of comet dust back to Earth. Mars rovers sent by the U.S. National Aeronautics and Space Administration (NASA) continued to crawl across the dusty surface of the Red Planet long after scientists had expected their mission to end. The Cassini spacecraft

sent back spectacular pictures of Saturn and its many moons, including one that appeared to show huge fountains of water spewing into the cold vacuum of space and freezing into ice and snow. A Japanese space probe touched down on an asteroid that formed from a collision between two of these rocky bodies, which circle the sun between the orbits of Mars and Jupiter. Finally, rockets on two continents launched spacecraft to three planets—Venus, Mars, and distant Pluto.

Shuttle returns to flight. The space shuttle Discovery roared into orbit from the Kennedy Space Center in Florida on July 26, 2005. Discovery carried a crew of seven on the first shuttle mission since Columbia disintegrated during its return from space on Feb. 1, 2003. Discovery's 2005 mission was part test flight—to check out repairs and other changes made to the shuttle to prevent another accident—and part supply mission to the ISS. As the shuttle ap-

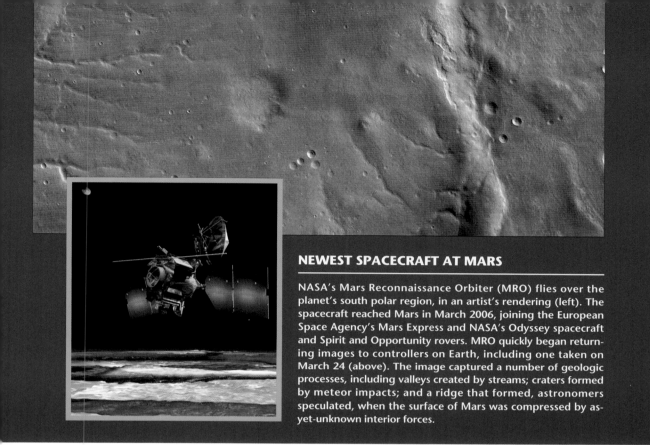

NEWEST SPACECRAFT AT MARS

NASA's Mars Reconnaissance Orbiter (MRO) flies over the planet's south polar region, in an artist's rendering (left). The spacecraft reached Mars in March 2006, joining the European Space Agency's Mars Express and NASA's Odyssey spacecraft and Spirit and Opportunity rovers. MRO quickly began returning images to controllers on Earth, including one taken on March 24 (above). The image captured a number of geologic processes, including valleys created by streams; craters formed by meteor impacts; and a ridge that formed, astronomers speculated, when the surface of Mars was compressed by as-yet-unknown interior forces.

proached the space station, Commander Eileen Collins fired control jets to perform a slow back flip so the station's two crew members could photograph Discovery from all angles with telephoto lenses. The crew was looking for any damage to Discovery's fragile thermal protection system.

Other cameras on the ground and on the shuttle's external tank also filmed and photographed the thermal tiles during the ascent to space. Cameras on Discovery photographed the tank as it fell away after liftoff, enabling scientists to confirm that the foam insulation that keeps ice from building up before launch was intact. Columbia and her crew of seven were lost when a piece of foam fell off and cracked the thermal protection system on the left wing. The damage allowed gas heated by the friction of reentry to enter the wing and melt the aluminum structure beneath.

Even before Discovery left the atmosphere, a video camera sent back live images of a piece of foam falling from the external tank toward the shuttle. Other pictures showed that the foam had caused no apparent damage. However, the loss of the foam even after design changes had been made to prevent such an occurrence meant that

tank engineers would have to go back to the drawing board to devise another method of attaching the insulation. NASA, in effect, grounded the shuttle again and was not expected to launch another mission until July 2006.

Work on the foam problem was temporarily delayed when Hurricane Katrina struck New Orleans on Aug. 29, 2005. The storm scattered the engineers and technicians who work on the external tank at the Michoud Assembly Facility near New Orleans. A "ride-out" crew had remained at the factory throughout the storm, keeping the pumps in operation and preventing the buildings from flooding. Nevertheless, the Michoud staff had to repair the facility before they could finish designing and installing the latest tank-safety modifications.

International Space Station. Although the return of the shuttle to space was short-lived, Discovery still had a mission to complete before it returned to Earth. Its crew pitched in with the two men on the station—Expedition 11 cosmonaut Sergei Krikalev and astronaut John Phillips—to unload supplies from Raffaello, a module in Discovery's cargo bay. They refilled the module with trash, scientific samples, and worn-out equipment for the trip back to Earth.

SPACE TECHNOLOGY continued

Astronauts Stephen Robinson of the United States and Soichi Noguchi of Japan donned spacesuits three times to work outside the docked shuttle and station. They practiced thermal protection system repair techniques developed to fix damage such as the crack that doomed Columbia. They also replaced a malfunctioning *gyroscope*, a piece of equipment that helps keep the ISS solar arrays facing the sun without burning precious fuel, and performed other repairs. Robinson rode on the end of a robot arm to pull out a strip of insulation used as a thermal protection system "gap filler" that a camera spotted protruding from Discovery's belly. Engineers were concerned that the material might cause higher-than-normal temperatures on the shuttle when it reentered Earth's atmosphere.

Robinson's impromptu repair highlighted the extra care NASA took on the mission, after the Columbia Accident Investigation Board found that the agency had been lax in addressing safety concerns. On Aug. 9,

2005, Discovery landed safely at Edwards Air Force Base in California. Had serious launch damage made it necessary, its crew was prepared to move into the ISS and wait for NASA to send a second shuttle on a rescue mission that had already been planned and practiced.

The added safety precautions were expensive. U.S. President George W. Bush had instructed NASA in 2004 to retire the three remaining shuttles by the end of 2010. The agency planned to switch to a simpler and presumably less expensive spaceship called the Crew Exploration Vehicle (CEV). The CEV was being designed to move human space exploration beyond the ISS, back to the surface of the moon and, eventually, to Mars and beyond.

However, the space station had not been completed when Columbia crashed. Many pieces of ISS *hardware* (modules and other construction components) awaiting launch in special facilities at the Kennedy Space Center were

STARDUST SAMPLES

NASA technicians dressed in clothing designed to prevent contamination (below, right) open the Stardust capsule, which fell to Earth in the Utah desert in January 2006. The capsule contained the first samples of comet dust ever returned to Earth. The capsule and the probe that carried it were launched in 1999 to collect samples from Comet Wild-2 (pronounced *vihlt-2*). Scientists believe that comets contain material left over from the formation of the solar system 4.6 billion years ago. As the Stardust probe flew past Wild-2, particles of dust and debris from the comet became trapped in a collection device filled with *aerogel* (above, right), a substance that is 99.8-percent air. The aerogel was encased in a canister inside the capsule. The probe dropped the capsule in the Utah desert as it flew past Earth in an orbit around the sun.

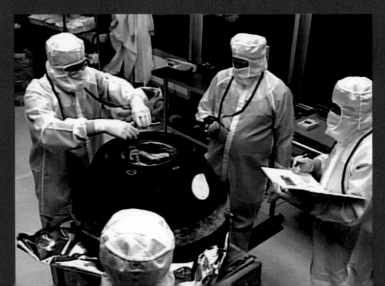

designed to fly only on the shuttle. On March 2, 2006, the heads of the U.S., Russian, Japanese, Canadian, and European space agencies agreed to fly 16 more shuttle missions to finish the station. If NASA's engineers believe that it is safe to do so, a shuttle will also make one final visit to the Earth-orbiting Hubble Space Telescope (HST). There, astronauts will install new instruments, perform maintenance to extend the HST's life span, and install a rocket motor to bring the massive telescope safely back to Earth in a controlled reentry after it finally wears out.

International exploration. The space station continued to play host to visitors from many nations during 2005 and 2006. In addition to U.S., Russian, and Japanese astronauts and cosmonauts, a spacefarer from Brazil and a U.S. millionaire spent time aboard the ISS. They rode up in the so-called "taxi seat" on Russia's three-seat Soyuz capsule, which has been available for astronauts from other nations and tourists, who can spend about a week in space while ISS crews are exchanged.

The third tourist to ever fly in space, U.S. scientist and entrepreneur Gregory Olson, paid Russia a reported $20 million to visit the ISS when Expedition 12—William McArthur of the United States and Valery Tokarev of Russia—were launched to the station on Sept. 30, 2005. Olson returned home on October 10 with Phillips and Krikalev. (Krikalev set a record for spending 803 days in space and became the first person to serve two tours aboard the ISS.)

Marcos Pontes, a pilot in Brazil's air force, occupied the taxi seat on March 29, 2006, when Russia's Pavel Vinogradov and Jeffrey Williams of the United States lifted off on Expedition 13 to replace McArthur and Tokarev. A German astronaut, Thomas Reiter, was scheduled to join Expedition 13 on the next shuttle mission in July, which would bring the station crew size to three, as it was before the Columbia accident. When the station is completed, six men and women will be able to live and work there.

China in space. As of mid-2006, China had not become a member of the international partnership that operates the space station. Nevertheless, the country continued to develop its own ability to fly astronauts into space. On Oct. 12, 2005, Fei Junlong and Nie Haisheng were launched into orbit atop a Long March 2F rocket from facilities in the Gobi, a desert in northern China and Mongolia. Their Shenzhou 6 spacecraft featured more than 100 modifications from Shenzhou 5, China's first piloted spacecraft, which flew in 2003. During China's

second human spaceflight, Fei and Nie tested all the equipment in their orbital module, including sleeping and kitchen facilities. They landed safely back in the Gobi on Oct. 17, 2005.

Deep Impact. Spacecraft that rely on robot technology and commands from the ground to operate without people aboard produced some of the most exciting scientific results of 2005 and 2006. Perhaps the most spectacular was a NASA spacecraft called Deep Impact. On July 4, 2005, it sent a probe into comet Tempel 1. Deep Impact then used its sensors to analyze the composition of the resulting debris cloud and to photograph the huge crater left by the high-speed impact. Analysis of the data was giving scientists new insights into these ancient snowballs, which carry material that remains unchanged since the formation of the solar system about 4.6 billion years ago.

Stardust. Scientists also got an actual sample of comet dust from Comet Wild-2 (pronounced *vihlt-2*, after Paul Wild, its Swiss discoverer). NASA's Stardust sample-return probe parachuted to a perfect landing in Utah on Jan. 15, 2006. Inside was a tennis-racket-shaped collector that had trapped dust from the comet in a bed of *aerogel*, a substance that is composed of 99.8-percent air and is the lightest known solid. Stardust was launched in 1999 and followed the comet on a 4.63-billion-kilometer (2.88-billion-mile) journey through the solar system.

Mars rovers. On Mars, the rovers named Spirit and Opportunity continued their unprecedented exploration of features on opposite sides of the planet. Originally sent to Mars for 90-day missions, the rovers worked so well after the 2004 extension of their mission that NASA granted them a second mission extension, to September 2006. Although both have suffered partial failures in the six-wheel mobility systems that move them around the surface, the systems' design has proved robust enough for controllers at the Jet Propulsion Laboratory in Pasadena, California, to keep them rolling. Along the way, the two solar-powered robotic geologists have added dramatically to the body of evidence that liquid water once flowed on the surface of Mars. The findings have nurtured hopes that evidence of extraterrestrial life—possibly tiny fossils—will be found there.

Cassini. Farther out in the solar system, the Cassini probe continued its tour around the ringed planet Saturn and its moons in 2006. Cassini is powered by a nuclear generator, which converts heat produced by the decay of radioactive plutonium into electric power. Because it

SPACE TECHNOLOGY continued

must travel so far from the sun, Cassini was not able to convert solar energy to electric power, as other spacecraft can.

One of the most exciting discoveries yet from the probe was evidence that fountains of water vapor and ice spewing from cracks at the south pole of the tiny moon Enceladus may originate in underground reservoirs of liquid water. Like the suspected ocean beneath the surface of Jupiter's moon Europa and the sands of Mars, liquid water on Enceladus could be another target for life-seeking robots and, eventually, human explorers.

Japanese scientists made an important discovery with their Hayabusa spacecraft, which landed on the asteroid Itokawa twice in November 2005. Hayabusa found that Itokawa formed as the result of an ancient collision between two asteroids. The asteroids themselves were so loosely assembled by gravity that scientists referred to them as "rubble piles." The spacecraft's cameras photographed a clear boundary between the two bodies, which apparently collided only about 10 million years ago. Hayabusa's managers hoped to bring the craft back to Earth and retrieve its sample-collecting capsule in 2010.

Three more missions. Scientists and engineers launched three new missions to three different planets in 2005 and 2006. On Aug. 12, 2005, NASA launched the Mars Reconnaissance Orbiter (MRO). The MRO reached Mars seven months later, and on March 10, 2006, entered its planned initial orbit. It will descend even closer to the surface of Mars in November 2006. There, the MRO will use information collected by NASA's Odyssey spacecraft and the European Space Agency's (ESA) Mars Express—launched in 2001 and 2003, respectively—to target certain areas in its search for water. During its planned four-year mission, the MRO will also collect information about the weather, climate, and geography of Mars in preparation for human exploration.

On Nov. 9, 2005, a Soyuz rocket lifted off from the Baikonur Cosmodrome in Kazakhstan carrying the ESA's new Venus Express. The craft is the first to visit Earth's nearest planetary neighbor since NASA's Magellan probe orbited the planet from 1990 to 1994. Venus Express arrived on April 11, 2006, to study Venus's atmosphere, which consists almost entirely of carbon dioxide.

On January 19, NASA launched the New Horizons probe, the first spacecraft sent to explore Pluto. New Horizons was to fly past Pluto and its largest moon, Charon, in July 2015. The spacecraft will send back information on the surface, geology, and atmosphere of Pluto and Charon. ■ Frank Morring, Jr.

See also **ASTRONOMY; ENGINEERING.**

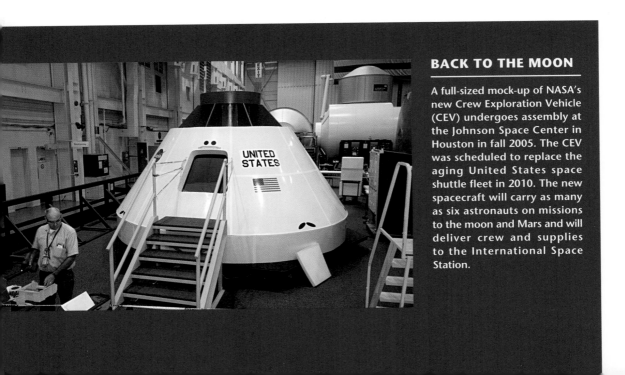

BACK TO THE MOON

A full-sized mock-up of NASA's new Crew Exploration Vehicle (CEV) undergoes assembly at the Johnson Space Center in Houston in fall 2005. The CEV was scheduled to replace the aging United States space shuttle fleet in 2010. The new spacecraft will carry as many as six astronauts on missions to the moon and Mars and will deliver crew and supplies to the International Space Station.

SUPPLEMENT

Seven new or revised articles reprinted from the
2006 edition of *The World Book Encyclopedia.*

Fossil ferns and a lump of coal, *left,* were both formed from the remains of plants that died many millions of years ago. While the plants lived, they stored up energy from the sun. The plants that became fossils gave up their store of energy in the process. Only the outline of their appearance remains. But the energy of the coal-forming plants is preserved in the coal. When the coal is burned, it releases this energy in the form of heat.

WORLD BOOK photo

Coal

Coal is a black or brown rock that can be ignited and burned. As coal burns, it produces useful energy in the form of heat. People use this heat to warm buildings and to make or process various products. Chemicals from coal are used to make dyes, fertilizer, pharmaceuticals, and other products. However, the main use of coal is the production of electric power. The heat from burning coal is used to make steam that drives engines called *turbines* in electric power plants. Coal-burning plants supply more than a third of the world's electric power. Much coal is also used in making steel. Coal accounts for about 25 percent of total world consumption of *commercial energy*—that is, energy produced by businesses and governments and sold to the public. The countries of the world use more than 5 billion tons (4.5 billion metric tons) of coal a year. The United States uses nearly 20 percent of this coal.

Coal formed over millions of years from partially decayed plant matter called *peat.* The chemical process that transforms peat to coal is called *coalification.* Coal consists primarily of the elements carbon, hydrogen, nitrogen, oxygen, and sulfur.

Coal was extremely important in fueling the Industrial Revolution, a period of rapid industrial development in the 1700's and 1800's. However, wood remained the principal fuel for many years. The change from wood to coal as the primary source of energy took place in different countries at different times, influenced by the availability and cost of these fuels. Some less developed countries still rely heavily on wood as their primary source of fuel. In much of the world, however, petroleum and natural gas have become the leading sources of energy.

People are rapidly using up the world's supplies of petroleum and natural gas that can be removed from the earth economically. If the present rates of use continue, little may remain of these supplies by about 2050. But the world's supply of coal can last for several hundred years at the present rate of use. Increased use of coal in place of natural gas and petroleum for electric power production could help relieve gas and oil shortages. But coal has long been a major cause of air pollution, and many people are concerned about the effects of coal combustion on the environment. Liquids derived from coal can substitute for gas or oil as transportation fuels. But such fuels are costly to produce.

In the past, few jobs were harder or more dangerous than that of an underground coal miner. During the 1800's, many miners had to work underground 10 or more hours a day, six days a week. Picks were almost the only equipment they had to break the coal loose. The miners shoveled the coal into wagons. In many cases, children as young as 10 years of age hauled the coal from the mines. Women worked as loaders and haulers. Over the years, thousands of men, women, and children were killed in mine accidents. Thousands more died of lung diseases from breathing coal dust.

James C. Cobb, the contributor of this article, is State Geologist and Director of the Kentucky Geological Survey.

Today, in industrialized countries, machines do most of the work in coal mines. Mine safety has been improved, work hours have been shortened, and child labor is prohibited. However, coal mining is still dangerous and poorly regulated in many of the less developed regions of the world.

This article discusses the composition of coal, the uses of coal, how coal was formed, where coal is found, and how it is mined. It also discusses the cleaning and shipping of coal, the coal industry, and the history of the use of coal.

The composition of coal

Coal is referred to as a *fossil fuel* because it formed from the remains of living things. Petroleum and natural gas are also fossil fuels. The composition of coal varies more than that of oil or gas. It varies from one coal *seam* (deposit) to another, and even within an individual seam.

Coal is a *sedimentary rock* that formed when *sediments* (particles of older rock and plant remains) settled to the bottom of bodies of water and hardened. It is often referred to as a mineral. But unlike a true mineral, coal has no fixed chemical formula. By weight, it contains at least 50 percent organic matter and no more than 50 percent inorganic matter. The principal chemical elements in coal are carbon, hydrogen, nitrogen, oxygen, and sulfur—elements found in all living organisms. Much of the hydrogen and carbon are combined in molecules called *hydrocarbons,* which give coal its ability to burn and produce useful heat. The inorganic matter, or *mineral matter,* in coal commonly includes calcite, clay, marcasite, pyrite, and quartz. When coal is burned, the minerals become ash, a powderlike residue. The coal industry refers to ash-producing substances in coal as "ash" even before the coal is burned.

Other components of coal are known as *major elements, minor elements,* and *trace elements.* They occur in small quantities primarily in the mineral matter. The major elements include aluminum, calcium, iron, and silicon. Minor elements may include chlorine, magnesium, potassium, sodium, and titanium. The trace elements include arsenic, lead, mercury, and selenium. The trace elements occur in extremely small amounts—usually measured in parts per thousand, parts per million, or parts per billion.

Coal also contains moisture and gases that accumulate in its pores and that stick to the coal itself. The gases are mainly carbon dioxide and the hydrocarbon methane.

Occasionally, petrified peat in the form of *coal balls* occurs in coal seams. Such peat was saturated with calcium carbonate that seeped into it dissolved in water before the peat was compacted and coalified. Coal balls preserve many parts of the original plants that make up the coal.

How composition affects value. The composition of coal is important in determining the coal's suitability for certain applications. It also determines the value of coal in the marketplace. *Heating value,* sulfur content, moisture content, and ash are measured when grading coal for commercial purposes. Heating value refers to the amount of heat that is produced by a given amount of coal when it is burned. It is usually reported in British thermal units (Btu's) per pound or in kilojoules per kilogram. A Btu is equal to 251.996 calories or 1,054.35 joules. In general, the higher the heating value, the more valuable the coal. Moisture and ash tend to reduce the value of coal, because moisture reduces heating value and ash may contain substances that corrode boilers or equipment, or build up inside them. It is also costly to dispose of ash. Government regulations in the United States and other countries control sulfur emissions from coal burning. For this reason, higher sulfur content may also render a coal less valuable.

Analyzing coal composition. Coal scientists use a variety of tests to measure the quantities of the components of coal to judge its suitability and value. The most basic tests are two types of chemical analysis—*proximate analysis* and *ultimate analysis.* Proximate analysis indicates the quantity of ash, moisture, *volatile matter,* and *fixed-carbon* in a coal sample. Volatile matter is material, other than moisture, that is given off when coal is heated to a specific high temperature in the absence of oxygen. The remaining solid material contains fixed-carbon, which will burn if heated in the presence of oxygen, and ash, which will not burn. Proximate analysis is a basic test to judge coal quality. Scientists use ultimate analysis, also called *elemental analysis,* to determine the amounts of carbon, hydrogen, oxygen, nitrogen, and sulfur in a sample of coal. Using the results of this analysis, they can calculate the heating value of the coal in the sample.

Some scientists use *coal petrography,* the observation of coal under a microscope, to study composition. For

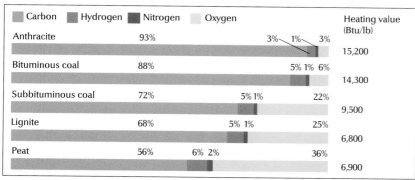

Carbon	Hydrogen	Nitrogen	Oxygen		Heating value (Btu/lb)
Anthracite	93%	3%	1%	3%	15,200
Bituminous coal	88%	5%	1%	6%	14,300
Subbituminous coal	72%	5%	1%	22%	9,500
Lignite	68%	5%	1%	25%	6,800
Peat	56%	6%	2%	36%	6,900

Figures may not add up to 100 percent due to rounding.
Source: U.S. Geological Survey

Composition of coal

This graph shows the typical chemical composition of the four types of coal and of peat, along with their *heating value.* The heating value is the amount of heat produced by a given amount of coal or peat when it is burned.

example, they may examine a polished pellet of finely ground coal under high magnification to observe structures in the organic material of coal. These structures, called *macerals,* are the "building blocks" of coal. There are three basic groups of macerals: (1) vitrinites, (2) liptinites, also called exinites, and (3) inertinites. *Vitrinites* developed from woody plant tissues; *liptinites,* from algae, spores, and the resinous and waxy parts of plants; and *inertinites,* from charred and decomposed plant material. Scientists may also view a thin slice of coal called a *thin section* under a high-magnification microscope. Such observation reveals many details about the organic and inorganic makeup of the coal.

Study of the mineral matter in coal helps scientists identify pollutants and predict how the ash will behave during combustion. Ash can foul furnaces, leading to damage and costly repairs. Laboratory scientists must carefully identify and measure the mineral in coal. They use such tools of physics as *electron microscopy* and *X-ray diffraction* to perform this work. The trace elements in coal can pollute the air or foul or corrode equipment during combustion. Geologists chemically test coal before it is used to identify and measure the quantities of these elements.

In another method of analysis, powdered coal is placed in a strong acid or alkali solution. Much of the coal dissolves, leaving a residue of microscopic fossilized pollen and spores that survived the coalification process. These fossils are valuable in identifying types of plants and geological periods associated with the coal.

Coal rank. Coal is usually assigned a *rank* based on how much the organic matter was altered during coalification. As rank increases, the carbon content and heating value of the coal also increases. The four ranks of coal, from the lowest to the highest, are (1) lignites, or brown coals; (2) subbituminous coals; (3) bituminous coals; and (4) anthracites. The lowest-ranking lignites have a carbon content of only about 30 percent. The highest-ranking anthracites contain about 98 percent carbon. Many properties of coal, including chemical and physical characteristics, change with rank.

The uses of coal

Coal as a fuel. Coal is a useful fuel because it is abundant and has a relatively high heating value. Sulfur, the main impurity in coal, limits coal's usefulness as a fuel. Most of the sulfur in burning coal combines with oxygen and forms a gas called *sulfur dioxide,* which contributes to air pollution. To prevent serious air pollution, the burning of *medium-* and *high-sulfur coals* in a power plant requires *scrubbers,* pollution control devices that absorb sulfur dioxide fumes as they pass through the plant's smokestacks. Scrubbers prevent dangerous amounts of sulfur dioxide and other pollutants from entering the atmosphere. *Low-sulfur coal* can be burned in power plants without the use of scrubbers.

Different systems are used in different countries to classify coal by sulfur content. The system used by the U.S. Environmental Protection Agency (EPA) classifies coal according to the weight of sulfur in a sample that can produce 1 million Btu's of heat. Such a sample is low-sulfur coal if it produces 0.60 pound (0.272 kilogram) or less of sulfur, medium-sulfur coal if its sulfur content is 0.61 to 1.67 pounds (0.277 to 0.757 kilogram), and high-

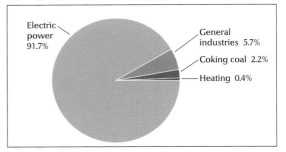

Figures are for 2002. Source: U.S. Energy Information Administration.

sulfur coal if it produces 1.68 pounds (0.762 kilogram) or more of sulfur.

Some of the ash produced by burning pulverized coal may escape into the air. This so-called *fly ash* also contributes to air pollution. Devices have been developed that trap fly ash in the power plant before it can escape into the air. In some regions, the amount of fly ash and other particles from power plants is strictly monitored and regulated to reduce air pollution.

Electric power production. Most electric power

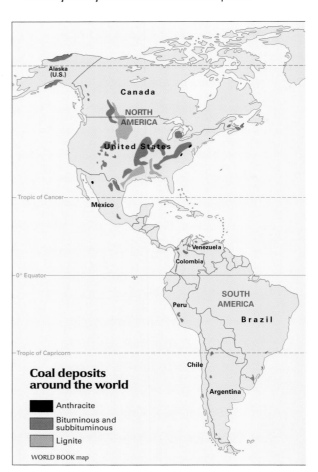

Coal deposits around the world

WORLD BOOK map

plants fueled by coal are steam-turbine plants. They use high-pressure steam to generate electric power. The steam turns the fanlike blades of turbines, which drive the generators that produce electric power.

Bituminous coal is the preferred fuel for electric power production in many countries, including the United States, because it is plentiful and has a high heating value. However, much bituminous coal has medium or high sulfur content. In many countries, more and more power plants are burning at least some low-sulfur subbituminous coal and lignite to meet government air pollution standards. These coals must be handled carefully because they are more likely than other types to ignite spontaneously in stockpiles, freeze in railcars, and dry out, break up, and produce much coal dust. Modern handling techniques have improved control over these problems.

Other uses of coal as a fuel. In parts of Asia and Europe, coal is widely used for heating homes and other buildings. In the United States and many other countries, natural gas and fuel oil have almost entirely replaced coal as a domestic heating fuel. However, the rising cost of oil and natural gas has caused some factories and other commercial buildings to turn to coal. Anthracite is the cleanest-burning coal, but it is expensive and in limited supply. And subbituminous coals

and lignites have such a low heating value that large amounts must be burned to heat effectively. Thus, bituminous coals are the kinds most widely used for domestic heating.

In the past, coal provided heat for the manufacture of a wide variety of products, including glass and canned foods. Since the early 1900's, manufacturers in many parts of the world have come to use natural gas in making most of these products. Coal is used mainly by the steel and cement industries. But some other industries have switched back to coal to reduce fuel costs.

Coal as a raw material. Many substances made from coal serve as raw materials in manufacturing. *Coke*—a hard, foamlike mass of nearly pure carbon—is the most widely used of these substances. It is made by heating bituminous coal to about 2000 °F (1100 °C) in an airtight oven. The lack of oxygen prevents the coal from burning. The heat changes some of the coal into gases. The remaining solid matter is coke. It takes about 1 ½ tons (1.4 metric tons) of bituminous coal to produce 1 ton (0.9 metric ton) of coke.

The coal used to make coke is called *coking coal*. It requires specific characteristics, such as low sulfur and ash content. Only certain types of bituminous coals have all the necessary characteristics.

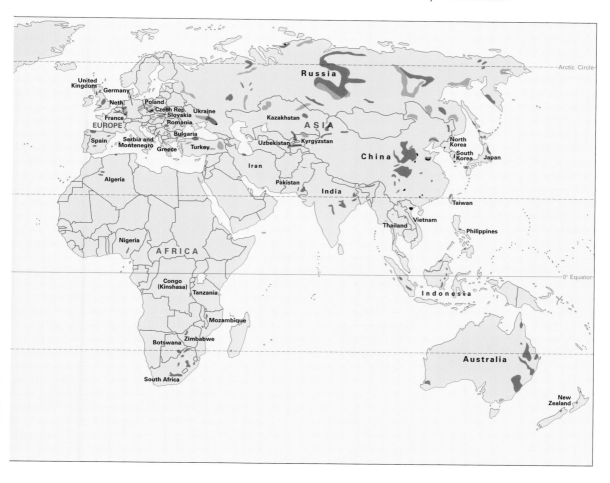

Coke is used primarily to make iron and steel. Most coking plants are a part of steel mills. The mills burn coke with iron ore and limestone to change the ore into the pig iron required to make steel. It takes about 900 pounds (410 kilograms) of coke to produce 1 ton (0.9 metric ton) of pig iron.

The coke-making process is called *carbonization*. Some of the gases produced during carbonization turn into liquid ammonia and coal tar as they cool. Through further processing, some of the remaining gases change into light oil. Manufacturers use the ammonia, coal tar, and light oil to make such products as dyes, fertilizers, and pharmaceuticals. Coal tar is also used for roofing and for road surfacing.

Some of the gas produced during carbonization does not become liquid. This *coal gas,* or *coke oven gas,* burns like natural gas. But coal gas has a lower heating value and, unlike natural gas, gives off large amounts of soot as it burns. Coal gas is used chiefly at the plants in which it is produced. Coal gas provides heat for the coke-making and steel-making processes.

Gas can be produced from coal directly, without carbonization, by various *gasification* methods. The simplest such method involves burning coal in the presence of forced air or steam. The resulting gas, like coke oven gas, has a low heating value and produces soot. It is used chiefly in certain manufacturing processes. Coal can be used to make high-energy gas, gasoline, and fuel oil. But the present methods of producing these fuels from coal are costly and complex. The section *The coal industry* discusses how researchers are working to develop cheaper and simpler methods.

Coal combustion products are coal components that remain after burning. Such products, including ash, scrubber sludge, and other materials, were once considered waste materials. They were often expensive to handle and took up valuable space in landfills. Today, millions of tons of these products are sold in the world each year. They are used to make asphalt, cement and concrete, road materials, soil stabilizers, wallboard, and other products. Industries that make these items often build facilities near power plants so that they can use the materials generated by coal combustion.

How coal was formed

The formation of coal is a geologic process that takes millions of years. It begins with the accumulation of dead plant material under extremely wet conditions. The ideal setting is a humid swamp. There, constant rainfall and a high water table keep the plant material water-logged, limiting normal decay. Under those conditions, thick deposits of dead plant material can build up. The partial decay of plant material produces peat.

The next step in the formation of coal is for the land surface upon which peat accumulates to gradually sink. Such sediments as sand and mud bury the peat as the surface sinks. The buried peat is preserved from decay and erosion, and so it can be transformed into coal.

The last step, coalification, changes peat by concentrating carbon and hydrogen and expelling such by-products as carbon dioxide, methane, and water. The rate at which peat changes to coal depends on tempera-

The development of coal

The formation of coal involved three main steps. (1) The remains of dead plants turned into a substance called *peat.* (2) The peat became buried. (3) The buried peat was subjected to heat and pressure. After thousands or millions of years under pressure, the peat turned into coal. Each of these steps is illustrated below.

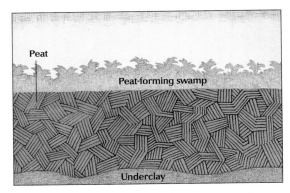

A thick layer of peat developed as plant matter accumulated and hardened on the floor of a swamp. The matter built up as plants that grew in the swamp died and sank to the bottom. Peat-forming swamps once covered much of the earth.

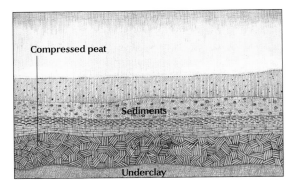

Deposits of loose mineral matter, called *sediments,* completely covered the peat bed. As these sediments continued to pile up over the bed, they compressed the peat.

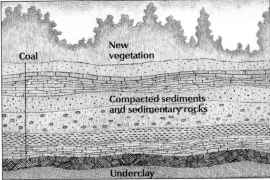

WORLD BOOK diagrams by Jean Helmer

Pressure on the peat increased as the sediments became more compact and heavier. Some sediments hardened into rock. The increasing pressure turned the peat into coal.

ture. Higher temperatures result in faster coalification, and lower temperatures produce slower coalification.

The interior of Earth is hot, and heat flows constantly from the interior to Earth's surface. Temperatures at the boiling point of water, 212 °F (100 °C), are found a few thousand feet or meters beneath the surface. This temperature is high enough to cause the transformation of peat to coal over a long time. It generally takes from 3 to 10 feet of peat to produce a 1-foot thickness of coal, or 3 to 10 meters of peat to produce 1 meter of coal.

Much coal formed during the Carboniferous Period, from about 360 million to 286 million years ago. Coal formed during this period is found in North America, Europe, and Asia. Various trees grew in Carboniferous swamps and produced huge amounts of peat-forming matter after they died. Another important time for the formation of coal ranged from the Cretaceous Period to the Miocene Epoch, from about 145 million to 24 million years ago. The thick subbituminous coals of the Powder River Basin in Wyoming and Montana and the lignite coal of the U.S. Gulf Coast were formed during this period. The swamps that formed these coals had plants similar to modern cypress and tupelo trees.

Peat is accumulating on Earth today in conditions similar to those that produced coal. High-quality peat deposits as thick as 50 feet (15 meters) occur in the swamps of Indonesia. These deposits cover thousands of square miles or square kilometers, and receive 5 to 10 feet (1.5 to 3 meters) of rainfall each year. This peat could eventually develop into coal, but the process could take as long as 1 million years.

The coal seams, or *coal beds,* mined today range in thickness from a few inches or centimeters to more than 400 feet (120 meters). The thickest seams are of subbituminous coal and lignite. Many coal deposits consist of two or more seams separated by layers of rock. These formations were produced as new coal-forming swamps developed over buried ones. Each new swamp became buried and developed into a separate seam.

Most coal beds lie nearly parallel to Earth's surface. But some have been tilted by movements of underground rock layers and lie at an angle to the surface. Most of the deepest beds consist of anthracites or bituminous coals. Over a long period, geologic forces can push coal upward from deep beneath the surface. In addition, erosion can expose coal at the surface.

Where coal is found

Coal is found on every continent. Deposits occur as far north as the Arctic and as far south as Antarctica. Some coal deposits occur off ocean coastlines. However, underwater deposits are not considered valuable because they are difficult to mine.

Coal is typically found in bowl-shaped geologic features known as *coal basins* or *sedimentary basins.* Coal basins vary greatly in size, depth, age, number, and thickness of seams. The Appalachian Basin in the eastern United States, for example, is about 1,000 miles (1,600 kilometers) long and 150 miles (240 kilometers) wide. Its coal seams range from *outcroppings* that occur at the surface to seams that reach depths exceeding 5,000 feet (1,500 meters). The quality and rank of the coal varies within a coal basin.

Coal deposits that can be mined profitably are called *coal reserves.* In most cases, a coal seam must be at least 24 inches (61 centimeters) thick for mining engineers to class it as a reserve. Most estimates of coal reserves include only tested deposits.

Estimating coal reserves. To estimate coal reserves, mining engineers drill into the ground in suspected coal-bearing areas. Using a special drill bit, they cut cylindrical samples of sediment and rock called *cores.* After a drill cuts through a coal seam, geologists and engineers can use the resulting cores to determine the depth and thickness of the seam. Laboratory scientists can analyze core samples to determine the coal's composition and suitability for use.

Mining engineers use a formula based on coal rank and quality to calculate the tons of coal reserves beneath a given land area. According to the formula used for bituminous coal, for example, a bed yields about 1,800 tons for each foot of coal thickness per acre, or about 13,240 metric tons per meter of thickness per hectare. By drilling a number of test holes, and applying the proper formula, geologists and engineers can estimate the size and potential value of a particular deposit. A large area of tested reserves is called a *coal field.*

The world's coal reserves. No complete estimate exists for the total amount of coal that lies beneath Earth's surface. The world's proved recoverable reserves of coal total over 1.1 trillion tons (1 trillion metric tons). This figure represents the amount of coal that can be recovered from known deposits with current technology. The major proved recoverable reserves are in Australia, China, Germany, India, Indonesia, Poland, Russia, South Africa, and the United States. Australia, China, and Russia have large deposits of bituminous, subbituminous, and lignite coals. Australia's largest deposits lie in the eastern part of the country. Central China has the country's largest deposits. Russia's main deposits lie in the central and western parts of the country. South Africa has large deposits of bituminous coal in the northeastern part of the country.

About half of all U.S. coal reserves lie in the eastern half of the nation, from the Appalachian Highlands to the eastern edge of the Great Plains. The rest are in the western part of the country, especially in the Rocky Mountain States and the northern Great Plains. Many geologists believe that Alaska may have huge coal reserves. But the state's potential reserves have not been fully assessed. The reserves in the eastern United States include nearly all the nation's anthracite deposits and more than 80 percent of its bituminous coal deposits. The western reserves include almost all the subbituminous coal and lignite. Canada's coal reserves consist chiefly of bituminous coal in eastern Canada. Canada also possesses large deposits of subbituminous coal and lignite. The largest reserves are in Alberta and British Columbia.

Surface mining

About 60 percent of the coal mined in the world, and in the United States, comes from *surface mines.* Surface mining involves removing the *overburden,* the soil and rock that lie over a coal deposit. After this material has been removed, the coal can easily be dug up and hauled away. Surface mining is usually limited to deposits within 100 to 200 feet (30 to 61 meters) of the surface. The more overburden that must be removed, the more difficult and costly surface mining becomes.

Nearly all surface mining is *strip mining*—that is, mining by first stripping away the overburden. Many coal seams are exposed on the sides of hills or mountains. In some cases, these seams can be mined without removing any overburden. Miners use machines called *augers* to dig out the coal. This method of surface mining is known as *auger mining.*

Surface mining requires fewer miners than does underground mining. Strip miners are chiefly heavy-machine operators. Unlike underground miners, they have little need for traditional mining skills.

Strip mining depends on powerful machines that dig up the overburden and pile it out of the line of work. The dug-up overburden is called *spoil.* In time, a strip mine and its spoil may cover an enormous area. The digging up of vast areas of land has raised serious environmental concerns. As a result, some governments require that all new strip-mined land be *reclaimed*—that is, returned as close as possible to its original condition. Strip mining thus involves methods of (1) mining the coal and (2) reclaiming the land.

Mining the coal. Most strip mines follow the same basic steps to produce coal. First, bulldozers or loaders clear and level the vegetation and soil above the mining area. Many small holes are then drilled through the rocky overburden to the coal bed. Each hole is loaded with explosives. The explosives are set off, shattering the rock in the overburden. Giant power shovels or other earthmoving machines then clear away the broken rock. Some of these earthmovers are as tall as a 20-story building and can remove more than 3,500 tons (3,180 metric tons) of overburden per hour. After a fairly large area of coal is exposed, explosives may be used to loosen the coal itself. Coal-digging machines then scoop up the coal and load it into trucks. The trucks carry the coal from the mine.

Although most strip mines follow the same basic steps, strip-mining methods vary according to whether the land is flat or hilly. Strip mining can thus be classed as either (1) area mining or (2) contour mining. Area mining is practiced where the land is fairly level. Contour mining is practiced in hilly or mountainous country. It involves mining on the *contour*—that is, around hillsides.

In area mining, an earthmover digs up all the broken overburden from a long, narrow strip of land along the edge of the coal field. The resulting deep ditch is referred to as a *cut.* As the earthmover digs the cut, it piles the spoils along the side of the cut that is away from the mining area. The spoil pile forms a ridge called a *spoil bank.* After the cut is completed, the coal is dug, loaded into trucks, and hauled away. The earthmover then digs an identical cut alongside the first one. It piles the spoils from this cut into the first cut. This process is repeated over and over across the width of the coal field until all of the coal has been mined. The spoil banks form a series of long, parallel ridges on the land that can later be leveled.

If a coal seam lies near the top of a hill, an earthmover may simply remove the hilltop and so expose the coal. This type of mining is called *mountain top removal.* It is practiced primarily in the eastern United States. It can be efficient because it may enable the removal of several

Bucyrus-Erie Company

Strip mining depends on giant earthmoving machines like the one at the top of this picture. The earthmover strips away the soil and rock that lie over a coal deposit. A coal-digging machine, *center,* then scoops up the coal and loads it into a truck.

coal seams at once. Some people oppose its use because it produces dramatic changes to the landscape.

In contour mining, an earthmover removes the overburden immediately above the point where a seam *outcrops* (is exposed) all around a hill. The resulting cut forms a wide ledge on the hillside. The spoils may be stored temporarily on the hillside or used to fill in the cuts. After the exposed coal has been mined and hauled away, the earthmover may advance up the slope and dig another cut immediately above the first one. But the depth of the overburden increases sharply with the rise of the slope. After one or two cuts, the overburden may be too great for a coal company to remove profitably. But if the seam is thick enough, a company may dig an underground mine to remove the rest of the coal.

Reclaiming the land. The chief environmental problems that strip mining can cause result from burying fertile soil under piles of rock. The rocks tend to give off acids when exposed to moisture. Rainwater runs down the bare slopes, carrying acids and mud with it. The runoff from the slopes may wash away fertile soil in surrounding areas and pollute streams and rivers with acids and mud.

The first step in reclaiming strip-mined land is to reduce the steep slopes formed by the spoils. The spoil banks created by area mining can be leveled by bulldozing. The spoils from contour mining can be used to fill in the cuts in the hillsides. As much topsoil as possible should then be returned to its original position so that the area can be replanted. Much reclaimed land is turned into farms or recreation areas.

Auger mining. A coal auger is shaped like an enormous corkscrew. It bores into the side of a coal outcrop on a slope and twists out the coal in chunks. Contour mines often use augers when the overburden in a slope is too great to remove. An auger can penetrate the outcrop and recover coal that could not otherwise be mined. Some augers can bore 200 feet (61 meters) or more into a hillside.

Companies often employ auger mining to mine outcrops of high-quality coal that cannot be mined economically by other methods. However, auger mining can recover only a small portion—as little as 15 percent—of the coal in a seam. The method is best used in combination with contour mining.

Underground mining

Underground mining involves digging tunnels into a coal deposit. Miners must go into the tunnels to remove the coal. Most coal deposits deeper than 200 feet (61 meters) are mined underground. Underground mining is more hazardous to workers than surface mining. The miners may be injured or killed by cave-ins, falling rocks, explosions, and poisonous gases. To prevent such disasters, every step in underground coal mining must be designed to safeguard the workers.

Underground mining generally requires more human labor than surface mining does. But even so, underground mines are highly mechanized. Machines do all the digging, loading, and hauling in nearly all the mines. In industrialized nations, few mines are nonmechanized.

In most cases, miners begin an underground mine by digging two access passages from the surface to the coal bed. One passage will serve as an entrance and exit for the miners and their equipment. The other will be used to haul out the coal. Both passages will also serve to circulate air in and out of the mine. As the mining progresses, the workers dig tunnels from the access passages into the coal seam.

Underground mines can be divided into three main groups according to the angle at which the access passages are dug into the ground. The three groups are (1) shaft mines, (2) slope mines, and (3) drift mines. Some mines have two or all three types of passages.

In a shaft mine, the access passages run straight down from the surface to the coal seam. The entrance and exit shaft must have a hoist. Most mines under more than 700 feet (210 meters) of cover are shaft mines. In a slope mine, the access passages are dug on a slant. They may follow a slanting seam or slant down through the cover to reach the seam. Drift mines are used to mine seams of coal that outcrop in hills or mountains. The access passages for these mines are dug into a seam where the coal bed outcrops on a slope.

Two main systems of underground mining are used: (1) the room-and-pillar system and (2) the longwall system. Each system has its own set of mining techniques. Either system may be used in a shaft, slope, or drift mine. The room-and-pillar system is by far the more common system of underground mining in the United States. The longwall system is more widely used elsewhere, especially in European countries.

The room-and-pillar system involves initially leaving pillars of coal standing in a mine to support the overburden. Miners may begin a room-and-pillar mine by digging three or more long, parallel tunnels into the coal seam from the access passages. These tunnels are called *main entries.* In most cases, the *ribs* (walls) of coal separating the main entries are 40 to 80 feet (12 to 24 meters) wide. Cuts are made through each wall every 40 to 80 feet. The cuts form square or rectangular pillars of coal that measure 40 to 80 feet on each side. The coal dug in building the entries is hauled to the surface.

The pillars help support the overburden in the main entries. But in addition, the entry roofs must be bolted to hold them in place. To bolt the roof, the miners first drill holes 3 to 6 feet (0.9 to 1.8 meters) or more into the roof. They then anchor a long metal bolt into each hole and fasten the free end of each bolt to the roof. The bolts bind together the separate layers of rock just above the roof to help prevent them from falling. The miners must also support the roof in all other parts of the mine as they are developed. For safety, miners can only work under a supported roof.

A conveyor belt or a railroad track is built in one of the main entries to carry the coal to the access passages. A railroad may also transport the miners along the main

Kinds of underground mines There are three main kinds of underground mines: (1) shaft mines, (2) slope mines, and (3) drift mines. In a shaft mine, the entrance and exit passages are vertical. In a slope mine, they are dug on a slant. In a drift mine, the passages are dug into the side of a coal bed exposed on a slope.

WORLD BOOK diagram

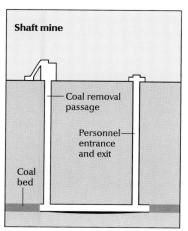

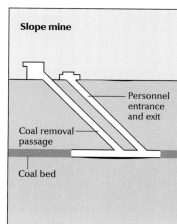

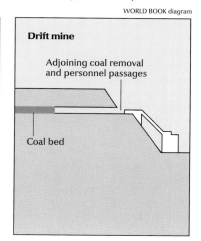

entries. At least two main entries serve chiefly to circulate air through the mine. The mine may also need such facilities as water drainage ditches, gas drainage pipes, compressed air pipes, water pipes, and electric power cables. These facilities are built into the main entries and later extended to other parts of the mine.

After the main entries have been constructed, the miners dig sets of *subentries* at right angles from the main entries into the coal seam. Each set of subentries consists of three or more parallel tunnels, which serve the same purposes as the main entries. Cuts are made through the walls separating these tunnels, forming pillars like those between the main entries. At various

The room-and-pillar system

Most underground mines in the United States use the *room-and-pillar system* of mining. First, the miners dig tunnels called *main entries* into the coal bed from the entrance and exit passages. They then dig sets of *subentries* into the bed from the main entries and sets of *room entries* into the bed from the subentries. Pillars of coal are left standing in all the entries to support the mine roof. As the room entries are extended, they create large *panels* of coal. The miners eventually dig *rooms* into the panels to recover as much coal as possible from the bed. This floor plan of a room-and-pillar mine shows how the entries are developed.

WORLD BOOK diagram by Linda Kinnaman

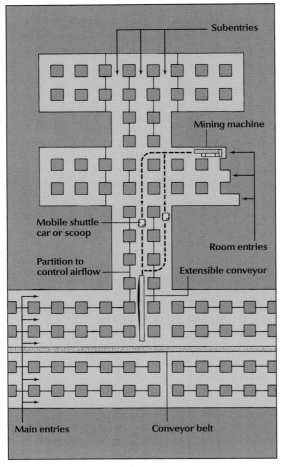

Subentries

Mining machine

Mobile shuttle car or scoop

Room entries

Partition to control airflow

Extensible conveyor

Main entries   Conveyor belt

points along each set of subentries, the miners dig *room entries* at right angles into the seam. They then begin to dig *rooms* into the seam from the room entries.

As the miners enlarge a room, they leave pillars of coal to support the overburden. A room is mined only a certain distance into the seam. When this distance is reached, the miners may remove the pillars. The room roof collapses as the pillars are removed, and so they must be removed in *retreat*—that is, from the back of the room toward the front. The miners' exit from the room thus remains open as the roof falls. Pillars are also sometimes removed from entries. Like room pillars, they must only be removed in retreat to protect the miners.

All room-and-pillar mining involves leaving some pillars in place. Room-and-pillar mines differ, however, in their mining methods. Mechanized room-and-pillar mines use two main methods: (1) continuous mining, and (2) the conventional mechanized method.

Continuous mining uses machines called *continuous miners* to gouge the coal from the *coal face,* the coal exposed on the surface of a wall. One worker operating a continuous miner can produce about 2 tons (1.8 metric tons) of coal per hour. The machine automatically loads the coal onto shuttle cars or a conveyor belt, which carries it to the railroad or conveyor in the main entries.

A continuous miner can usually dig and load coal much faster than the coal can be hauled out of a mine. The machine can work faster than the haulage, roof-bolting, ventilation, construction, and drainage systems can be completed. As a result, a continuous miner must frequently be stopped to allow the other mine systems to catch up.

The conventional mechanized method was more widely practiced during the 1930's and 1940's than it is today. During the 1930's, it largely replaced the earlier method of digging coal by hand. Since about 1950, continuous mining has increasingly replaced the conventional method.

The conventional mechanized method involves five steps. (1) A machine that resembles a chain saw cuts a long, deep slit, usually along the base of the coal face. (2) Another machine drills a number of holes into the face. (3) Each hole is loaded with explosives. The explosives are set off, shattering the coal. The deep slit along the base causes the broken coal to fall to the floor. (4) A machine loads the coal onto shuttle cars, scoops, or a conveyor. (5) Miners bolt the roof exposed by the blast.

A separate crew of miners carries out each of the five steps. After a crew has completed its job on a particular face, the next crew moves in. The miners can thus work five faces of coal at a time. But there are frequent pauses in production as the crews change places.

The longwall system of underground mining involves digging main tunnels or entries like those in a room-and-pillar mine. However, the coal is mined from one long face, called a *longwall,* rather than from many short faces in a number of rooms.

A longwall face is about 300 to 700 feet (91 to 210 meters) long. The miners move a powerful cutting machine back and forth across the face, plowing or shearing off the coal. The coal falls onto a conveyor belt. Movable steel props support the roof over the length of the immediate work area. As the miners work the machine farther into the seam, the roof supporters are advanced.

Types of underground-mining equipment

The type of equipment that an underground mine requires depends on the method of mining it uses. Mechanized mines use three main methods: (1) the conventional method, (2) continuous mining, and (3) longwall mining. Each of the three methods calls for a different type of equipment.

WORLD BOOK illustrations by Robert Addison

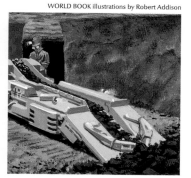

Conventional-mining equipment. The conventional method of mining involves a series of steps, three of which require special machinery. First, a cutting machine, *left,* cuts a deep slit along the base of the coal *face* (coal exposed on the surface of a mine wall). Another machine, *center,* drills holes into the face. Miners load the holes with explosives and then set the explosives off. The undercutting along the bottom of the face causes the shattered coal to fall to the floor. A loading machine, *right,* gathers the coal onto a conveyor belt.

Continuous-mining equipment eliminates the series of steps in mining a face. A continuous-mining machine, *right,* gouges out the coal and loads it onto a shuttle car in one operation.

Longwall-mining equipment. Longwall mining differs from the other methods of underground mining in its system of roof support. The other methods are used only in room-and-pillar mines, where pillars of coal are left to support the mine roof. In the longwall method, movable steel props support the roof over one long coal face. The miners move a cutting machine back and forth across the face, shearing off coal. The coal falls onto a conveyor. As the miners advance the cutter into the bed, the roof supports are moved forward. The roof behind the miners is allowed to fall.

Consolidation Coal Company

A worker operating a continuous miner can produce about 2 tons (1.8 metric tons) of coal per hour. These machines dig about 60 percent of the coal mined underground in the United States.

The roof behind the miners is allowed to fall. After a face has been dug out 4,000 to 6,000 feet (1,200 to 1,800 meters) into the seam, a new face is developed and mined. This process is repeated until as much coal as possible has been removed from the seam.

The longwall system originated in Europe. Underground mines in Europe are much deeper, on the average, than those in the United States. The pressure of the overburden becomes intense in an extremely deep mine. Longwall mining relieves the pressure by allowing the roof to cave in throughout most of a mine. In a European longwall mine, the roof remains in place only over the main entries, over the longwall face, and over two tunnels leading to the face. The mines can thus recover up to 90 percent of the coal in a seam.

Mine safety laws in some countries, including the United States, require longwall mines to have fully developed subentries as well as main entries. Such longwall mines include some of the main features of room-and-pillar mines. One kind of longwall mining is called the *retreating longwall system.* This type of mining uses the room-and-pillar system to reach and expose the long coal face. Longwall equipment then mines the coal. These mines are much more productive than room-and-pillar mines because less coal is left in place.

Some mines have adopted a variation of the longwall method called *shortwall mining.* A shortwall face is only about 150 to 200 feet (46 to 61 meters) long, and it is mined with continuous-mining machines rather than with longwall equipment. This system, which was developed in Australia, is suited to coal seams whose structure prevents them from being divided into long faces.

Cleaning and shipping coal

Some coal is shipped to buyers exactly as it comes from the mine, without any processing. In the coal industry, such coal is called *run-of-mine coal.* It ranges in size from fine particles to large chunks.

The two largest users of coal, the electric power and coking industries, have definite quality requirements for the coal they buy. Much run-of-mine coal does not meet these requirements because it is the incorrect size or contains unacceptable amounts of impurities. As a result, mining companies sort their coal according to size. They also clean the coal to remove impurities. The companies sort about 50 percent of their coal without cleaning it. They both sort and clean about 40 percent of the coal.

Cleaning coal. Mining companies clean impurities from coal in *preparation plants.* Most large coal mines have a preparation plant on the mine property. Machines and other equipment are used to remove the impurities. They perform work referred to as *coal processing, beneficiation,* or *coal cleaning.*

Ash and sulfur are the chief impurities in coal. Run-of-mine coal may also contain pieces of rock or clay. These materials must be removed in addition to the impurities. Preparation plants rely on the principle of *specific gravity* to remove impurities. According to this principle, if two solid substances are placed in a solution, the heavier substance will settle to the bottom first. Most mineral impurities in coal are heavier than pure coal. As a result, they can be separated from run-of-mine coal placed in a solution. The entire coal-cleaning process involves three main steps: (1) sorting, (2) washing, and (3) dewatering.

Sorting. Large pieces of pure coal may settle to the bottom of a solution faster than small pieces that have many impurities. Therefore, the pieces must first be sorted according to size. In many preparation plants, a screening device sorts the coal into three sizes—coarse, medium, and fine. Large chunks are crushed and then sorted into the three main batches according to size.

Washing. The typical preparation plant uses water or dense fluids called *heavy liquids* as solutions for separating the impurities from coal. Each batch of sorted coal is piped into a separate washing device, where it is mixed with the solution. The devices separate the impurities by means of specific gravity. The heaviest pieces—those containing the largest amounts of impurities—drop into a refuse bin. Washing removes much ash and sulfur from coal. But only small amounts of *organic sulfur,* which is closely bound in the coal, can be removed.

Dewatering. The washing leaves the coal dripping wet. If this excess moisture is not removed, the heating value of the coal will be greatly reduced. Preparation plants use vibrators, spinning devices called *centrifuges,* and hot-air blowers to dewater coal after it is washed.

In most cases, the separate batches of coal are mixed together again either before or after dewatering. The resulting mixture of various sizes of coal is shipped chiefly to electric power companies and coking plants. All coking plants and many power companies grind coal to a powder before they use it. They therefore accept shipments of mixed sizes. Some coal users require coal of a uniform size. Preparation plants that supply these users leave the cleaned coal in separate batches.

Shipping coal. Most coal shipments within a country are carried by rail, barge, or truck. A particular shipment may travel by two or all three of these means. Huge cargo ships transport coal across oceans, between coastal ports, and on large inland waterways.

Barges provide the cheapest way of shipping coal within a country. But they can operate only between river or coastal ports. Trucks are the least costly means of moving small shipments of coal short distances by land. Much coal, however, must be shipped long distances over land to reach buyers. Railroads offer the most economical means of making such shipments.

Many large shipments of coal are delivered to electric power companies and coking plants by *unit trains.* A unit train normally carries only one kind of freight and travels nonstop from its loading point to its destination. A 100-car unit train may carry 10,000 tons (9,100 metric tons) or more of coal.

Coal can even flow through pipes from mines to power plants. In the United States, for example, a 273-mile (439-kilometer) underground pipeline carries coal from a mine in Arizona to a power plant in Nevada. The coal is crushed and mixed with water to form a *slurry* (soupy substance) that can be pumped through the pipeline.

In the past, nearly all coal shipments consisted of anthracite, bituminous coal, or subbituminous coal. It costs

How impurities are removed from coal

Mining companies remove impurities from coal by a process called *cleaning.* The process involves three main steps. (1) A screening device sorts the coal into batches of three sizes. (2) Each batch is piped into a separate washing device and mixed with water or solutions called *heavy liquids.* The impurities are heavier than pure coal. As a result, the first pieces of coal to settle to the bottom of each solution are those that contain the most impurities. Any loose pieces of rock or clay mixed in with the coal also settle. All the waste pieces are discarded. (3) The clean pieces are dewatered with vibrators, spinning devices, or hot-air blowers. The coal is then ready for shipment to buyers.

WORLD BOOK diagram

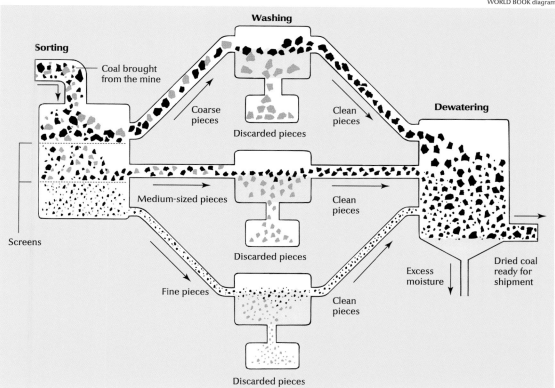

Sorting

Washing

Coal brought from the mine

Coarse pieces

Discarded pieces

Clean pieces

Dewatering

Screens

Medium-sized pieces

Discarded pieces

Clean pieces

Fine pieces

Clean pieces

Excess moisture

Dried coal ready for shipment

Discarded pieces

Clean pieces

Discarded pieces

Unit trains carry most large overland shipments of coal in the United States. A unit train normally carries only one kind of freight and travels nonstop from its loading point to its destination.

Dennis Brack, Black Star

the same to ship a given amount of any rank of coal. Lignite, having a low heating value, could not formerly compete with the higher-ranking coals in distant markets. It was used chiefly by power plants built in the lignite fields. Conveyor belts or small railways carried the coal from the mines to the plants. Today, a growing need for low-sulfur coal and improvements in coal preparation technology have increased demand for lignite. Some lignite is shipped by rail from mines to power plants hundreds of miles or kilometers away.

The coal industry

In most countries, the central government owns all or nearly all the coal mines. The major exceptions are Australia, Canada, Germany, South Africa, and the United States. In these countries, all or nearly all the coal mines are privately owned. But the central government regulates certain aspects of the coal industry in these nations.

Australia and the United States are the leading coal exporters in the world. The other leading exporters include Canada, China, Indonesia, Poland, and South Africa. Japan purchases approximately 30 percent of the world's coal exports—far more than any other country.

Coal producers. Some coal mines are owned and operated by large coal-mining companies. Others are owned by oil companies, railroads, ore-mining firms, steel companies, or electric utilities. In addition, there are many small firms that own and operate one or two mines. However, the small corporations together produce a much smaller amount of coal than the large corporations do. Steel companies and electric utilities that own coal mines produce coal chiefly for their own use. Their mines are known as *captive mines.*

In some countries, the coal industry is represented by a coal union or organization. These organizations seek to increase efficiency within the industry, to encourage favorable legislation, and to inform the public about the industry. Some organizations representing the coal industry also represent the petroleum industry or other

heavy industries. In the United States, the National Mining Association (NMA) works to promote the interests of coal producers. The NMA is jointly sponsored by operators of coal, metal, and mineral mines, and the firms that supply them with equipment, technical advice, and transportation.

Mineworkers. Most large coal-mining companies have a full-time staff of professional workers, including engineers, geologists, lawyers, and business experts. They also employ electricians, mechanics, and construction workers. Skilled miners, however, provide the labor on which the industry depends. Underground mining requires more miners than does surface mining.

National Coal Association

Coal miners provide the labor on which the coal industry depends. These miners have just finished their day's work in an underground mine. The train will carry them to the mine exit.

Mechanization has helped miners become more productive. For example, each coal miner in the United States produced, on the average, about 7 tons (6.4 metric tons) of coal daily in 1950. Today, the production rate averages about 32 tons (29 metric tons) per miner per day. On the average, a strip miner produces more than twice as much coal as does an underground miner.

Increased mechanization has also made miners' jobs more specialized. The job of most miners is to operate a certain type of machine, such as a continuous miner or a power shovel. A beginning miner must work as an apprentice for a specified period to qualify for a particular job. In some countries, mine supervisors must be licensed by the government. In the United States, for example, state departments of mining issue licenses. Generally, the licenses are granted to miners who have two to five years' experience and who pass a written examination. Most mining engineering jobs call for a college degree in engineering. If the job is directly related to mine safety, it may also require an engineering license. Some mining engineering jobs require only an engineering license.

Until the late 1800's, coal miners lived and worked under miserable conditions. The mines were dangerously unsafe, and the miners earned barely enough to live on. Many miners banded together in labor unions that called strikes in protest. In the United States, for example, the United Mine Workers of America (UMW), a major industrial trade union, was organized in 1890. Since the early 1900's, mineworkers' lives have greatly improved in the industrialized countries.

Mine safety. In the early days of underground coal mining, accidents killed or disabled thousands of miners every year. Governments have since introduced regulations that set minimum health and safety standards for employers and employees alike. Death rates have fallen dramatically. Many coal companies give every new miner a course in mine safety. In some countries,

the government requires a safety course. Mine safety involves four main types of problems: (1) accidents involving machinery, (2) roof and rib failures, (3) accumulations of gases, and (4) concentrations of coal dust.

Accidents involving machinery kill or injure more coal miners in a typical year than any other kind of mining accident. Most strip mine accidents involve machinery. The machines in underground mines must often operate in cramped, dimly lit spaces. Thus, the miners must be doubly alert to prevent accidents.

Roof and rib failures can be prevented in many cases if a mining company carries out a scientific roof support plan. In some countries, a government organization must approve this plan before mining can begin. Mining engineers make a roof support plan after studying all the rock formations surrounding the coal bed. The plan deals with such matters as the number of pillars that must be left standing, entry widths, mine geometry, and the number of roof bolts that must be used.

Accumulations of gases. Certain gases that occur in underground coal mines can become a serious hazard if they accumulate. Methane and carbon monoxide are especially dangerous. Methane is an explosive gas that occurs naturally in coal seams. It is harmless in small amounts. However, a mixture of 5 to 15 percent methane in the air can cause a violent explosion. Carbon monoxide is a poisonous gas produced by the combustion of such fuels as coal and oil. Blasting in an underground mine may produce dangerous levels of carbon monoxide if the mine is not properly ventilated.

The air vents in a mine normally prevent harmful gases from accumulating. A powerful fan at the surface circulates fresh air through the mine. The circulating air forces polluted air to the surface. As an added precaution against methane, some countries have laws that require all underground mines to have automatic methane detectors. A mine is required to shut down temporarily if a detector shows a methane accumulation above a specific level.

Concentrations of coal dust. Anyone who breathes large amounts of coal dust over a period of years may develop a disease called *pneumoconiosis* or *black lung* (see **Black lung**). The disease interferes with breathing and may eventually cause death. Thousands of coal miners have been victims of the disease. In addition, high concentrations of coal dust are explosive. A mixture of coal dust and methane is especially dangerous.

Proper ventilation removes much of the coal dust from the air in a mine. But mines also use other dust control measures. In the United States, for example, federal law requires that underground mines be *rockdusted.* In this process, the miners spray powdered limestone on all exposed surfaces in the mine entries. The limestone dilutes and coats the coal dust and so lessens the chance of an explosion. Mines use water sprays to hold down the dust along a face that is being mined.

Government regulation. National or regional governments typically set and enforce safety standards for coal mines. Government agencies regulate mine ventilation, coal dust concentrations, roof supports, and mining equipment. The regulation of coal dust has helped reduce the occurrence of black lung among miners. In some nations, benefits programs provide financial and medical benefits to miners disabled by black lung. Other

Leo Touchet, The Photo Circle

Bolting the roof is an essential safety practice in an underground mine. Roof bolts are long metal rods that are inserted into the mine roof. After a bolt is fastened to the roof, *above,* it helps prevent the rock layers immediately overhead from falling.

How strip-mined land is reclaimed

The law requires mine owners in the United States to reclaim all the land they use for strip mining. The first step is to level the piles of dug-up soil and rock, *left.* The area may then be reseeded, *center.* The project is finally completed when the new vegetation is fully grown, *right.*

Bucyrus-Erie Company

Bucyrus-Erie Company

Bucyrus-Erie Company

government agencies may regulate environmental aspects of coal-mining activities. In some countries, dishonest individuals or companies illegally operate secret mines to avoid regulation. Such mines are typically far more dangerous for workers than those operating under government safety regulations.

Coal research has become increasingly important. Government agencies, environmental organizations, mining companies, and power companies often sponsor such research. The goals of most coal research are (1) to find ways to burn more coal without increasing air pollution and (2) to develop economical methods of converting coal into liquid fuels and natural gas.

Pollution control. Most industrialized nations regulate coal-burning power plants to reduce emissions of sulfur dioxide. In some countries, the burning of high- and

medium-sulfur coal is regulated. These coals are sometimes left in reserve, to be used when better sulfur dioxide pollution controls are developed, and low-sulfur coal burned instead. In the United States, the Environmental Protection Agency enforces air pollution control requirements set forth in the Clean Air Act of 1970 and its amendments. These requirements help protect public health and the environment by establishing air quality standards.

New processes for producing power can make coal use more efficient and safer for the environment. In one such process, called *fluidized-bed combustion,* crushed coal is burned in a bed of limestone. The limestone captures sulfur from the coal and so prevents sulfur dioxide from forming. The heat from the burning coal boils water that is circulated through the bed in metal coils. The

Leading coal-producing countries

Tons of coal mined in a year

China	●●●●●●●●●●●●●●
	1,459,000,000 tons (1,323,600,000 metric tons)
United States	●●●●●●●●●●
	1,121,300,000 tons (1,017,300,000 metric tons)
Australia	●●●●
	356,900,000 tons (323,700,000 metric tons)
India	●●●●
	338,600,000 tons (307,200,000 metric tons)
Russia	●●●
	299,500,000 tons (271,700,000 metric tons)
South Africa	●●●
	250,300,000 tons (227,100,000 metric tons)
Germany	●●●
	225,700,000 tons (204,700,000 metric tons)
Poland	●●
	178,900,000 tons (162,300,000 metric tons)
Korea, North	●
	105,300,000 tons (95,500,000 metric tons)
Indonesia	●
	99,600,000 tons (90,400,000 metric tons)

Figures are for 2001.
Source: U.S. Energy Information Administration.

Leading coal-producing states and provinces

Tons of coal mined in a year

Wyoming	●●●●●●●●●●●●●●
	373,200,000 tons (338,530,000 metric tons)
West Virginia	●●●●●●
	150,100,000 tons (136,150,000 metric tons)
Kentucky	●●●●●
	124,100,000 tons (112,620,000 metric tons)
Pennsylvania	●●●
	68,400,000 tons (62,050,000 metric tons)
Texas	●●
	45,200,000 tons (41,050,000 metric tons)
Alberta	●●
	37,700,000 tons (34,200,000 metric tons)
Montana	●●
	37,400,000 tons (33,920,000 metric tons)
Indiana	●●
	35,300,000 tons (32,060,000 metric tons)
Colorado	●●
	35,100,000 tons (31,850,000 metric tons)
Illinois	●●
	33,300,000 tons (30,220,000 metric tons)

Figures are for 2002.
Sources: U.S. Energy Information Administration; Statistics Canada.

boiling water produces steam, which may be used to produce electric power.

Coal conversion. To turn coal into a high-energy fuel, the hydrogen content of the coal must be increased. Bituminous coals have the highest hydrogen content of the four ranks of coal. On average, they consist of about 5 percent hydrogen. The hydrogen must be increased to about 12 percent to produce a high-energy liquid fuel and to about 25 percent to produce manufactured gas. The process of converting coal into a liquid fuel is called *coal hydrogenation* or *liquefaction.* In the most common coal hydrogenation method, a mixture of pulverized coal and oil is treated with hydrogen gas at high temperatures and under great pressure. The hydrogen gradually combines with the carbon molecules, forming a liquid fuel. This process can produce such high-energy fuels as gasoline and fuel oil.

Coal can easily be turned into low-energy gas by the carbonization and gasification methods described in the section *The uses of coal.* Low-energy gas can also be produced from unmined coal. This process, called *underground gasification,* involves digging two widely spaced wells from ground level to the base of a coal seam. The coal at the bottom of one well is ignited. Air is blown down the second well. The air seeps through pores in the seam, and the fire moves toward it. After a passage has been burned between the two wells, the air current forces the gases up the first well. Compared with natural gas, low-energy gas made from coal has limited uses. Low-energy gas must be enriched with hydrogen for its heating value to equal that of natural gas.

The present methods of obtaining high-energy fuels from coal cost too much for commercial use. Hydrogen is expensive to produce. In addition, most fuels made from coal contain unacceptable amounts of sulfur and ash. Researchers seek to develop cheaper methods of coal conversion. Coal research programs often rely heavily on government funding.

History of coal use

No one knows where or when people discovered that coal can be burned to provide heat. The discovery may have been made independently in various parts of the world during prehistoric times. The Chinese were the first people to develop a coal industry. By the A.D. 300's, they were mining coal from surface deposits and using it to heat buildings and smelt metals. Coal had become the leading fuel in China by the 1000's.

Commercial coal mining developed more slowly in Europe. During the 1200's, a number of commercial mines were started in England and in what is now Belgium. The coal was dug from open pits and was used mainly for smelting and forging metals. But most Europeans regarded coal as a dirty fuel and objected to its use. Wood, and charcoal made from wood, were the preferred fuels in Europe until the 1600's. During the 1600's, a severe shortage of wood occurred in western Europe. Many western European countries, but especially England, sharply increased their coal output to relieve the fuel shortage.

Developments in England. During the 1500's, English factories burned huge quantities of charcoal in making such products as bricks, glass, salt, and soap. The wood shortage in the 1600's forced most English facto-

ries to switch to coal. By the late 1600's, England produced about 80 percent of the world's annual coal output. It led in coal production for the next 200 years.

Charcoal had also been widely used in England as a fuel for drying *malt,* the chief ingredient in beer. Brewers tried using coal for this process. But the gases it produced were absorbed by the malt and so spoiled the flavor of the beer. The brewers found, however, that the undesirable gases could be eliminated if they preheated the coal in an airtight oven. They thus developed the process for making coke. About 1710, an English ironmaker named Abraham Darby succeeded in using coke to smelt iron. Coke then gradually replaced charcoal as the preferred fuel for ironmaking.

The spread of the new ironmaking process became part of a much larger development in England—the Industrial Revolution. The revolution consisted chiefly of a huge increase in factory production. The increase was made possible by the development of the steam engine in England during the 1700's. Steam engines provided the power to run factory machinery. But they required a plentiful supply of energy. Coal was the only fuel available to meet this need.

During the 1800's, the Industrial Revolution spread from England to other parts of the world. It succeeded chiefly in countries that had an abundance of coal. Coal thus played a key role in the growth of industry in Europe and North America.

Developments in North America. The North American Indians used coal long before the first European settlers arrived. For example, the Pueblo Indians in what is now the southwestern United States dug coal from hillsides and used it in baking pottery. European explorers and settlers discovered coal in eastern North America during the last half of the 1600's. In the 1700's, a few small coal mines opened in what are now Nova Scotia, Virginia, and Pennsylvania. The mines supplied coal

Bettmann Archive

A Pennsylvania mine of the late 1800's was like coal mines everywhere before mining became mechanized and child labor was abolished. Boys and mules provided much of the labor.

chiefly to blacksmiths and ironmakers. Most settlers saw no advantage in using coal as long as wood was plentiful. Wood and charcoal remained the chief fuels in America until about 1880.

The Industrial Revolution spread to the United States during the first half of the 1800's. By then, coal was essential not only to manufacturing but also to transportation. Steamships and steam-powered railroads were becoming the chief means of transportation, and they required huge amounts of coal to fire their boilers. As industry and transportation grew in the United States, so did the production and use of coal. By the late 1800's, the United States had replaced England as the world's leading coal producer.

The United States led in coal production until the mid-1900's. Its demand for coal then declined as the use of petroleum and natural gas increased. The Soviet Union surpassed the United States in coal production from the late 1950's through the late 1970's. From the 1980's to the early 2000's, China usually ranked first in coal production, and the United States, second.

Recent developments. Mining companies in most countries are required by law to reclaim strip-mined land. Much of this land is turned into farms and recreation areas. In 1977, for example, the U.S. Congress passed a law requiring mine owners to reclaim all the land they use for strip mining after 1978.

The growing scarcity of petroleum and natural gas has led to a sharp rise in the demand for coal. As a result, coal production since the 1970's has continued to increase. The increased output has been used mainly to produce electric power. Today, electric power can be produced more cheaply from coal than from either natural gas or petroleum.

A type of natural gas extracted from coal beds, called *coal bed methane,* has become the fastest-growing source of natural gas in the United States and some other regions. The presence of natural gas in coal beds has been known since the beginning of mining. Widespread production of coal bed methane began in the 1970's. Scientists and engineers have overcome a number of technical and environmental obstacles to enable clean and practical production of the gas.

In 2005, the Kyoto Protocol, an international agreement to decrease the rate at which carbon dioxide and five other gases are released into the atmosphere, went into effect. Many scientists believe that the amount of carbon dioxide in the world's atmosphere has increased about 30 percent since the start of the Industrial Revolution. Coal combustion accounts for about 35 percent of worldwide emissions of this gas. The basic purpose of the protocol is to limit climate change caused by *global warming,* an increase in the average temperature of Earth's surface. The protocol seeks to control the primary *greenhouse gases,* which, when present in the atmosphere, trap heat from the sun and so increase the temperature of Earth's surface.

Delegates from around the world adopted the Kyoto Protocol as a preliminary document in 1997. They then worked to complete the agreement in a series of meetings. More than 160 nations eventually agreed to the protocol, and it went into effect in 2005. However, some countries that are heavily dependent on coal, including Australia and the United States, refused to ratify the treaty.

In the 1990's, the United States established regulations to reduce nitrogen oxides. In 2005, the federal government issued regulations to reduce mercury emissions from coal-burning power plants. Today, researchers are seeking ways to capture carbon dioxide from emissions and place it into the ground to prevent it from entering the atmosphere.   James C. Cobb

Coal production in the United States since 1800*

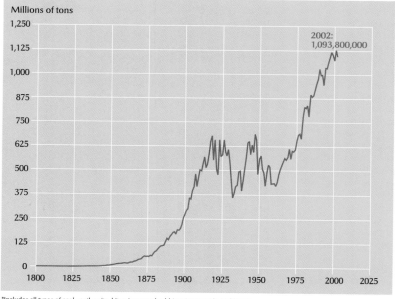

Millions of tons

2002: 1,093,800,000

Year	Tons	Metric tons
1800	108,000	98,000
1810	178,000	161,000
1820	334,000	303,000
1830	881,000	799,000
1840	2,474,000	2,244,000
1850	8,356,000	7,580,000
1860	20,041,000	18,181,000
1870	40,429,000	36,677,000
1880	79,407,000	72,037,000
1890	157,771,000	143,127,000
1900	269,684,000	244,653,000
1910	501,596,000	455,040,000
1920	658,265,000	597,168,000
1930	536,911,000	487,077,000
1940	512,256,000	464,711,000
1950	560,388,000	508,375,000
1960	434,330,000	394,018,000
1970	612,659,000	555,795,000
1980	829,700,000	752,691,000
1990	1,029,076,000	933,562,000
1995	1,032,974,000	937,098,000
2000	1,073,600,000	973,950,000
2002	1,093,800,000	992,280,000

*Includes all types of coal—anthracite, bituminous and subbituminous coals, and lignite.
Source: U.S. Energy Information Administration.

Colorful, delicious fruits make up an important food group in a healthful human diet. This fruit market in Barcelona, Spain, sells a variety of fruits from around the world, including apples, cherries, grapes, lemons, mangoes, melons, oranges, papayas, pears, and strawberries.

Fruit commonly refers to the edible tissue that surrounds the seed or seeds of many flowering plants. People around the world enjoy eating fruits as desserts or snacks. In fact, the word *fruit* comes from a Latin term meaning *enjoy*. Fruit growers worldwide produce hundreds of millions of tons of fruit annually. The most popular kinds include apples, bananas, grapes, oranges, peaches, pears, plums, and strawberries. Fruits make up an important food group in a healthful human diet. They provide rich sources of vitamins and *carbohydrates* (starches and sugars).

The term *fruit* has a somewhat different meaning for *botanists* (scientists who study plants) and for *horticulturists* (experts in growing plants). Botanists define fruit as the part of a flowering plant that contains the seeds. By this definition, fruits also can include acorns, cucumbers, and tomatoes.

Horticulturists describe fruit as an edible seed-bearing structure that (1) consists of fleshy tissue and (2) grows on a *perennial* (plant that lives for more than two growing seasons). This definition excludes nuts, which do not have a fleshy body, and vegetables, which typically grow on *annuals* (plants that live for only one growing season). It also excludes many foods that most people consider fruits. For example, watermelons meet both the botanical and common definitions of fruit. But horticulturists regard them as vegetables because they grow on annual vines. Many people also consider rhubarb a fruit because of its use as a dessert. But people eat the rhubarb leafstalk, not the seed-bearing structure. Thus horticulturists and botanists classify rhubarb as a vegetable.

This article describes the different categories of fruits in botany and in horticulture. It then discusses how people cultivate and market fruit and how they develop new fruit varieties called *cultivars.*

Types of fruits in botany

Fruits grow on flowering plants, which scientists call *angiosperms*. A fruit develops from the plant's *ovary,* the tissue surrounding the seed-bearing structure of the flower. Flowers may have one or more ovaries. Each ovary contains one or more seeds, depending on the plant. Fruits protect the seeds and help them to *disperse* (scatter) to form new plants. Many fruits have three layers after they mature: (1) an outer layer called the *exocarp,* (2) a middle layer called the *mesocarp,* and (3) an inner layer called the *endocarp.* Collectively, the three layers make up the *pericarp.*

Botanists classify fruits into two main groups, *simple fruits* and *compound fruits*. Simple fruits develop from a single ovary. Compound fruits develop from two or more ovaries.

Simple fruits make up by far the largest group of fruits. Many simple fruits have a fleshy pericarp, and others have a dry pericarp. There are three main kinds of fleshy simple fruits: (1) *true berries,* (2) *drupes* (pronounced *droopz)*, and (3) *pomes* (pronounced *pohmz)*.

True berries include bananas, blueberries, green peppers, grapes, oranges, tomatoes, and watermelons. Some of these fruits, known as *pepos (PEE pohz)*, have a firm exocarp. They include muskmelons and watermelons. Berries called *hesperidiums (HEHS puh RIHD ee uhmz)* possess a leathery exocarp. Citrus fruits rank

Simple fruits

Simple fruits are classified into two main groups, depending on whether their tissue is fleshy or dry. Fleshy simple fruits include most of the seed-bearing structures that are commonly called fruits. They are divided into three main types: (1) berries, (2) drupes, and (3) pomes. The drawings below show some examples of each of these types and of several dry simple fruits.

WORLD BOOK illustrations by James Teason

Berries consist entirely of fleshy tissue, and most species have many seeds. The seeds are embedded in the flesh. This group includes only a few of the fruits that are commonly known as berries.

Orange   Grapes   Watermelon

Drupes are fleshy fruits that have a hard inner stone or pit and a single seed. The pit encloses the seed.

Peach   Cherries   Plum

Pomes have a fleshy outer layer, a paperlike core, and more than one seed. The seeds are enclosed in the core.

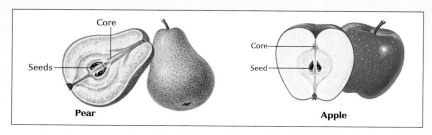

Pear   Apple

Dry simple fruits are produced by many kinds of trees, shrubs, garden plants, and weeds. The seed-bearing structures of nearly all members of the grass family, including corn and wheat, belong to this group.

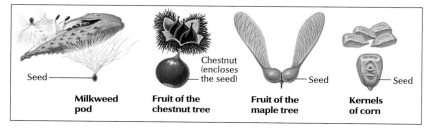

Milkweed pod   Fruit of the chestnut tree   Fruit of the maple tree   Kernels of corn

Compound fruits

A compound fruit consists of a cluster of seed-bearing structures, each of which is a complete fruit. Compound fruits are divided into two groups, (1) aggregate fruits and (2) multiple fruits.

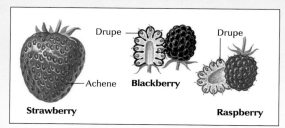

Strawberry   Blackberry   Raspberry

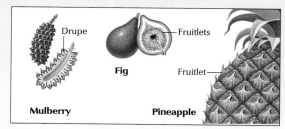

Mulberry   Fig   Pineapple

Aggregate fruits include most of the fruits that are commonly called berries. Each fruitlet of a blackberry or raspberry is a small drupe. Each "seed" of a strawberry is a dry fruit called an achene.

Multiple fruits include mulberries, figs, and pineapples. Mulberry fruitlets are small drupes. Each "seed" in a fig and each segment of a pineapple is a fruitlet.

as the best known hesperidiums. Many fruits that have the word *berry* in their common name, such as blackberries, raspberries, and strawberries, are not in fact true berries. Scientists classify them as compound fruits.

Drupes possess a fleshy mesocarp surrounding a hard endocarp called a *stone* or *pit*. A thin exocarp forms the skin. Examples of drupes include apricots, cherries, olives, peaches, and plums.

Pomes have a distinctive core. The core consists of a thin, paperlike endocarp surrounding hollow, seed-bearing cavities. Apples and pears rank among the most popular types of pomes.

Dry simple fruits include beans, milkweed, peas, rice, wheat grains, and true nuts. Botanists regard true nuts as single-seeded fruits with a hard pericarp called a *shell*. People eat the seeds of these plants but not the pericarps. True nuts include acorns, chestnuts, and hazelnuts. Some so-called nuts, almonds for example, are actually the seeds of drupes. In fact, food processors frequently substitute apricot seeds for almonds in processed foods.

Compound fruits consist of a cluster of ripened ovaries. There are two main types of compound fruits, *aggregate fruits* and *multiple fruits*.

Aggregate fruits develop from single flowers, each of which has many ovaries. The strawberry represents an unusual type of aggregate fruit. Each so-called seed on a strawberry is a true fruit. Botanists call these seed-like fruits *achenes (ay KEENZ)*. The edible fleshy part surrounding the achenes develops from the base of the flower rather than from the ovaries. Other aggregate fruits include blackberries and raspberries.

Multiple fruits grow from a cluster of flowers on a single stem. Figs, mulberries, and pineapples are multiple fruits. Botanists also consider an ear of corn to be a multiple fruit. Each kernel forms a single fruit called a *caryopsis (KAR ee AHP sihs)*.

Types of fruits in horticulture

Farmers have long cultivated various fruits far outside the areas where the plants originally grew. Peaches, for example, once grew only in China but now thrive in many parts of the world. Horticulturists classify fruits into three groups, based on their temperature requirements for growth: (1) temperate fruits, (2) subtropical fruits, and (3) tropical fruits.

Temperate fruits need an annual cold season to grow properly. Farmers raise them chiefly in the *temperate* regions between the tropics and the polar areas. Most temperate fruits thrive in Europe and North America. They also grow in Asia, Australia, and New Zealand. Such temperate fruits as apples, apricots, cherries, peaches, pears, and plums grow on trees. Temperate areas also produce many fruits that grow on plants smaller than trees, including blueberries, cranberries, grapes, kiwis, raspberries, and strawberries.

Subtropical fruits require warm or mild temperatures throughout the year but can survive an occasional light frost. This type of climate characterizes subtropical regions. The most widely grown subtropical fruits, the citrus fruits, consist primarily of grapefruits, lemons, limes, and oranges. Brazil, Israel, Italy, Mexico, Spain, Turkey, and the United States all have important citrus-

growing regions. Other subtropical fruits include avocados, figs, and olives.

Tropical fruits cannot tolerate even a light frost. Bananas and pineapples rank as the best-known tropical fruits. Growers cultivate them throughout the tropics, mostly for export. Other tropical fruits include acerolas, cherimoyas, litchis, mangoes, and papayas.

Fruit production

Many fruit *species* (kinds) grow on trees or other long-lived woody plants. Tree fruits include apples, mangoes, and the major citrus fruits. Grapes develop on woody vines, while many other small fruits grow on bushes. Bananas and strawberries develop on plants that have nonwoody stems.

Unlike most other crop plants, fruits are not grown from seeds. Instead, growers develop them from such plant tissues as stems, buds, and roots in a process called *vegetative reproduction*. Plants grown from seeds may vary in many ways from generation to generation. But plants grown vegetatively display similar growth habits and yield fruit of similar quality. Fruit growers strive to *propagate* (reproduce) plants in which such traits remain as consistent as possible over time.

Growers propagate fruit plants in three main ways: (1) by grafting, (2) from cuttings, and (3) from specialized plant structures. Most fruit trees require the grafting method. In this process, the grower joins a bud or piece of stem from a desirable cultivar to a *rootstock* from another plant. A rootstock is a root or a root plus its stem. The resulting tree will produce the desired fruit. Moreover, the rootstock may influence such tree characteristics as size, productivity, and disease resistance.

Farmers propagate some plants by rooted cuttings or specialized structures, such as modified stems called *runners*. Rooted cuttings consist of pieces of stem that have grown roots when placed in water or moist soil. Mature strawberry plants send out long, thin runners that grow along the surface of the soil. Where the runners touch the ground, modified buds called *nodes* form roots that produce *plantlets* (new leaves and stems). These plantlets are actually part of the parent plant but can develop into new plants if separated from the parent.

Most growers buy fruit plants from nurseries that specialize in propagating them. Nurseries produce plants under controlled conditions to reduce or eliminate diseases and insects. They often sell their plants with a guarantee that the plants are pest-free.

A branch of horticulture called *pomology* deals with growing fruit. Pomologists have developed efficient methods of planting, tending, and harvesting fruit.

Planting. Because fruit plants are perennials, growers need not replant them annually. Trees and vines may remain productive for 30 to 50 years or longer. Small fruit plants, including strawberries and raspberries, have a productive life of only a few years. Farmers in mild climates typically plant trees, bushes, and vines in the fall. In cold climates, planting often occurs in spring.

In the past, farmers almost always grew full-sized, free-standing fruit trees. They generally planted the trees from about 20 to 40 feet (6 to 12 meters) apart to allow room for growth. Today, however, most growers prefer specially propagated dwarf trees. Growers space these trees closer together, from about 4 to 20 feet (1.2 to 6 meters) apart. Closer spacing produces a larger crop in the same area. Smaller trees also enable growers to care for and harvest the crop more easily.

Caring for the crop. Most fruit growers use machinery to fertilize, cultivate, and irrigate the plantings. Farmers must fertilize fruit plants at least once a year. Some fertilizers are applied to the soil, while others are sprayed on the plants. Many growers cultivate the soil around young fruit plants periodically. This practice encourages crop growth by controlling weeds and improving the circulation of air and water through the soil. Most fruit plants require considerable moisture. Only a few fruits, such as dates and olives, can grow in dry regions without irrigation. If irrigation is needed, growers use ditches or sprinklers to distribute the water.

Some fruit plants, including blueberry bushes, are free-standing. But growers must train other types, such as grapevines, raspberry bushes, and many young fruit trees, to grow on trellises or other supports. Some trees may even need their trunks propped up so that the trees develop a uniform shape and sturdy structure. Fruits grown on supports receive maximum sunlight, producing a more uniform and better quality product. Supports also make harvesting easier.

Nearly all fruit plants need pruning at least annually. Growers must prune the plants to rid them of unproductive or diseased branches. Most growers also remove some of the crop from trees during the early stages of fruit growth. This practice, called thinning, helps increase the size and quality of the remaining fruit.

Fruit growers often use a system of *integrated pest management* (IPM), which combines natural and chemical controls to fight pests. Farmers usually apply chemical pesticides with tractor-pulled sprayers or specially equipped light airplanes or helicopters.

Sudden spring frosts can endanger fruit crops in temperate or subtropical regions. Growers may use

---

How horticulturists classify fruits

Any seed-bearing structure produced by a flowering plant is a fruit. But the word *fruit* has a more limited meaning in common usage and in horticulture, the branch of agriculture that includes fruit growing. Thus, the word usually refers to the edible sweet or tart fruits that are popular foods and widely grown farm crops. Horticulturists classify these fruits into three groups, based on temperature requirements for growth: (1) temperate fruits, (2) subtropical fruits, and (3) tropical fruits. Some examples of each of these types are shown below.

WORLD BOOK illustrations by James Teason and Wildlife Art Ltd.

Temperate fruits must have an annual cold season. They are raised mainly in the temperate zones, the regions between the tropics and the polar areas.

Subtropical fruits need warm or mild temperatures throughout the year but can survive occasional light frosts. They are grown chiefly in subtropical regions.

Tropical fruits cannot stand frost. They are raised mainly in the tropics. Large quantities of some species, especially bananas and pineapples, are exported.

water distributed by sprinklers to protect plants from frost damage. Water releases heat as it freezes. If sprinkled onto the crops continuously, the water protects flowers and young fruit from freezing. Another method of frost protection uses large fans on towers. The fans can mix naturally occurring warm air 30 to 50 feet (9 to 15 meters) above the ground with the colder air at plant level.

Harvesting. Most fruits ripen quickly after reaching their mature size. Harvesting occurs at different stages of the growth process, depending on the type of fruit and its intended use. Some fruit crops require harvesting when still immature. They include the gooseberries and cherries used to make artificial coloring. For apples, bananas, peaches, and pears, commercial harvesting occurs when the fruits reach full size, but before they ripen. Most fruits taste best if left to ripen on the plant, so home gardeners typically harvest their crops when ripe. Citrus fruits do not go through a distinct ripening process. Thus, harvesting can take place over a long period after they mature.

Fruits bruise more easily than do most other crops, so growers must harvest them with care. Workers pick most fruit crops by hand. However, the increasing cost of hand labor has encouraged the use of fruit-harvesting machines. Some of these machines have arms that shake the fruit loose from the plants. The loosened fruit drops onto outstretched cloths. Other harvesting machines have a rubber fingerlike structure that gently "combs" fruit from the plants.

Once the fruit has been harvested, farmers usually deliver it immediately to cold storage or controlled-atmosphere storage facilities. There the fruit can finish ripening under controlled conditions. When the fruit becomes ready for market, workers wash it, sort it, and pack it into containers. They then ship the fruit to such buyers as retail stores. Some fruits, including apples, can remain fresh for about a year if stored at temperatures near freezing. But most small fruits or tropical fruits remain fresh for only a few days or weeks in storage. Farmers ship much fruit directly from farms to food processing plants. These plants preserve fruit by such methods as canning, drying, and freezing. Processors often use imperfectly formed fruit in jams, preserves, juices, and other products.

Marketing fruit

The worldwide marketing of fruits has expanded dramatically since the late 1900's. Because of improved transportation and storage techniques, most major food markets now offer fresh fruit grown and shipped from thousands of miles or kilometers away. Such technology enables markets to offer fruits out of season. For example, apples ripen in the fall. But because the Northern Hemisphere and Southern Hemisphere experience fall at opposite times of the year, markets in one hemisphere can sell fresh apples out of season by importing them from the other hemisphere. Thus, markets in the United States can sell apples in spring that they have imported from New Zealand.

Technology also has made more types of fruits available in many countries. Such fruits as carambolas (also called star fruit), durians, and mangosteens once rarely appeared outside the tropics. Today, however, they have become available in much of the world. Human immigration helps further increase the variety of fruit in a particular country. Many immigrants bring unusual fruits with them from their old countries to their new homes. Therefore, nations with large immigrant populations often have especially rich and varied fruit markets.

Developing new fruit cultivars

Horticulturists have modified numerous original fruit species to create improved cultivars. Scientists use several methods to develop new cultivars. In one traditional method, they crossbreed two or more existing cultivars that have different desirable characteristics. Such crossbreeding creates a single new cultivar that exhibits desirable traits from both of its parents. Crossbreeding programs best suit small fruit plants that have short life cycles. Plants with longer life cycles make the process too time-consuming.

Another method for developing cultivars involves using *sports* or *chance seedlings* with desirable characteristics. Sports are *mutations* (random changes) that typically occur on individual branches or buds of plants. In chance seedlings, entire young plants exhibit unusual traits. Fruit plants propagated from sports or chance seedlings can inherit their desirable traits. Horticulturists employed this method to create many of the best-known fruit cultivars, including Delicious apples and varieties of navel oranges.

Since the late 1900's, scientists have increasingly used genetic engineering techniques to develop new cultivars. Genetic engineering involves altering a plant's *genes* (units of heredity) to give the plant certain desirable traits. Horticulturists often call such techniques *transgenic technology.*

Transgenic technology enables breeders to focus on precisely the characteristics they want to improve in a cultivar. For example, the first *genetically modified* (GM) food to reach the commercial market was a tomato. Breeders incorporated a gene into this tomato that slowed the ripening process. This genetic modification permitted growers to harvest the fruit at a riper, more flavorful stage. It also enabled the fruit to remain in good condition in the market longer. Other genetic modifications have incorporated vitamins and other nutrients into plants. Still others have improved the plants' resistance to diseases, weed killers, and drought.

Genes used in transgenic technology may come from organisms other than plants. For example, scientists have inserted a gene from a bacterium into several types of crop plants. This gene enables the plants to produce a bacterial protein that kills certain insects when they feed on the plants.

Despite the potential benefits of transgenic technology, there are also potential problems. For example, some people fear that this new technology may produce unintended consequences that cause harm to the environment. Some also fear that the patenting of GM cultivars may give corporations unreasonable monopolies over the production and marketing of such crops. Mainly for these reasons, various governments are considering bans on the sale of GM foods. James E. Pollard

Gas provides high-energy, clean-burning fuel for many home and industrial uses, such as this industrial furnace used to harden gears in an automobile factory.

Gas

Gas is one of our most important resources used as both a fuel and a raw material. We burn gas to provide heat and to produce energy to run machinery. Millions of people and businesses use gas to heat buildings and water, to cook meals, to dry laundry, and to cool the air. The chemical industry uses the chemicals in gas to make detergents, drugs, plastics, and many other products. Gas produces little air pollution when it is burned.

People sometimes confuse gas with gasoline, which is often called simply *gas*. But gasoline is a liquid. On the other hand, gas fuel—like air and steam—is a *gaseous* form of matter. That is, it does not occupy a fixed amount of space, as liquids and solids do.

There are two kinds of gas—*natural gas* and *manufactured gas*. Most of the gas used in the world is natural gas. Most scientists believe that natural gas has been forming beneath Earth's surface for hundreds of millions of years. The natural processes that created gas also created petroleum. As a result, natural gas is often found with or near oil deposits. The same meth-

ods are used to explore and drill for both fuels. Manufactured gas is produced chiefly from coal or petroleum, using heat and chemical processes. Manufactured gas costs more than natural gas and is used in regions where large quantities of the natural fuel are not available.

The gas industry consists of five main activities: (1) exploring for natural gas; (2) producing gas, either by drilling natural gas wells or by manufacturing gas; (3) transmitting gas, usually by pipeline, to large market areas; (4) distributing gas to the user; and (5) storing gas for transmission and distribution at a later time. Each part of the gas industry requires its own special skills and equipment. Some gas companies conduct all five activities, but most companies handle only one or two.

The modern natural gas industry began in the United States. The industry started to expand rapidly in the late 1920's with the development of improved pipe for transmitting gas economically over great distances. Today, long-distance gas pipelines serve much of the world.

The composition of natural gas

Pure natural gas is made up of chemical compounds of the elements hydrogen and carbon. These compounds are called *hydrocarbons*. Some hydrocarbons are naturally gaseous, some are liquid, and some are

Michael A. Adewumi, the contributor of this article, is Professor of Petroleum and Natural Gas Engineering at Pennsylvania State University.

An offshore drilling platform, such as this one in the North Sea east of Scotland, enables a gas producer to obtain natural gas from deposits beneath the ocean floor.

At a coal gasification plant, coal is converted to a manufactured gas. One of the most important of such gases is coke oven gas, which is used for certain manufacturing processes.

solid. A hydrocarbon's form depends on the number and arrangement of the hydrogen and carbon atoms in the hydrocarbon molecule.

Natural gas is composed mainly of methane, the lightest hydrocarbon. In a molecule of methane (CH_4), one atom of carbon is bound together with four atoms of hydrogen. Other gaseous hydrocarbons usually found in natural gas include ethane (C_2H_6), propane (C_3H_8), and butane (C_4H_{10}). Natural gas that is impure may contain such gases as carbon dioxide, helium, and nitrogen.

When natural gas burns, the hydrocarbon molecules break up into atoms of carbon and hydrogen. The atoms combine with oxygen in the air and form new substances. The carbon and oxygen form carbon dioxide (CO_2), an odorless, colorless gas. The hydrogen and oxygen produce water vapor (H_2O). As the molecules break up and recombine, heat is released. Heat is measured in Btu's (British thermal units) in the inch-pound system of measurement customarily used in the United States, and in *calories* or in *joules* in the metric system. One cubic foot (28,316 cubic centimeters) of burning gas releases about 1,000 Btu's, or about 252,000 calories or 1,050 *kilojoules.* A kilojoule equals 1,000 joules.

Uses of gas

Gas as a fuel. About 40 percent of the gas consumed in the industrialized countries is used in residences or by such businesses and institutions as offices, hotels, restaurants, stores, hospitals, and schools.

And industry and electric power generation typically consume about 30 percent each. In many less developed countries, most natural gas is used to generate electric power. In both industrialized and less developed countries, small amounts of natural gas are used as transportation fuel and to operate grain dryers, irrigation equipment, and other farm equipment.

For the industrialized countries as a whole, gas provides about 20 percent of the total energy needs. However, many countries of the world lack either large deposits of natural gas or the systems needed to produce and transport it. These countries consume only small amounts of gas, chiefly manufactured gas.

In the home. Wherever large quantities of natural gas are available, gas is the most popular cooking fuel. One reason for its popularity is that it costs less than most other fuels. In addition, gas can provide the desired amount of heat instantly, be controlled easily and even automatically, and be shut off instantly.

Many residential consumers also use gas to heat their homes and water, dry laundry, and operate air conditioners. Many people cook outdoors on gas grills. Some homes have gas fireplaces.

Many people who live in mobile homes or in farm areas or other places far from gas pipelines burn *liquefied petroleum gas* (LPG) for cooking and heating. LPG is also called *LP gas, propane, butane,* or *bottled gas.* It is produced from gaseous compounds in petroleum. These compounds become liquid when they are put under pressure. The liquid takes up much less space than the original gas and is easily transported in small pressurized containers. As the fuel is used, normal air pressure changes the liquid back to gas. In some countries, liquefied petroleum gas is used as a fuel for automobiles.

In industry. Gas has many uses in industry. Companies use gas flames or gas heat in coating, cutting, and shaping metals and other materials. Gas heat is used to harden the nose cones of spacecraft so they do not burn up from the intense heat generated by atmospheric friction. Meat packers use gas flames to remove the bristles from hogs. Manufacturers use gas to produce or process brick and tile, cement, ceramics, foods, glass, iron and steel, paper, textiles, and countless other products. Burning gas produces heat for industrial processes ranging in temperature from about 350 °F (177 °C) for baking automobile finishes to about 3000 °F (1600 °C) for making steel.

Many modern factories have *complete heat and power* (CHP) systems, which supply all their power needs. In such systems, gas is the only outside source of energy. It powers a turbine or engine that drives a generator to produce electric power. The exhaust heat from the turbine or engine is used for heating and cooling.

Industry also uses *gas infrared heaters.* Infrared rays from such heaters heat only the objects that they strike, not the air. These heaters are especially useful for keeping people warm in large warehouses or other buildings that are difficult to heat.

Gas as a raw material. Natural gas is an important source of *petrochemicals* (chemicals made from natural gas or petroleum). Petrochemicals serve as building blocks for the manufacture of many products, including fertilizers, paints, plastics, and synthetic rubber.

Petrochemical production is based on compounds of hydrogen and carbon found in gas and oil. These compounds include methane, ethane, and propane. They can be removed from the raw material and used alone, or broken apart and restructured to produce compounds not present in the raw material. The compounds or their parts are combined with other chemicals to make detergents, drugs, and other products.

Some *components* (parts) of natural gas are convert-

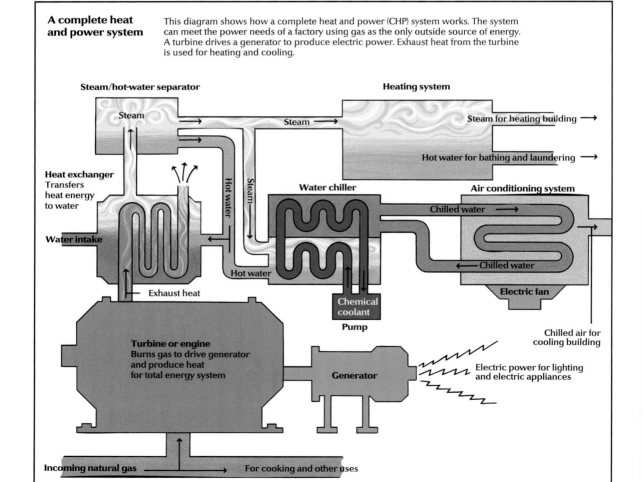

A complete heat and power system

This diagram shows how a complete heat and power (CHP) system works. The system can meet the power needs of a factory using gas as the only outside source of energy. A turbine drives a generator to produce electric power. Exhaust heat from the turbine is used for heating and cooling.

How natural gas was formed

Ages ago, the remains of tiny marine organisms sank to the sea floors and were buried by sediments, *left.* The decaying matter became gas and oil trapped in porous rock under nonporous rock, *center.* Later, Earth's crust shifted, and dry land appeared over many deposits, *right.*

WORLD BOOK diagram by George Suyeoka

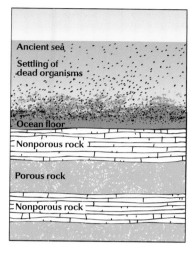

Ancient sea
Settling of dead organisms
Ocean floor
Nonporous rock
Porous rock
Nonporous rock

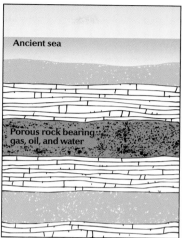

Ancient sea
Porous rock bearing gas, oil, and water

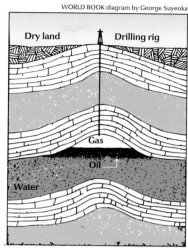

Dry land   Drilling rig
Gas
Oil
Water

ed into liquid form, called *natural gas liquids* (NGL). This group of chemical compounds includes ethane, propane, and butane. For many years, gas producers have made NGL's by trapping gas in its natural state and storing it at extremely low temperatures or by pressurizing gases that are formed early in the process of refining crude oil. Modern *gas-to-liquid* (GTL) technologies use chemical processes to convert natural gas to liquids. These technologies can produce NGL's at lower cost than the traditional methods can. Many NGL plants and petrochemical factories operate near gas fields to be close to their sources of supply. For additional information about the chemistry of gas, see the section *The composition of natural gas* in this article.

How natural gas was formed

Most scientists believe that natural gas was formed millions of years ago, when water covered much more of Earth's surface than it does today. Down through the ages, tiny marine organisms called *plankton* died and settled on the ocean floors. There, fine sand and mud drifted down over the plankton. Layer upon layer of these deposits piled up. The great weight of the deposits, plus bacteria, heat, and other natural forces, changed the chemical compounds in the plankton into natural gas and petroleum. The gas and oil flowed into openings in limestone, sandstone, and other kinds of rocks that are *porous* (full of tiny holes). Layers of solid rocks formed over the porous rocks and sealed the gas and oil beneath them. Later, movements in Earth's crust caused the ancient seas to draw back, and dry land covered many gas and oil deposits.

Natural gas is referred to as a *fossil fuel* because it formed millions of years ago from the remains of living things. Petroleum and coal are also fossil fuels.

From well to user

Natural gas is found underground in various types of *reservoir rock.* This type of rock includes limestone, sandstone, carbonate, and other porous rocks. The pores are interconnected, and so gas can move through the rock. A dome of nonporous rock forms a cap over the reservoir rock, trapping the gas. The gas cannot escape unless well drillers open a hole through the solid rock or unless Earth's surface shifts and cracks the cap. Natural gas is often found on top of oil deposits or dissolved in them, because the same natural processes formed both fuels.

Natural gas can also be also found in *gas hydrates,* solids that resemble wet snow. Gas hydrates, also called *methane hydrates,* are formed when water freezes in the presence of methane in the ocean depths. Under such conditions of extreme cold and pressure, the ice crystals form "cages" with gas molecules trapped inside. Gas hydrates also form readily in the extreme cold that occurs in the Arctic. The volume of gas stored in gas hydrates may be as great as 5,000 times the amount of conventional gas reserves known to exist in the world today. But retrieving the deposits is difficult because gas hydrates are not stable at normal surface temperatures and pressures. Without special equipment, the gas escapes from the ice when it reaches the surface. No one has yet found a cost-effective way to extract large quantities of fuel from gas hydrates.

Exploring for gas. Modern exploration methods constantly uncover new reserves of natural gas. These methods show where there are geological formations that could hold gas. But they cannot indicate the actual presence of gas. The only sure way to find out if an area contains deposits of natural gas is to drill a well.

In *proved areas,* where gas or petroleum has already been found, about 75 percent of new wells produce

Finding and processing natural gas

Many processes are involved in delivering clean-burning gas to industries, businesses, and homes. The pictures on this page show how gas is found, transported, and processed.

Geo Data Corporation

Exploring for gas may begin with *thumper trucks*. A thumper truck uses a huge vibrating pad to send sound waves into the earth that aid in locating underground gas deposits.

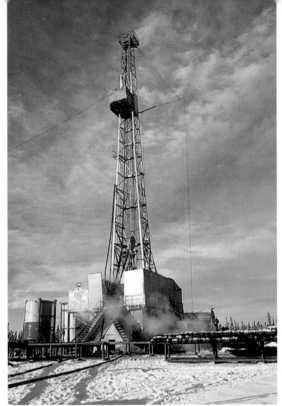

© Ted Czocowski, Image Finders

Drilling a well is the only sure way to find out if an area has gas. Tall derricks hold the well-drilling equipment. The same methods are used in drilling for gas as in drilling for petroleum.

TransCanada PipeLines Ltd.

Laying underground pipelines requires digging trenches. Pipelines carry gas from producing fields to local distribution systems that may be thousands of miles or kilometers away.

ConocoPhillips

A natural gas processing plant removes unnecessary components from the fuel. This plant in Canada stores gas in aboveground tanks to ensure a constant supply to its customers.

one of the two fuels. Drilling in unproved areas is called *wildcatting.* A wildcat well is drilled wherever a prospector believes that gas or oil may be present, in areas far from producing wells. About 70 percent of the wildcat wells produce some gas or oil. But many wildcat wells fail to yield enough fuel to pay for the cost of drilling them. Prospectors continue to drill in unproved areas because they keep part of the rights to any petroleum and gas deposits they discover. One successful wildcat well can more than make up for the cost of the unsuccessful ones.

In exploring for gas in unproved areas, prospectors rely on studies made by earth scientists called *geologists* and *geophysicists.* These studies involve maps, drilling records, and *seismic* measurements—that is, measurements of vibrations in the ground.

After selecting a promising site for a well, a geologist studies a detailed map of the features above and below Earth's surface. The map enables the geologist to locate underground formations called *traps,* where natural gas and petroleum can accumulate. In addition, geologists try to determine whether there is reservoir rock underground.

Geologists may also study *well logs.* A well log is a record of the rock formations encountered during the drilling of a well. Well logs measure such characteristics as the *porosity* (presence of pores) and fluid content of the rock. By comparing well logs, geologists can determine how rocks differ from one area to another.

Geophysicists commonly use an exploration technique known as *reflection seismology.* In this technique, a loud noise, such as an explosion, is produced at or just below the surface. The sound waves that result travel into the ground and are reflected back to the surface by underground rock layers. In populated areas, a *vibroseis truck,* also called a *thumper truck,* may be used to produce the sound waves. A thumper truck has a huge vibrating pad that repeatedly strikes the ground. In offshore areas, sound waves are produced by sending a compressed-air discharge or an electronic pulse from a ship into the water.

Groups of *geophones,* which are similar to microphones, pick up the reflected sound waves. The pattern of the sound waves is recorded on an instrument called a *seismograph.* Sound waves change in *amplitude* (height) when they are reflected from rocks that contain gas. These changes appear as irregularities, called *bright spots,* on the seismograph record.

Producing gas. Drilling for gas involves the same methods as those used in drilling for oil. The most common method is *rotary drilling.* It is much like making a hole in wood with a carpenter's drill. Another method, *cable-tool drilling,* is used chiefly to make shallow holes in soft rock. It is similar to punching a hole in wood with a hammer and a nail.

Offshore wells are drilled in water that may be more than 10,000 feet (3,050 meters) deep. The North Sea, the Gulf Coast waters of the United States, and the waters off the western coast of Africa are among the richest offshore producing areas. Offshore drilling is usually more productive than drilling on land, mainly because much less gas and oil have been taken from beneath the sea. But offshore drilling costs several times more.

Instead of simply drilling down from land, offshore drillers must work from a barge, a movable rig, or a fixed platform.

Transmitting and distributing gas. The raw natural gas that flows from a well must be cleaned and treated before it is distributed. A pipe called a *gathering line* carries the gas from the well to an *extraction unit,* which removes such impurities as sand, sulfur, and water. The gas may then flow to nearby processing plants. The plants remove components that are not needed in the fuel and that can be easily *condensed* (converted to liquid), such as propane and gasoline. The processed natural gas is then fed into long-distance *transmission pipelines,* which carry it to communities along their routes. These pipelines are typically buried to minimize the risk of vandalism or accidental damage.

Gas is sent through transmission pipelines under high pressures—usually about 1,000 pounds per square inch (70 kilograms per square centimeter). The pressure drops along the route because of the friction of the gas against the pipe walls. The pressure also falls when communities remove gas. *Compressor stations* along the line restore high pressure and push the gas on to its farthest destination. Many lines have automatically operated stations that increase or decrease the pressure to meet the demands of various communities.

Gas usually travels through pipelines at about 15 miles (25 kilometers) per hour. Gas being pumped through pipelines from a well to a local distribution system thousands of miles or kilometers away can take several days to reach its destination.

Inspectors on foot and in airplanes check continually for conditions that might damage pipelines. After floods and heavy rains, for example, inspectors ensure that the pipelines remain covered with earth. In addition, instruments installed along the pipelines automatically report leaks and other faulty conditions.

In cities and towns, *distribution lines* carry the gas to consumers. There are two kinds of distribution lines—*mains* and *individual service lines.* Mains are large pipes connected to the transmission pipelines. Service lines are smaller pipes that branch out from the mains. The service lines carry the fuel sold by gas utility companies to homes, factories, restaurants, and other buildings.

Pure natural gas is odorless and would not be noticed if it leaked out. For this reason, gas utility companies add *mercaptans,* chemicals that contain sulfur, to the gas to give it a smell.

Storing gas. Consumers use much more gas in winter than in summer. Pipelines cannot carry enough gas to meet the demand for fuel on the coldest days. As a result, gas must be stored when the demand is low for use when the demand is high.

During the summer, many gas companies pump great quantities of natural gas back into the ground. Most underground storage areas are old gas or oil fields that are no longer productive, or other porous rock formations. Ideal storage areas lie near pipelines, compressor stations, and—most important of all—large market areas.

If a gas company selects a nonproductive gas or oil field for storage, it must prepare the site to receive and hold gas. The company may have to repair and clean,

Natural gas regions of the world

This map shows the world's major natural gas producing areas on land and offshore. Modern methods of exploring for natural gas continue to increase the world's known supplies.

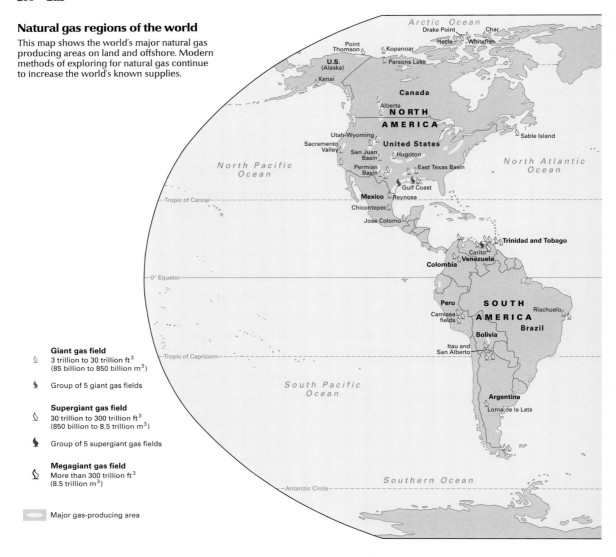

Giant gas field
⚶ 3 trillion to 30 trillion ft³
(85 billion to 850 billion m³)

⚶ Group of 5 giant gas fields

Supergiant gas field
⚶ 30 trillion to 300 trillion ft³
(850 billion to 8.5 trillion m³)

⚶ Group of 5 supergiant gas fields

Megagiant gas field
⚶ More than 300 trillion ft³
(8.5 trillion m³)

▢ Major gas-producing area

or even replace, the well *casings* (large pipes put down the well to prevent it from caving in). The firm also may have to redrill the old well or drill a new one.

To find and prepare new underground storage sites, geologists and engineers use methods like those used in exploring and drilling for gas or oil. After a new storage field has been prepared and tested, huge machines pump in the gas under high pressure. When the company removes the stored gas to meet heavy demand in cold weather, it cleans and treats the fuel before sending it to the consumers.

Underground storage reservoirs are also important in the conservation of natural gas. Before the wide use of reservoirs, oil drillers often *flared* (burned) the natural gas found in an oil well to get rid of it during periods of low demand. Some of the great oil-producing countries in the Middle East and Africa still waste a large amount of gas in this way.

Natural gas is also stored through *liquefaction*—by changing it into a liquid. Methane becomes liquid when its temperature is lowered to about −260 °F (−162 °C). Raising the temperature returns the fuel to its gaseous form. Liquid natural gas (LNG) requires only about 1/600 of the storage space needed to store an equal quantity of natural gas. LNG can also be shipped overseas. For use in large volumes, LNG is more practical than LPG or other liquid gases because it has the same chemical makeup as natural gas. As a result, suppliers can easily switch between LNG and natural gas.

How gas is manufactured

Gas is manufactured for its chemical by-products and for use as fuel. There are several types of manufactured gas. The most important are coke oven gas, also called coal gas, and acetylene.

Coke oven gas is made by roasting coal. As the coal

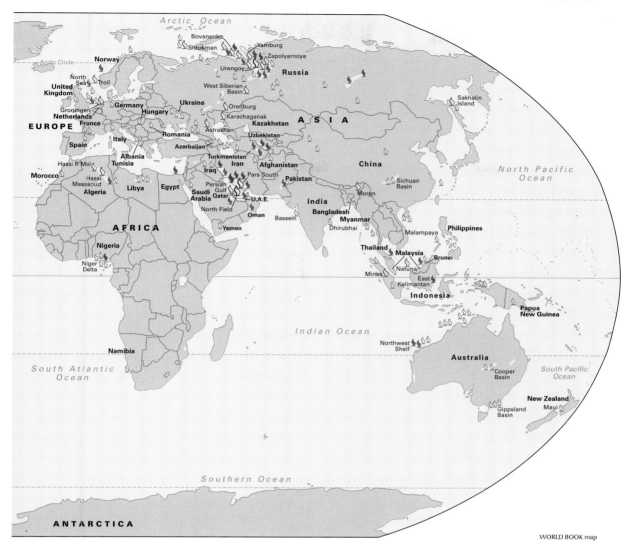

WORLD BOOK map

turns into coke, vapors consisting of many chemicals escape from the coal. The vapors are sent through water, which absorbs some of the unwanted chemicals. The rest of the gas bubbles up through the water. This gas may be further purified by various processes that remove chemical by-products. Coke oven gas has a much lower *heating value* than natural gas. Heating value refers to the amount of heat produced by a given amount of fuel when it is burned.

Acetylene is produced chiefly by creating a reaction between water and calcium carbide, a compound of calcium and carbon. Gas producers can also make acetylene by heating methane molecules to break them apart. Acetylene has a higher heating value than natural gas. It produces an extremely hot flame. Acetylene is used in welding and cutting metals.

Other manufactured gases include oil gas, producer gas, and water gas. *Oil gas* is made by spraying oil

onto hot bricks to break apart the petroleum molecules. *Producer gas* is made by sending air slowly through a bed of hot coal or coke. The oxygen in the air combines with carbon in the coal, forming carbon monoxide. *Water gas* is made by forcing steam through a hot bed of coal or coke, forming carbon monoxide and hydrogen.

Gas and the environment

Harmful gases may be generated as by-products of the burning of natural gas. But burning other fossil fuels, and some alternative fuels, produces greater quantities of these gases, along with harmful particles of matter. Thus, natural gas burns relatively cleaner than these other fuels. Partly because of this environmental advantage, the demand for natural gas continues to increase.

Leaks in natural gas pipelines can release methane

into the atmosphere, and burning natural gas produces carbon dioxide. Methane and carbon dioxide are often called *greenhouse gases,* because they trap heat from the sun and hold it near Earth's surface. A build-up of greenhouse gases may result in *global warming,* an increase in the average temperature of Earth's surface.

History of the gas industry

Early uses of natural gas. The ancient Chinese were the first people known to use natural gas for industrial purposes. Thousands of years ago, they discovered natural gas deposits and learned to pipe the fuel through bamboo poles. They burned gas to boil away salty water and collect the salt that remained.

As early as the A.D. 500's, temples with "eternal" fires were built in western Asia, near what is now Baku, Azerbaijan. Worshipers came from as far as Persia and India to see the mysterious, continuous fires and wonder at the power of the temple priests. Secret pipes carried natural gas into the shrines from nearby rock fractures.

First uses of manufactured gas. In 1609, Jan Baptista van Helmont, a Belgian chemist and physician, discovered manufactured gas. He found that a "spirit," which he named *gas,* escaped from heated coal. In the late 1600's, John Clayton, an English clergyman, roasted coal and collected the gas in animal bladders. He then punctured the bladders and lit the escaping gas.

In 1792, William Murdock, a British engineer, lighted his home with gas he made from coal. He installed gas lighting in his employers' factory and foundry in 1803. By 1804, Murdock had installed 900 gaslights in cotton mills. He became known as the father of the gas industry. The work of Murdock and other experimenters interested Frederick Albert Winsor, a German businessman. Winsor decided to manufacture gas on a large scale. He learned the process from Murdock and obtained a British patent for manufacturing gas in 1804. In 1807, Winsor and his partners staged the first public street lighting with gas—along London's Pall Mall. They formed the first gas company in 1812. The first gas company in the United States was established in 1817 in Baltimore to light that city's streets.

Development of the natural gas industry began

Natural gas in the United States

This graph shows proved reserves and marketed production of natural gas in the United States since the early 1900's. Reserves rose rapidly from the 1920's to the 1960's as new deposits were found. But since about 1970, few deposits have been found and reserves have fallen. The production of gas in the United States grew rapidly from the mid-1950's to the mid-1970's.

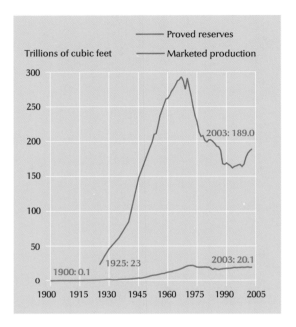

in the United States. The earliest known discoveries of natural gas in the country occurred in 1775. That year, French missionaries in the Ohio Valley reported seeing "pillars of fire," which probably were caused by seeping gas accidentally set on fire. Also in 1775, the American colonial leader George Washington saw a "burning spring"—flames rising from water—near what is now Charleston, West Virginia.

In 1821, mysterious bubbles appeared in a well be-

Leading natural gas producing countries

Marketed production of gas in a year

United States	●●●●●●●●●●●●●●●●
	20,630,000,000,000 ft³ (584,180,000,000 m³)
Russia	●●●●●●●●●●●●●●●●
	20,511,000,000,000 ft³ (580,810,000,000 m³)
Canada	●●●●◖
	7,224,000,000,000 ft³ (204,560,000,000 m³)
United Kingdom	●●◖
	3,928,000,000,000 ft³ (111,230,000,000 m³)
Algeria	●●
	2,972,000,000,000 ft³ (84,160,000,000 m³)
Netherlands	●◖
	2,747,000,000,000 ft³ (77,790,000,000 m³)

Figures are for 2001.
Source: U.S. Energy Information Administration.

Leading natural gas producing states and provinces

Marketed production of gas in a year

Texas*	●●●●●●●●●●●●●●
	6,442,000,000,000 ft³ (182,431,000,000 m³)
Louisiana*	●●●●●●●●●●●◖
	5,249,000,000,000 ft³ (148,625,000,000 m³)
Alberta	●●●●●●●●●◖
	4,685,000,000,000 ft³ (132,676,000,000 m³)
New Mexico	●●●◖
	1,689,000,000,000 ft³ (47,831,000,000 m³)
Oklahoma	●●●◖
	1,615,000,000,000 ft³ (45,743,000,000 m³)
Wyoming	●●◖
	1,364,00,000,000 ft³ (38,621,000,000 m³)

*Includes federal offshore production.
Figures are for 2001. Sources: U.S. Energy Information Administration; Statistics Canada.

World production and consumption of natural gas

This graph shows the amount of natural gas produced and used in various regions of the world. Most regions produce about the same amount of natural gas that they use. Africa produces nearly two times the natural gas that it uses.

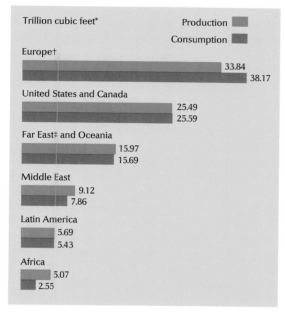

Trillion cubic feet*

Production
Consumption

Europe†
33.84
38.17

United States and Canada
25.49
25.59

Far East‡ and Oceania
15.97
15.69

Middle East
9.12
7.86

Latin America
5.69
5.43

Africa
5.07
2.55

*One cubic foot equals 0.028 cubic meter.
†Includes both European and Asian parts of Russia.
‡Excludes Russia.
Figures are for 2003. Source: U.S. Energy Information Administration.

ing drilled for water at Fredonia, New York. The driller gave up his efforts. Soon afterward, on the same site, a gunsmith named William Aaron Hart completed the first natural gas well in the United States. It was 27 feet (8 meters) deep. Hart piped the gas to nearby buildings, and it was burned for lighting. Another shallow natural gas well was drilled near Westfield, New York, in 1826.

The first company known to have distributed natural gas was formed in Fredonia in 1865. By then, about 300 U.S. companies were distributing manufactured gas. Oil was discovered near Titusville, Pennsylvania, in 1859, and natural gas development was nearly forgotten in the oil rush that followed. The gas found in the oil fields lacked both markets and pipeline systems.

The first "long-distance" pipeline was completed in 1872. This 25-mile (40-kilometer) wooden pipeline carried natural gas to consumers in Rochester, New York. Also in 1872, the first iron pipeline for natural gas began carrying the fuel 5 ½ miles (9 kilometers) to Titusville. This pipeline delivered 4 million cubic feet (110,000 cubic meters) of gas daily to about 250 consumers.

In 1879, the American inventor Thomas A. Edison developed the first practical incandescent lamp. This development and the introduction of electric lighting nearly destroyed the gas industry. However, the industry started to grow again as more and more consumers turned to manufactured gas for cooking and water heating. Meanwhile, natural gas development re-

mained at a standstill. However, during the early 1900's, huge gas reserves were discovered in Texas, Louisiana, and Oklahoma. From 1906 to 1920, natural gas production in the United States more than doubled, to 800 billion cubic feet (23 billion cubic meters) a year. The lack of long-distance pipelines held back further growth in the industry. By 1925, there were about 3 ½ million natural gas consumers, but all were within a few hundred miles or kilometers of the gas fields.

The natural gas industry began to expand rapidly in the late 1920's with the introduction of seamless, electrically welded steel pipe. This pipe was stronger than earlier pipe. It could carry gas under higher pressures and, therefore, in greater quantities. With the new pipe, companies could profitably build lines over 1,000 miles (1,600 kilometers) long. The earliest of such lines delivered gas from Texas fields to cities in the midwestern United States.

Until the 1960's, large quantities of natural gas were not available in many industrialized countries, and manufactured gas was used widely. In the 1960's, the development of newly discovered gas fields led to the rapid expansion of Europe's natural gas industry. This expansion was especially fast in the Soviet Union and the Netherlands. The world's largest known gas field was discovered in the Soviet Union in 1966. The United Kingdom began to produce much natural gas from deposits found under the North Sea in the mid-1960's.

The modern gas industry. Modern methods of exploring for natural gas have led to the greatest world supply of gas in history. A number of nations that had depended mostly or entirely on manufactured gas have discovered large deposits of natural gas and are switching to the cheaper fuel. Producing countries are exporting an increasing amount of natural gas by new pipelines or in liquid form by tanker ships.

The largest known natural gas reserves are in the Middle East and Russia. The United States and Russia lead the nations, by far, in natural gas production. Canada ranks third in yearly production, followed by the United Kingdom and Algeria. The Netherlands, a leading European producer, exports much of its gas to Germany, France, and Belgium. Texas and Louisiana account for more than half of all production in the United States. Alberta leads the Canadian provinces in gas production.

The gas industry is developing more efficient ways to use natural gas. See the *Gas in industry* section of this article for information on these advances. A device under development is a *gas fuel cell*. It produces electric power chemically, using methane from natural gas.

The problem of air pollution has created interest in natural gas as a transportation fuel. Natural gas produces lower amounts of dangerous emissions when burned than gasoline and diesel fuel do. Gas is being used, mostly on an experimental basis, to power some automobiles, trucks, and ships.

In some regions, natural gas consumption exceeds the amount made available from the discovery of new reserves. Some people fear that the demand for gas might soon exceed the world's supply. The gas industry has been exploring for additional sources of natural gas. It has also sought new and better ways of producing clean-burning gas from coal.   Michael A. Adewumi

Jupiter's colorful appearance comes from bands of clouds in the planet's lower atmosphere. The large oval-shaped mark, called the Great Red Spot, is a giant cloud of swirling gas.

Jupiter is the largest planet in our solar system. It has a *mass* (amount of matter) that is greater than the masses of all the other planets in the solar system added together. Astronomers call Jupiter a *gas giant* because the planet consists mostly of hydrogen and helium, and has no solid surface. Jupiter is named for the king of the gods in Roman mythology.

Jupiter ranks among the brightest objects in the night sky. Usually, only the moon and Venus appear brighter. When seen with the unaided eye, Jupiter has a pale orange color. Viewed through a telescope or in images taken by spacecraft, Jupiter appears as a globe covered with swirling, brightly colored clouds of brown, orange, white, and yellow.

Jupiter lies at the center of a system of cosmic objects so vast and diverse that it resembles a miniature solar system. The planet has 16 moons that measure at least 6 miles (10 kilometers) in diameter and dozens of smaller satellites. Four faint rings of dust particles encircle the planet. Jupiter also has a strong *magnetic field*. A magnetic field is the area around a magnet in which its influence can be detected. Jupiter's magnetic field extends beyond the planet throughout a huge region of space called the *magnetosphere*. Astronomers sometimes refer to the planet together with its rings, satellites, and magnetosphere as the *Jovian system*.

Roger V. Yelle, the contributor of this article, is Professor of Planetary Science at the University of Arizona.

Astronomers have made detailed observations of Jupiter for centuries. It was one of the first planets studied by the famous Italian astronomer Galileo in the early 1600's. Beginning in the 1970's, several spacecraft have explored the Jovian system in great detail.

Characteristics of Jupiter

Astronomers determine Jupiter's characteristics using observations made through telescopes and by spacecraft. They study these characteristics to learn about the planet's structure and origin.

Orbit and rotation. Like all the planets in our solar system, Jupiter travels around the sun in an *elliptical* (oval-shaped) orbit. Jupiter's orbit is nearly circular. It lies about five times as far from the sun as Earth's orbit does. The average distance between the sun and Jupiter measures around 484 million miles (779 million kilometers). Jupiter's orbit is tilted by 1.3 degrees from the *ecliptic plane,* the imaginary plane that contains Earth's orbit. Jupiter takes nearly 12 Earth years to orbit the sun.

Jupiter rotates faster than any other planet, taking about 9 hours 55 minutes to turn completely on its axis. This is the length of a day on Jupiter. The planet's rapid spinning causes it to bulge slightly at the equator. Its diameter at the equator measures 88,846 miles (142,984 kilometers), while the distance between its *geographic poles*—the ends of its axis—measures only 83,082 miles (133,708 kilometers).

Like Earth's axis, Jupiter's axis is not perpendicular to the planet's *orbital plane,* the imaginary plane that contains its orbit. Jupiter's axis lies tilted from the perpendicular by 3.1 degrees. As a result of this tilt and its motion about the sun, Jupiter, like Earth, has seasons.

Mass and density. Jupiter's mass is about 318 times as great as the mass of Earth, but about 1,000 times smaller than that of the sun. Jupiter's average density is about 1.3 times as great as the density of water at room temperature. This density is much lower than the density of Earth. Jupiter's low density indicates that it is composed mostly of light elements rather than rock.

Chemical composition. Jupiter's elemental composition resembles that of the sun. The planet consists mostly of hydrogen (chemical symbol, H) and helium (He). It also contains small amounts of heavier elements, including oxygen (O), carbon (C), nitrogen (N), sulfur (S), and many others. In general, Jupiter has a higher concentration of heavy elements than does the sun.

Most of the elements in Jupiter's atmosphere consist of atoms linked together in molecules. Molecules that have been detected in Jupiter's atmosphere include molecular hydrogen (H_2), water (H_2O), ammonia (NH_3), methane (CH_4), and hydrogen sulfide (H_2S). Smaller amounts of other molecules form in chemical reactions in the atmosphere. These include ethane (C_2H_6), acetylene (C_2H_2), ethylene (C_2H_4), hydrogen cyanide (HCN), and other compounds. Helium exists as individual atoms in Jupiter's atmosphere.

Temperature. Because of its great distance from the sun, Jupiter is much colder than Earth. To determine a planet's temperature, astronomers measure the energy that the planet *radiates* (gives off in the form of heat). The amount of energy delivered in a certain peri-

Jupiter at a glance

The orbit of Jupiter lies between those of Mars and Saturn. The ancient symbol for Jupiter, *right,* represents a lightning bolt.

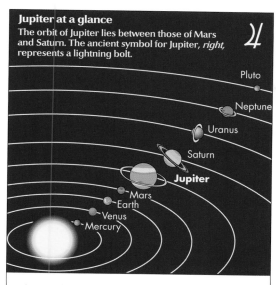

Distance from the sun:
  Shortest—460,140,000 mi (740,520,000 km);
  Greatest—507,420,000 mi (816,620,000 km);
  Average—483,780,000 mi (778,570,000 km).
Distance from Earth:
  Shortest—365,700,000 mi (588,500,000 km);
  Greatest—601,500,000 mi (968,100,000 km).
Diameter at equator: 88,846 mi (142,984 km).
Length of year: 12 Earth years.
Rotation period: 9 hours 55 minutes.
Temperature: –236 °F (–149 °C).
Atmosphere: Mostly hydrogen and helium; some methane, ammonia, water vapor, hydrogen sulfide, and traces of other gases.
Number of satellites: At least 63.

Major satellites of Jupiter*

Name	Mean distance from Jupiter In miles	In kilometers	Diameter of satellite In miles	In kilometers	Year of discovery
Metis	79,540	128,000	27	43	1979
Adrastea	80,160	129,000	10	16	1979
Amalthea	112,700	181,400	104	167	1892
Thebe	137,900	221,900	62	99	1980
Io	262,100	421,800	2,264	3,643	1610
Europa	417,000	671,100	1,940	3,122	1610
Ganymede	664,900	1,070,000	3,270	5,262	1610
Callisto	1,170,000	1,883,000	2,996	4,821	1610
Leda	6,941,000	11,170,000	6	10	1974
Himalia	7,121,000	11,460,000	106	170	1904
Lysithea	7,282,000	11,720,000	22	36	1938
Elara	7,295,000	11,740,000	53	86	1905
Ananke	13,220,000	21,280,000	17	28	1951
Carme	14,540,000	23,400,000	29	46	1938
Pasiphae	14,680,000	23,620,000	37	60	1908
Sinope	14,880,000	23,940,000	24	38	1914

*Satellites smaller than 6 miles (10 kilometers) in diameter are not listed.

od is called *power.* The power that Earth radiates equals the power it absorbs from the sun. Jupiter, however, radiates about twice as much power as it absorbs from the sun. This indicates that some of Jupiter's energy comes from a source other than the sun. The energy may be heat left over from Jupiter's formation. It might also come from heat created as the planet slowly contracts under the influence of gravity. Although Jupiter gives off some of its own energy, astronomers do not consider it a star because no nuclear reactions occur in its interior.

By measuring the total power a planet radiates, astronomers determine its *effective temperature.* The effective temperature of Jupiter is –236 °F (–149 °C), about 234 Fahrenheit degrees (130 Celsius degrees) lower than the effective temperature of Earth. Jupiter's effective temperature corresponds roughly to the temperature of its atmosphere about 30 miles (50 kilometers) above the cloud tops. The actual temperature of the planet varies with altitude and location.

Magnetic field. Jupiter's magnetic field is mainly *dipolar*—that is, Jupiter has a magnetic north and south pole like the poles on a bar magnet. Physicists describe the overall strength of such a field using a measure called the *magnetic dipole moment.* Jupiter's magnetic dipole moment measures about 20,000 times as strong

as that of Earth. Like Earth's magnetic field, Jupiter's field lies tilted by about 10 degrees from the planet's axis of rotation. Jupiter's magnetic poles are aligned opposite those of Earth. A compass needle that points north on Earth would point south on Jupiter.

Jupiter's magnetic field traps electrically charged particles, such as electrons and *ions* (charged atoms or groups of atoms). As a result, the magnetosphere contains a hot, low-density *plasma,* a form of matter made up of charged particles. The plasma is concentrated in a thin disk near the planet's equator. It comes from Jupiter's moons, especially Io. Io has active volcanoes that eject much sulfur and oxygen into the magnetosphere. The hot plasma can damage the optics and electronics of spacecraft operating in the Jovian system. The magnetosphere also deflects the *solar wind* around the Jovian system. The solar wind is a continuous flow of charged particles from the sun.

Telescopes on Earth can detect the glowing ions in Jupiter's magnetosphere. The ions also produce visible effects when they enter Jupiter's atmosphere near the poles. There, collisions between the charged particles and the atmosphere create bands and streamers of light called *auroras.* Jupiter's auroras glow brighter than those of any other planet in the solar system.

Radio emissions. Astronomers discovered Jupiter's magnetic field in 1955 when they detected radio waves *emitted* (given off) by the planet. The radio emissions result from the movement of electrons in Jupiter's magnetic field. Some electrons travel through the field in a spiral path at high speeds. Electrons that move in this way emit radio waves in a process called *synchrotron radiation.* Observations show that synchrotron radiation creates some of Jupiter's radio emissions. Other emissions result from electrons moving between Io and Jupiter and from electrons moving within the atoms of Jupiter's atmosphere.

Jupiter's radio emissions vary in strength in a pattern that repeats about every 9 hours 55 minutes. Astronomers think the planet's magnetic field takes this long to complete one rotation. They also use this value for Jupiter's rotation period because the planet lacks solid features that can be used to measure its rotation.

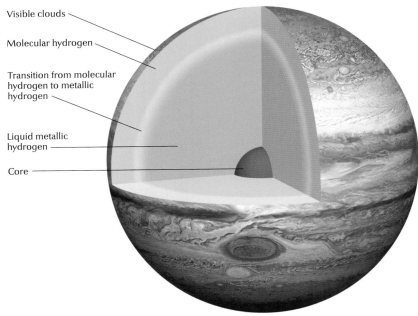

Visible clouds

Molecular hydrogen

Transition from molecular hydrogen to metallic hydrogen

Liquid metallic hydrogen

Core

Jupiter's interior consists primarily of the element hydrogen. Beneath the planet's clouds, molecular hydrogen (H_2) blends gradually from a gas to a liquid. At a depth of about 6,000 miles (10,000 kilometers), the pressure begins to break down hydrogen atoms. Their separated nuclei and electrons form a state called *liquid metallic hydrogen* that makes up most of the planet's mass. Most scientific models suggest that Jupiter has a dense core composed of heavier elements, such as carbon, iron, nitrogen, oxygen, and silicon.

WORLD BOOK illustration by Precision Graphics

Structure of Jupiter

Astronomers know more about Jupiter's atmosphere than they do about the planet's interior because the atmosphere is the part of the planet that we can see. The pressure of the atmosphere grows steadily as altitude decreases until the atmosphere merges gradually into the interior. Astronomers use observations of the planet and their knowledge of chemistry and physics to determine what its interior might be like.

Atmosphere. Temperatures in the uppermost parts of Jupiter's atmosphere measure about 1500 °F (800 °C). Throughout the upper atmosphere, temperatures drop as altitude decreases. Temperatures reach their lowest in a region where the atmospheric pressure equals about one-fifth the pressure at the surface of Earth. This area, called the *tropopause,* separates Jupiter's upper atmosphere from its lower atmosphere. Below the tropopause, temperatures begin to increase approaching the planet's interior.

Jupiter's colorfully swirled appearance comes from clouds in its lower atmosphere. Different compounds there condense to form clouds at different altitudes, creating layers of clouds with various chemical compositions. The uppermost clouds consist primarily of ammonia ice. These clouds make up most of what we see when we look at Jupiter. Pure ammonia ice is colorless. The clouds' colors result from tiny amounts of impurities called *chromophores.* Astronomers do not know for sure what chromophores are. They may include *organic* (carbon-based) molecules, sulfur compounds, or phosphorous compounds.

Astronomers think that below the ammonia clouds there is a second cloud layer made up of ammonium hydrosulfide, which forms when ammonia and hydrogen sulfide condense together. A cloud layer com-

posed of water ice may lie deeper still. At greater pressures, far beyond where light can penetrate, the atmosphere may contain clouds of iron or *silicates,* compounds of metals, silicon, and oxygen that form rocks on Earth.

Zones and belts. Alternating bands of light and dark clouds cover Jupiter's atmosphere. Astronomers refer to the wider, brighter bands as *zones* and to the darker, narrower bands as *belts.* The zones and belts result from wind patterns in Jupiter's lower atmosphere.

In the zones, winds blow from the west at speeds that reach up to 400 miles (650 kilometers) per hour near the equator. Winds in the belts blow from the east at slightly lower speeds. The zones appear bright because they contain high-altitude clouds that reflect much sunlight. The clouds in the belts lie at somewhat lower altitudes. Their darker appearance probably results from a higher concentration of chromophores.

Jupiter's alternating east and west winds result from *convection currents.* These currents are movements of the atmosphere created by the rising of warm gases and the falling of cooler gases. Because Jupiter's internal energy heats the atmosphere unevenly, warm gases rise in certain places and cool gases descend in others. This rise and fall creates convection currents.

Jupiter's rapid rotation bends the convection currents into patterns that stretch east and west around the entire planet. This effect of rotation, known as the *Coriolis effect,* also creates wind patterns on Earth. On Earth, the patterns vary because oceans, continents, and mountain ranges interfere with the circulation of the atmosphere. On Jupiter, which lacks a solid surface or other obstacles, the east and west winds remain remarkably stable. As a result, while individual bands continuously undergo small changes, the overall pattern of zones and belts has remained unchanged in the hundreds of years since people first observed Jupiter through telescopes.

Ovals. Jupiter's atmosphere displays many features that are oval or circular in shape. The most prominent of these is a vast, reddish oval called the Great Red Spot.

The Great Red Spot extends about 7,450 miles (12,000 kilometers) from north to south. The spot's width from east to west, which is slowly shrinking, measured about 10,500 miles (17,000 kilometers) in the early 2000's. The spot's circumference is larger than that of Earth. The spot travels around Jupiter with the wind at about 22 degrees of latitude south of the equator. As with the other clouds on Jupiter, astronomers do not know exactly what causes the spot's reddish color.

The English scientist Robert Hooke first observed a large spot in Jupiter's atmosphere in 1664. Astronomers first recorded the Great Red Spot's precise form and position in 1831. Since then, the spot has remained near the same latitude. Images taken by the two Voyager spacecraft in 1979 revealed that the spot is a swirling cloud of gas that takes about seven days to complete one full rotation. Wind speeds at the outer edges of the spot reach up to 425 miles (685 kilometers) per hour.

Unlike the winds in hurricanes on Earth, which swirl around a low-pressure region, winds in the Great Red Spot and in other long-lived ovals in Jupiter's atmosphere swirl around areas of high pressure. Scientists call these weather systems *anticyclones.*

Other oval features in Jupiter's atmosphere include *white ovals.* White ovals are much smaller and less stable than the Great Red Spot. They tend to lie on the edges of zones but frequently move around the planet.

Weather. In addition to longer-lived atmospheric patterns, such as zones, belts, and ovals, Jupiter has storms and other active weather. Astronomers first saw flashes of lightning on Jupiter's night side in images from the Voyager spacecraft. Later, the Galileo craft observed numerous lightning flashes and tracked the movement of clouds from night to day. Astronomers determined that the lightning flashes originate in small cloud plumes that resemble thunderheads on Earth. The lightning flashes on Jupiter are much more powerful than those on Earth.

Interior. Below the clouds, Jupiter's pressure, temperature, and density increase until the atmosphere gradually blends into the fluid interior. Eventually, the hydrogen and helium that make up most of the planet become more like a liquid than a gas.

About 6,000 miles (10,000 kilometers) below the clouds, the pressure becomes 1 million times as great as the atmospheric pressure at Earth's surface. At this depth, hydrogen atoms begin to break down, with the electrons becoming separated from their nuclei. The separated nuclei and electrons compose an unusual form of hydrogen called *liquid metallic hydrogen* that can conduct electric current like an ordinary metal. Liquid metallic hydrogen makes up most of Jupiter's mass. Scientists believe that electric currents flowing through this hydrogen generate the planet's magnetic field.

Core. The region near Jupiter's center is difficult to probe. The pressure there equals about 70 million times the pressure at Earth's surface. Astronomers estimate the temperature of Jupiter's center to be around 45,000 °F (25,000 °C).

Most scientific models suggest that Jupiter has a dense core made up of substances that, under less severe conditions, would form rock and ice. They estimate the mass of the core to be about 10 to 15 times the mass of Earth. The rock-forming material in the core may include iron and silicates. The ice-forming material may include oxygen, carbon, and nitrogen. Other models indicate that Jupiter has no distinct core. Instead, they suggest that liquid metallic hydrogen merges gradually with heavier elements near the planet's center.

Satellites and rings

Astronomers have identified at least 63 satellites of Jupiter, but the planet has more small moons that have yet to be discovered. Jupiter's satellites can be divided into three groups: (1) the Galilean satellites; (2) the inner satellites; and (3) the outer satellites. The inner satellites are closely related to Jupiter's system of rings.

Galilean satellites. Astronomers call Jupiter's four largest moons the *Galilean satellites* because Galileo discovered them. In order of increasing distance from the planet, they are Io, Europa, Ganymede, and Callisto.

Io, the innermost Galilean satellite, ranks as the most geologically active body in the solar system. It has many volcanoes that frequently erupt sulfur dioxide gas. Most of the gas condenses and falls back to Io as ice, but some sulfur and oxygen ions escape into Jupiter's magnetosphere. Io's eruptions result from the gravitational pulls of Jupiter, Europa, and Ganymede. These forces pull Io in different directions, squeezing the moon's interior and causing it to heat up.

Europa also shows evidence of geological activity caused by the same forces that squeeze Io. Europa's icy surface features a broken network of *fissures* (narrow cracks). The fissures may indicate that the moon's icy crust rests atop deep oceans of liquid water or slushy water ice.

Ganymede ranks as Jupiter's largest moon and the largest moon in the solar system. Ganymede is also the only moon in the solar system known to generate its own magnetic field. Its surface has large light and dark regions. The dark regions contain more impact craters than the light regions do. The light regions typically exhibit many grooves and ridges.

Callisto, the outermost Galilean satellite, is one of the most heavily cratered bodies in the solar system. Impact craters uniformly cover its surface.

Inner satellites. The moons with orbits that lie inside those of the Galilean satellites are known as Jupiter's *inner satellites.* In order of increasing distance from the planet, they are Metis, Adrastea, Amalthea, and Thebe.

Compared to the large, round Galilean satellites, the inner satellites are small and irregular in shape. They range from 10 to 104 miles (16 to 167 kilometers) in average diameter. Amalthea is the largest, followed by Thebe, Metis, and Adrastea. Their surfaces all appear dark and red and have many craters from collisions. Astronomers have measured the average density of

Amalthea, which is about the same as the density of water on Earth's surface. Amalthea must be remarkably *porous* (filled with tiny holes) to have such a low density.

Outer satellites. Jupiter also has a large number of small, irregular satellites orbiting well beyond the Galilean satellites. Astronomers have discovered dozens of these *outer satellites,* but their actual number is probably higher. Jupiter's gravitation can also trap other bodies for a time, making them *temporary satellites.*

Himalia ranks as the largest of Jupiter's outer satellites, followed by Elara, Pasiphae, Carme, Sinope, Lysithea, Ananke, and Leda. Ananke, Carme, Pasiphae, and Sinope have *retrograde* orbits—that is, they orbit in a direction opposite to that of the other satellites and the direction of Jupiter's rotation. Elara, Himalia, Leda, and Lysithea orbit in the same direction that Jupiter rotates. Unlike the orbits of the inner satellites and the Galilean satellites, the orbits of some outer satellites are tilted by many degrees from Jupiter's *equatorial plane,* the imaginary plane that contains the planet's equator.

Rings. Jupiter's four rings consist of fine dust particles, all circling the planet on individual orbits. The rings all lie close to the planet's equator and are sometimes mistaken for a single ring. Compared to Saturn's rings, Jupiter's rings are smaller and fainter, and contain much less mass. In fact, astronomers could not confirm that Jupiter's rings existed until the two Voyager spacecraft observed them close-up in 1979. Astronomers call the brightest ring the *main ring.* Its outer edge corresponds to the orbit of Adrastea. A fainter ring called the *halo ring* lies inside the orbit of Metis. Two faint rings called the *gossamer rings* lie outside the main ring. Their outer edges correspond to the orbits of Amalthea and Thebe.

Astronomers think that the rings result from collisions between the inner satellites and tiny particles called *micrometeoroids.* These collisions eject some dust into the space around the moons. The dust particles orbit Jupiter as they fall toward the planet, forming the rings.

Formation of Jupiter

As the largest planet in our solar system, Jupiter plays a central role in our ideas about how the solar system formed. Jupiter and the sun share a similar composition. They likely formed at the same time from the *solar nebula.* The solar nebula was a giant rotating cloud of gas and dust.

The solar system began to take shape as the solar nebula collapsed under the influence of gravity. As the nebula contracted, its central region heated up while the outer regions remained cool. Around what is now Jupiter's orbit, temperatures became cold enough for water vapor to freeze into ice crystals.

According to the most widely held theory, ice and other solid material slowly gathered together to form what is now the core of Jupiter. The core grew as it attracted more material from nearby regions. As the core gained mass, its gravitational pull became stronger. Eventually, the core's gravitational pull became strong enough to capture hydrogen and helium, which were abundant in the solar nebula. For this reason, Jupiter today consists mostly of hydrogen and helium.

Astronomers use the term *accretion* to refer to the process by which tiny particles accumulate to form giant planets. When two particles collide, they may stick together to form a larger particle. As the process continues, larger and larger bodies collide. This process happens in different locations, creating many large objects called *planetesimals.* Eventually, many of the planetesimals near what is now Jupiter's orbit combined to form the planet. Some of the other planetesimals may have formed some of Jupiter's satellites, while still others may have escaped to great distances and become comets.

NASA/Cornell University

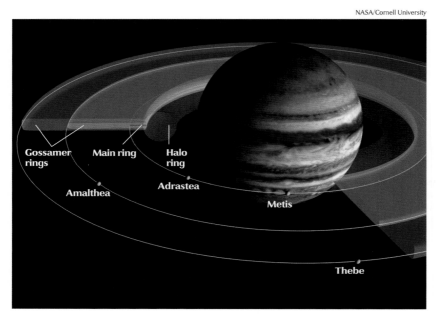

Gossamer rings

Main ring

Halo ring

Amalthea

Adrastea

Metis

Thebe

Jupiter's rings consist of dust circling the planet near the orbits of its inner satellites. The outer edge of the bright *main ring* lies along the orbit of Adrastea. The *halo ring* lies inside Metis's orbit, and the *gossamer rings* stretch to the orbits of Amalthea and Thebe. Astronomers think the dust comes from collisions between these moons and particles called *micrometeoroids.*

History of Jupiter study

Jupiter was known to the ancient astronomers, who tracked the planet's motion across the night sky. Astronomers first studied Jupiter through telescopes in the early 1600's. In 1610, Galileo discovered what later became known as the Galilean satellites. At the time, many people believed that every cosmic body revolved around Earth. The discovery of moons orbiting another planet helped convince Galileo and others that Earth was not at the center of the universe.

In the late 1960's, astronomers chose Jupiter as the target of the first spacecraft mission to the outer solar system. They launched the craft, called Pioneer 10, in 1972. On Dec. 3, 1973, Pioneer 10 became the first spacecraft to visit Jupiter when it passed within about 81,000 miles (130,000 kilometers) of the planet's cloud tops. Pioneer 11 (later renamed Pioneer-Saturn) flew by Jupiter on Dec. 2, 1974. The two Pioneer spacecraft captured images of Jupiter and its moons. The spacecraft sent back data on the planet's gravitational pull, magnetic field, radiation belts, atmosphere, and the plasma in its magnetosphere.

Astronomers used data from the Pioneer missions to design the more ambitious Voyager missions. The spacecraft Voyager 1 flew by Jupiter at a distance of about 174,000 miles (280,000 kilometers) on March 5, 1979. Voyager 2 passed within about 449,000 miles (722,000 kilometers) of the planet on July 9, 1979. The Voyager missions revolutionized our understanding of Jupiter. They discovered Jupiter's intense aurora and numerous features of the planet's atmosphere. They also confirmed the existence of Jupiter's rings. The two Voyager craft captured the first close-up views of many of the planet's moons, revealing them to be much more active and varied than astronomers had expected.

The Ulysses spacecraft, designed primarily to study the sun, passed within about 235,000 miles (378,000 kilometers) of Jupiter on Feb. 8, 1992. Its instruments measured Jupiter's radio emissions as well as the plasma, dust, and other particles in the Jovian system. Ulysses gathered more data on Jupiter's magnetosphere when it revisited the planet from late 2003 to early 2004.

In 1993, the American astronomers Carolyn and Eugene Shoemaker and the Canadian-born astronomer David H. Levy discovered a comet passing near Jupiter. Astronomers soon realized that the comet, called Shoemaker-Levy 9, had been captured by Jupiter's gravity and broken into 21 or more fragments. Astronomers from around the world watched as most of the fragments collided with Jupiter over a period of several days in July 1994.

The Galileo spacecraft became the first craft to orbit Jupiter when it arrived at the planet on Dec. 7, 1995. The same day, a probe Galileo had released months earlier entered Jupiter's atmosphere. The probe made the first precise measurements of the atmosphere's helium, ammonia, and many other substances. It also recorded the speed of the winds below the cloud tops. The probe surprisingly detected few clouds and little water vapor, but astronomers think that the probe entered an area that was not typical of the rest of the atmosphere.

The main Galileo spacecraft continued to orbit Jupiter for eight years, making numerous important discoveries about the Jovian system. It detected massive thunderstorms in Jupiter's atmosphere. It discovered that Ganymede has a magnetic field and found evidence that an ocean of water or soft ice lies beneath Europa's surface. Galileo also studied volcanoes on Io, determining that their eruptions were powered by hot, magnesium-rich silicates from deep in Io's interior.

The Cassini spacecraft, designed to study Saturn, flew by Jupiter at a distance of nearly 6 million miles (10 million kilometers) from the planet's cloud tops in December 2000. The Galileo spacecraft was still in orbit around Jupiter at that time, allowing astronomers to conduct coordinated observations of the planet from two different locations. The data gathered were used to study Jupiter's moons, its magnetosphere, and the weather systems in its atmosphere.   Roger V. Yelle

Stem cell is a cell that has the ability to develop into any of the different cell types that make up the tissues and organs of the body. The original cells from which an entire organism develops are stem cells. These cells are also found in adult organs. Stem cells have the ability to divide endlessly, producing more stem cells or other types of cells.

In 1998, scientists succeeded in isolating and growing stem cells from a human embryo in a laboratory. Such stem cells are called *embryonic stem cells.* Scientists think that stem cells can be used to replace damaged tissues and treat diseases, such as Parkinson disease and diabetes, in people.

Early in development, a human embryo is made up of a hollow ball of cells called a *blastocyst.* Blastocyst cells divide and eventually develop into all of the tissues and organs of a human being, a process called *differentiation.*

Embryonic stem cells can be grown in the laboratory from blastocysts and made to differentiate into nerve, liver, muscle, blood, and other cells. Scientists hope to control the differentiation of the cells to replace cells in diseased or damaged organs in human beings. Learning how to control the differentiation of stem cells will help scientists understand how human tissues and organs develop. It may also lead to new treatments for many diseases, such as cancer. Embryonic stem cells can also be used to test the effects of new drugs without harming animals or people.

In adults, stem cells are found in many places in the body, including the skin, liver, bone marrow, and muscles. In these organs, stem cells remain inactive until they are needed. The stem cells supply each organ with the cells needed to replace damaged or dead cells. Some stem cells in the bone marrow may produce new bone and cartilage cells when needed.

Other bone marrow stem cells divide to produce more stem cells, additional cells called *precursor cells,* and all of the different cells that make up the blood and immune system. Precursor cells have the ability to form many different types of cells, but they cannot produce more stem cells. Scientists can isolate bone marrow stem cells to use as donor cells in transplants. Adult stem cells, however, are rare and more difficult to detect and isolate. Scientists have also found that damaged stem cells may play a role in the development of certain cancers.

The discovery and isolation of embryonic stem cells has led to debate over whether it is right to use cells taken from human embryos for research. People have expressed concern about using human embryos and collecting some of their cells. Some people consider embryos already to be human beings. The embryos are destroyed in the process of isolating the stem cells. Many people consider it wrong to destroy human embryos, but other people believe that the potential medical benefits of stem cells justify their use.

By 2004, scientists in the United Kingdom and South Korea began producing cloned human embryos to produce stem cells. This process first involves destroying the nucleus of a human egg cell. A nucleus is then removed from another human cell and injected into the egg cell. The egg, with its new nucleus, develops into an embryo with the same genetic makeup as the donor. In 2005, scientists in South Korea announced that they were able to clone human embryos to produce a collection of stem cells that were genetically identical to several sick and injured patients. This kind of cloning is known as *therapeutic cloning* because doctors might be able to use the stem cells to replace damaged human tissues and treat diseases.

In the United States, the National Institutes of Health (NIH) sets the standards for medical research that can receive federal funding. The NIH forbids laboratories that receive federal funding from isolating human embryonic stem cells. In 2001, President George W. Bush allowed federal funds to support scientific research on existing supplies of stem cells that had been isolated previously in privately funded laboratories. A law passed by the California Legislature in 2002 allows laboratories that received funding from that state to isolate human embryonic stem cells.

In 2004, California voters approved the use of state funds for stem cell research. Other states, including Massachusetts and New Jersey, have passed similar laws. Studies of new embryonic stem cells remain ineligible for federal funds.   Meri T. Firpo

Tsunami, *tsoo NAH mee,* is a series of powerful ocean waves produced by an earthquake, landslide, volcanic eruption, or asteroid impact. Tsunami waves can travel great distances and still retain much of their strength. They differ from common ocean waves, which are caused by wind. The word *tsunami* is a combination of Japanese words meaning *harbor* and *wave.*

Tsunami waves are much longer than common ocean waves. In the open ocean, the water may take from 5 minutes to over 1 hour to reach its highest level and fall back again as a tsunami wave passes. By contrast, a common ocean wave causes the water level to rise and fall in 5 to 20 seconds. Tsunami waves in the open ocean usually raise and lower the water level by 3 feet (1 meter) or less. Because the change happens gradually, tsunamis frequently go undetected by ships.

The deeper the water is, the faster a tsunami wave travels. In the Pacific Ocean basin, where depths average about 13,000 feet (4,000 meters), tsunami waves can travel up to 600 miles (970 kilometers) per hour, as fast as a jet aircraft. As a tsunami wave approaches land, its speed drops to about 20 to 30 miles (30 to 50 kilometers) per hour. As the wave's speed decreases, its height usually grows by at least three times. The resulting flood of water can surge more than ⅔ mile (1 kilometer) inland and pile up in certain places to reach elevations higher than 100 feet (30 meters) above sea level.

Scientists strive to predict tsunamis so that endangered coastal areas can be evacuated. One method uses devices called *seismographs* to measure *seismic waves* (waves of vibration generated by earthquakes). By analyzing seismic waves, scientists can determine when and where an undersea earthquake has occurred and calculate its strength. As a result, they can then estimate the size of a possible tsunami and the time its waves will reach land. Seismic waves travel through the ground much faster than tsunami waves travel through the water. For this reason, scientists can sometimes warn people several hours before tsunami waves strike.

Other forecasting methods use *pressure sensors* placed on the ocean floor. When a sensor detects pressure from a large tsunami, it relays the information to a buoy. The buoy then transmits the data to a warning center. Scientists are also developing methods to detect tsunamis using radar signals from satellites.

In December 2004, an enormous undersea earthquake west of the Indonesian island of Sumatra created a tsunami that pounded low-lying coastlines throughout the Indian Ocean, killing more than 175,000 people and causing billions of dollars in property damage. The greatest number of deaths occurred in Indonesia, Sri Lanka, India, and Thailand. Deaths were reported as far as away as Somalia, about 3,000 miles (4,800 kilometers) from the tsunami's origin.   Steven Ward

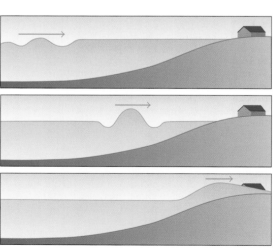

WORLD BOOK illustration

A tsunami wave on the open ocean is long and low, *top.* As the wave enters shallow water, it slows, causing its height to grow dramatically, *middle.* The wave then surges onto land, *bottom.*

A volcanic eruption at Kilauea, in Hawaii, shoots fountains of glowing lava into the air. As a volcano erupts, ash, gases, and molten rock from deep underground pour from an opening called a *vent.* Erupted rock and ash can build up around the vent to form the body of the volcano.

Volcano

Volcano is a place where ash, gases, and molten rock from deep underground erupt onto the surface. The word *volcano* also refers to the mountain of erupted rock and ash that often accumulates at such a place.

Volcanic eruptions result from *magma* (molten rock below the ground). Magma usually forms 30 to 120 miles (50 to 200 kilometers) beneath Earth's surface. It rises because it is less dense than the surrounding rock. Rising magma can collect below or inside a volcano in a region called a *magma chamber.* As the magma accumulates, the pressure inside the chamber increases. When the pressure becomes too great, the chamber breaks open, and magma rises in the volcano. If magma reaches the surface, an eruption occurs. The hole through which the magma erupts is called a *vent.*

Scott K. Rowland, the contributor of this article, is Associate Professor of Geology and Geophysics at the University of Hawaii in Honolulu.

If magma accumulates at a high enough rate, the volcano erupts almost continuously. With magma that accumulates more slowly, the eruption may halt for periods while new magma replaces that which has erupted.

The violence of an eruption depends largely on the amount of gas dissolved in the magma and the magma's *viscosity* (resistance to flow). Magmas with little gas produce relatively calm eruptions in which lava flows quietly onto the surface. Magmas with much gas can shoot violent jets of gas and ash high into the air. *Viscous* (thick and sticky) magmas tend to erupt more violently than runnier, more fluid magmas. Water mixing with the erupting magma can make any eruption more explosive.

Volcanoes can create many dangers. Hot ash, gas, lava, and mud can bury or burn people and buildings near an erupting volcano. The most violent eruptions launch large clouds of ash and gas high into the atmosphere, causing problems far from the volcano itself.

Volcanoes also provide benefits. Erupted materials

contain many nutrients and can break down to form fertile soils. Volcanic activity provides an important source of *geothermal energy,* energy from Earth's interior heat. Geothermal energy can power electric generators and heat water and buildings. Undersea volcanoes have built up over time to form islands on which millions of people live. Volcanoes also have inspired myths and legends in many cultures. The word *volcano* comes from Vulcan, the Roman god of fire. Scientists who study volcanoes are called *volcanologists.*

This article discusses volcanic eruptions, the dangers of volcanoes, where volcanoes form, the types of volcanoes, and how scientists study volcanoes. Most of the article deals with volcanoes on Earth. For information on volcanoes elsewhere, see the section *Volcanoes in the solar system* at the end of this article.

Volcanic eruptions

Volcanic eruptions differ in their violence and in the materials they produce. Some eruptions involve calm outpourings of lava. Other eruptions produce powerful explosions and large volumes of rock, ash, and gas.

Products of eruption. During an eruption, a variety of materials can come from a volcanic vent. These include lava, *pyroclasts* (rock fragments), and gases.

Lava is magma that flows onto Earth's surface. As lava spreads from a vent, parts of it begin to cool and harden. The resulting stream of lava and rock is called a *lava flow.* Lava flows vary greatly in appearance depending on their viscosity, temperature, and rate of advance.

Fluid lava flows spread easily from the vent. Two common types of fluid lava flows are *pahoehoe* (pronounced *PAH hoy hoy* or *pah HOY hoy)* and *aa* (pronounced *ah AH* or *ah ah).* Pahoehoe flows have smooth, glassy surfaces and wavy, ropelike ridges. They form when hot, fluid lava advances relatively slowly. Aa flows have rough, broken surfaces. They form when less fluid lava advances rapidly. *Pahoehoe* and *aa* are Hawaiian terms adopted by most volcanologists.

Highly viscous lavas cannot flow easily. They pile up around the vent to form thick mounds called *lava domes* or short, stubby flows with rugged surfaces. These domes and flows advance extremely slowly.

Pyroclasts, also called *pyroclastics,* form when fragments of magma are thrown into the air by expanding gas. More explosive eruptions tend to produce finer pyroclasts. Pyroclasts that settle to the ground can cement together to form a rock called *tuff.*

Volcanologists often classify pyroclasts by their size. The finest pyroclasts, dust-sized and sand-sized grains, make up *volcanic ash.* Volcanologists call pebble-sized pyroclasts *lapilli.* Rock-sized and boulder-sized fragments are known as *volcanic bombs.*

Volcanologists also classify pyroclasts by texture. *Pumice,* a lightweight pyroclast, contains many tiny cavities left behind by gas bubbles in the magma. The cavities trap air, enabling some pumice to float on water. *Scoria* (cinder), another pyroclast, also has many tiny cavities, but it does not float on water. Pumice and scoria come from vigorous eruptions that hurl magma fragments high into the air. They solidify before landing, often forming a loose pile around the vent called a *scoria cone* or *pumice cone.*

Spatter, a fluid pyroclast, comes from less vigorous eruptions. Blobs of spatter do not fly high, and they land while still molten. Spatter collects around vents in steep structures called *spatter cones* and *spatter ramparts.*

Gases from volcanic eruptions include water vapor, carbon dioxide, and sulfur dioxide. Deep underground, the gases are dissolved in the magma. As magma rises, the pressure it is under decreases. The gases come out of solution to form bubbles and may eventually escape.

The violence of an eruption depends on the amount of gas dissolved in the magma and the magma's viscosity. Magmas rich in gas develop many bubbles as they rise to the vent. The bubbles increase the pressure in the vent, causing a more explosive eruption. Viscous magmas resist the expansion of bubbles, leading to a buildup of pressure in the magma. When the pressure of the bubbles finally overcomes the magma's viscosity, an explosive eruption occurs. In more fluid magmas, the bubbles expand without building up excess pressure. The resulting eruptions are relatively mild.

When external water, such as groundwater or sea water, mixes with magma, the water rapidly turns to steam, expanding in the process. This increases the violence of an eruption. Some volcanologists call eruptions involving external water *hydromagmatic eruptions.*

Volcano terms

Caldera is a depression that forms when part of the ground above a magma chamber collapses.
Divergent boundary is a line where two plates pull apart.
Flood basalt is a huge deposit of hardened lava that covers hundreds of thousands of square miles.
Hot spot is an area where hot rock rises through the mantle far from the boundaries between plates.
Lahar is a volcanic mudflow.
Lava is molten rock that flows onto Earth's surface.
Lava flow is a moving stream of lava and rock.
Magma is molten rock beneath the ground.
Magma chamber is an area below or inside a volcano where magma collects.
Mantle is the rocky layer beneath Earth's crust where magma forms.
Mid-ocean ridge is a place on the ocean floor where magma erupts as two plates pull apart.
Monogenetic field is a large field of separate volcanic vents that share a common magma chamber.
Plates are the rigid pieces that make up Earth's rocky outer shell.
Pyroclastic flows and surges are clouds of hot ash and gas that move rapidly along the ground.
Pyroclasts are fragments of magma tossed into the air by expanding gas.
Shield volcano is a broad, gently sloping volcano composed of hardened lava flows.
Silicic caldera complex is a volcano that consists of a vast caldera above a huge magma chamber.
Stratovolcano is a steep-sided volcano made up of layers of pyroclasts and hardened lava.
Subduction zone is a boundary where two plates collide, forcing one plate to sink beneath the other.
Vent is an opening through which magma erupts along with the material that builds up around it.

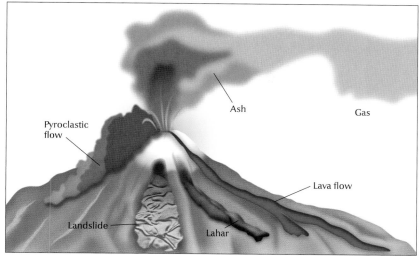

Ash

Gas

Pyroclastic flow

Lava flow

Landslide

Lahar

WORLD BOOK illustration by Adam Weiskind

Volcanic hazards can harm people, wildlife, and property. The greatest danger is posed by *pyroclastic flows,* fast-moving clouds of hot gas and ash that can burn, bury, choke, or poison. Lava flows can damage property but usually move too slowly to threaten people. Landslides and *lahars* (volcanic mudflows) can devastate areas near an erupting volcano. Ash and gas can travel great distances on the wind, causing problems far from the eruption.

As gases and pyroclasts erupt from a volcano, they draw in and heat some of the surrounding air. The heated air, gases, and pyroclasts form an *eruption column.* If an eruption draws in enough air and heats it sufficiently, the eruption column becomes lighter than the surrounding air. As a result, the column floats upward in a process called *convective rise.* Convective rise can carry erupted gas and ash high into the atmosphere. Some eruptions do not draw in enough air or heat it sufficiently to produce convective rise. In these eruptions, erupted materials remain closer to the ground.

The dangers of volcanoes

Volcanoes can endanger people, wildlife, and property. Volcanic disasters are much more difficult to avoid once an eruption begins. Instead, volcanologists and disaster planners strive to identify and evacuate dangerous areas before eruptions occur. Most damage results from (1) lava flows, (2) pyroclastic hazards, (3) lahars, (4) dangerous gases, (5) avalanches and landslides, and (6) tsunamis.

Lava flows. Many people fear being buried by lava, but lava flows rank as the least dangerous volcanic hazard. Lava usually advances at less than 6 miles (10 kilometers) per hour, slow enough for people and animals to escape. Unusually fluid lavas can flow fast enough to be dangerous, but this rarely occurs. However, lava flows can burn and bury buildings, roads, and other structures. Because lava hardens into solid rock, repairing buried areas can prove slow and costly.

Pyroclastic hazards. Flying volcanic bombs pose relatively little danger because they usually fall near the vent. Ash and lapilli, however, can travel on the wind for tens or even hundreds of miles or kilometers. Falling

AP/Wide World

Volcanic ash can travel on the wind and fall to the ground far from the erupting volcano. This man works to remove a crust of ash that accumulated on his automobile following an eruption.

ash and lapilli can contaminate water supplies, damage crops, and collect in great masses on roofs, causing them to collapse. Falling ash also blocks sunlight and reduces visibility, complicating evacuations. Ash can clog the engines of jet aircraft and parts of other machines, causing them to fail. Dispersed ash can remain in the stratosphere for years, where it can cool the atmosphere by blocking some sunlight.

The most dangerous volcanic hazards are *pyroclastic flows and surges,* clouds of hot ash and gas that travel mostly along the ground. They can choke or poison people with gases, bury people with debris, and burn them with temperatures of up to 1100 °F (600° C). They advance tens or hundreds of feet or meters per second and can even cross such natural barriers as rivers and ridges. People and animals cannot outrun pyroclastic flows and surges. In 1902, pyroclastic flows and surges from Mount Pelée on the island of Martinique swept through the city of St. Pierre, killing about 28,000 people.

Some pyroclastic flows occur when the column of ash and gas rising from a vent becomes too heavy to be supported by rising air. The column collapses and flows away from the vent. Other pyroclastic flows occur when the edge of a steep-sided lava dome or flow collapses, releasing pressurized gases mixed with lava fragments.

Lahars are mudflows that occur when loose ash on a steep volcano mixes with rain or melting snow. Eruptions or earthquakes, which often accompany volcanic activity, can dislodge the resulting mud. As lahars travel downhill, they pick up trees, boulders, and other debris. They usually follow river and stream valleys, where there may be many towns. Lahars leave a thick layer of mud that can harden like cement. Lahars from Nevado del Ruiz volcano in Colombia destroyed the town of Armero in 1985, killing more than 23,000 people.

Dangerous gases may be nearly invisible. Carbon dioxide, the most dangerous volcanic gas, causes suffocation in high concentrations. Where air circulation is poor, carbon dioxide can collect in low areas. People in these areas can be overcome before they realize the danger. Clouds of erupted gas can occasionally descend into populated areas, causing widespread suffocation.

Avalanches and landslides often occur on volcanoes because many are steep-sided and covered with loose ash and fractured lava flows. Many volcanoes also produce tremors or explosions that can dislodge snow and rubble. A landslide or avalanche may stem from an eruption or may happen when no eruption is occurring.

Tsunamis. A tsunami is a series of large ocean waves that can devastate coastal areas. Volcanic tsunamis can begin when a pyroclastic flow or avalanche enters the ocean, when a volcano in or near the ocean collapses, or when an underwater volcano erupts. During the 1883 eruption of Krakatau in Indonesia, the summit of the volcano collapsed, causing a tsunami that killed more than 30,000 people on nearby coasts.

Where volcanoes form

Volcanoes form above regions where magma is produced in the *mantle,* the rocky layer beneath Earth's crust. The theory of *plate tectonics* helps explain why volcanoes form where they do. According to the theory, Earth's outer shell consists of rigid pieces called *plates* that slowly move against one another. Nearly all volcanoes form along the edges of plates at *subduction zones* and *divergent boundaries.* Some volcanoes

Where volcanoes occur This map shows the location of many volcanoes. It also shows the large, rigid plates that make up Earth's rocky outer shell. Volcanoes usually occur along the boundaries between plates.

Kinds of volcanoes Most volcanoes fit into one of the following groups: (1) shield volcanoes, (2) stratovolcanoes, (3) silicic caldera complexes, (4) monogenetic fields, (5) mid-ocean ridges, and (6) flood basalts.

WORLD BOOK illustrations by Adam Weiskind

A shield volcano has a broad profile that resembles the shallow curve of a warrior's shield. Its gently sloping sides build up over time from the eruption of large amounts of fluid lava.

A monogenetic field is an area where magma from one source erupts through many vents. What appear to be individual volcanoes are actually vents in the larger field.

A stratovolcano is a steep-sided volcano formed by violent eruptions. The word *stratovolcano* refers to the *strata* (layers) of erupted rock fragments and hardened lava that make up its cone.

A mid-ocean ridge develops where two *plates* (rocky pieces of Earth's outer shell) spread apart on the ocean floor. As the plates separate, magma erupts between them and cools to form new plate material.

A silicic caldera complex is a broad depression that forms when the ground above a huge deposit of *magma* (molten rock) collapses. These volcanoes produce the most violent eruptions.

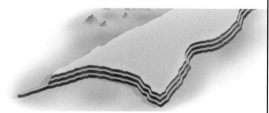

A flood basalt is a layered deposit of hardened lava that can cover hundreds of thousands of square miles. Scientists are not sure how such lavas erupted and spread so far.

appear above locations called *hot spots* that can be far from plate boundaries.

Subduction zones are boundaries where two plates push up against each other, forcing an oceanic plate to *subduct* (sink) beneath another plate. The subducting plate moves downward, carrying water trapped in sediments and rock into the mantle. When the plate reaches a depth of about 60 to 90 miles (100 to 150 kilometers), Earth's heat causes the water to boil. The boiling water rises into the overlying mantle. There, the water lowers the melting point of the rock, which turns to magma in a process called *hydration melting*. The magma can then rise through the overlying plate to erupt at the surface.

Subduction zones produce viscous magmas that contain much water vapor and other gases. For this reason, subduction zone volcanoes tend to erupt explosively.

Bands of subduction zone volcanoes often form along the edges of continents where an oceanic plate subducts beneath a continental plate. These include the volcanoes of the Andes Mountains in South America and the Cascade Range in North America. Other bands of volcanoes occur where two oceanic plates meet and one subducts beneath the other. These include the volcanoes of Japan and Indonesia. Subduction zones border most of the Pacific Ocean, creating a region of volcano and earthquake activity called the *Ring of Fire*.

Divergent boundaries are areas where two plates are pulling apart. As the plates separate, hot rock from the mantle rises to fill the space between them. The pressure on the rock decreases as the rock rises, causing it to melt in a process called *decompression melting*. The resulting magma erupts along the fracture between the plates and cools to form new plate material.

Divergent boundaries usually produce fluid magmas that contain few gases. Accordingly, eruptions at divergent boundaries tend to be less violent than eruptions at subduction zones. Most divergent boundaries involve two oceanic plates. For this reason, most eruptions at divergent boundaries occur underwater.

Hot spots are areas where columns of hot, solid rock rise slowly through the mantle. At the top of the column, decompression melting turns the rock into relatively fluid magma with little gas. This magma rises to erupt through the overlying plate. Many hot spot volcanoes, such as those of the Hawaiian Islands, occur far from plate boundaries. However, some hot spot volcanoes, such as those of Iceland, lie on or near plate boundaries.

Types of volcanoes

Many schemes exist for classifying volcanoes. Most volcanoes fit into one of these types: (1) shield volcanoes, (2) stratovolcanoes, (3) silicic caldera complexes, (4) monogenetic fields, (5) mid-ocean ridges, and (6) flood basalts.

Shield volcanoes build up over time from hundreds of thousands of lava flows. They erupt large volumes of fluid lava and few pyroclasts. The lava spreads far from the vent, creating a broad volcano with gently sloping sides. The word *shield* refers to the volcano's profile, which resembles the shallow curve of a warrior's shield.

Shield volcanoes include some of the largest volcanoes on Earth. Mauna Loa, a shield volcano on the island of Hawaii, rises about 30,000 feet (9,000 meters) from the ocean floor to its summit.

Most shield volcanoes form at hot spots. These include the volcanoes of Comoros, the Galapagos Islands, Hawaii, and many volcanoes in Iceland. Some of them, such as Westdahl Peak in Alaska, occur at subduction

zones. Other shield volcanoes, including Erta Ale in Ethiopia and Nyamuragira in Congo (Kinshasa), appear at divergent boundaries between continental plates.

Stratovolcanoes form from explosive eruptions that produce viscous lava and a large volume of pyroclasts. These materials pile up around the vent to form a steep-sided volcano. Most stratovolcanoes are smaller than shield volcanoes and erupt less often. The name *stratovolcano* refers to the *strata* (layers) of pyroclasts and hardened lava that make up the volcano's cone.

Stratovolcanoes are among the most common volcanoes on Earth. They include many famous historical volcanoes, such as Krakatau in Indonesia, Mount Pinatubo in the Philippines, and Vesuvius in Italy.

Most stratovolcanoes occur along subduction zones. Nyiragongo in Congo (Kinshasa), however, is a stratovolcano that lies on a divergent continental boundary.

Silicic caldera complexes produce the most violent volcanic eruptions. A silicic caldera complex can be difficult to recognize as a volcano because it consists of a broad, low-lying depression rather than a mountain.

Silicic caldera complexes form when a huge volume of magma collects below the surface in a giant magma chamber. The magma eventually erupts explosively, throwing ash high into the atmosphere and producing pyroclastic flows that damage vast areas. During the eruption, the ground above the chamber usually collapses, producing a large depression called a *caldera.*

A caldera can also form when part of a shield volcano or stratovolcano collapses into an opening left behind by erupting magma. These calderas are much smaller than silicic caldera complexes and are considered part of the larger volcano.

No silicic caldera complex has produced a major eruption in recent history. However, geologists have

How a volcano erupts   *Magma* (molten rock) forms deep underground and rises toward the surface, *left,* collecting in a *magma chamber.* As pressure builds, the chamber breaks open and magma rises through a *conduit, right.* At openings called *vents,* the magma erupts as gas, lava, and *pyroclasts* (rock and ash). Layers of erupted lava and pyroclasts make up the body of a *stratovolcano, shown here.*

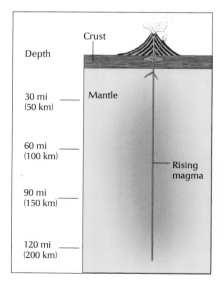

Crust

Depth

30 mi (50 km)   Mantle

60 mi (100 km)

Rising magma

90 mi (150 km)

120 mi (200 km)

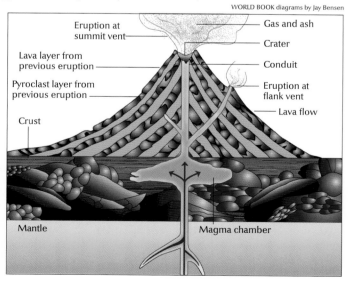

Eruption at summit vent

Lava layer from previous eruption

Pyroclast layer from previous eruption

Crust

Mantle

Gas and ash

Crater

Conduit

Eruption at flank vent

Lava flow

Magma chamber

Some famous volcanoes

Name	Location	Height above sea level in feet	in meters	Interesting facts
Aconcagua	Argentina	22,835	6,960	Highest mountain in Western Hemisphere; volcano extinct.
Cotopaxi	Ecuador	19,347	5,897	Eruption in 1877 produced *lahar* (mudflow) that traveled about 150 miles (241 kilometers) and killed about 1,000 people.
El Chichón	Mexico	3,478	1,060	Eruption in 1982 killed 187 people and released a cloud of dust and sulfur dioxide gas high into the atmosphere.
Krakatau	Indonesia	2,667	813	Great eruption in 1883 heard about 3,000 miles (4,800 kilometers) away; produced sea waves almost 130 feet (40 meters) high that drowned over 30,000 people on nearby coasts.
Lassen Peak	California	10,457	3,187	One of several volcanoes in the Cascade Range; last erupted in 1921.
Mauna Loa	Hawaii	13,677	4,169	World's largest volcano; rises about 30,000 feet (9,100 meters) from ocean floor.
Mount Etna	Sicily	10,902	3,323	About 20,000 people killed in 1669 eruption.
Mount Pelée	Martinique	4,583	1,397	*Pyroclastic flows and surges* (rushing clouds of hot ash and gas) from 1902 eruption destroyed city of St.-Pierre, killing about 28,000 people in minutes.
Mount Pinatubo	Philippines	4,875	1,486	Eruption in 1991, perhaps the largest of the 1900's, spewed about 20 million tons (18 million metric tons) of sulfur dioxide gas into the atmosphere.
Mount St. Helens	Washington	8,364	2,549	In 1980, violent eruptions released large amounts of molten rock and hot ash; killed 57 people.
Mount Tambora	Indonesia	9,350	2,850	In 1815, eruption released 6 million times more energy than the explosion of an atomic bomb; killed about 92,000 people.
Nevado del Ruiz	Colombia	17,717	5,400	Eruption in 1985 triggered lahars and floods; destroyed city of Armero and killed over 23,000 people.
Paricutín	Mexico	9,213	2,808	Newest volcanic mountain in the Western Hemisphere. Began in farmer's field in 1943; built cinder cone over 500 feet (150 meters) high in six days.
Stromboli	Mediterranean Sea	3,031	924	Active since ancient times; erupts constantly for months or even years.
Surtsey	North Atlantic Ocean	568	173	In 1963, underwater eruption began forming island of Surtsey; after last eruption of lava in 1967, island covered more than 1 square mile (2.6 square kilometers).
Thera[†]	Mediterranean Sea	1,850	564	Eruption in about 1500 B.C. may have destroyed Minoan civilization on Crete; legend of lost continent of Atlantis may be based on this eruption.
Vesuvius	Italy	4,190	1,277	In A.D. 79, produced history's most famous eruption, which destroyed towns of Herculaneum, Pompeii, and Stabiae.

†Name in ancient times. Now called Thira, or Santorini.

found evidence of such eruptions in Earth's past. These eruptions occur rarely because the large volume of magma they require takes a long time to accumulate. For example, a huge silicic caldera complex at what is now Yellowstone National Park in Wyoming produced three massive eruptions roughly 600,000 years apart.

Silicic caldera complexes are sometimes called *resurgent calderas*. This is because scientists think the caldera floor can *resurge* (rise again) as magma accumulates in the magma chamber before an eruption.

Silicic caldera complexes usually form on land near a hot spot or subduction zone. Famous examples include La Primavera in Mexico, Taupo in New Zealand, Toba in Indonesia, and the Yellowstone caldera.

Monogenetic fields form when magma from a single source flows to the surface through many different vents. Each vent forms during a single eruption. The word *monogenetic* means *of one origin*. Casual observers may not realize that what appears to be a single volcano is actually one vent in a large monogenetic field.

Monogenetic fields usually occur near subduction zones and hot spots. The best known is the Michoacan-Guanajuato field in Mexico. The field's newest vent, a large cone called Paricutín, formed in a thinly populated area during an eruption that lasted from 1943 to 1952. Monogenetic fields also occur in and around the cities of Auckland, New Zealand, and Flagstaff, Arizona.

Mid-ocean ridges are places at divergent boundaries where erupting magma creates new oceanic plate material. The mid-ocean ridge system forms the longest mountain chain on Earth. Estimates of its total length range from 30,000 to 50,000 miles (50,000 to 80,000 kilometers). Many geologists consider smaller segments of the ridge system to be individual volcanoes. At mid-ocean ridges, the material built up by eruptions spreads out as the plates pull apart. Ridges that spread rapidly, such as the East Pacific Rise, are broad and low. Ridges that spread slowly, such as the Mid-Atlantic Ridge, are narrow and steep.

Flood basalts, also known as *plateau basalts,* consist of layers of a dark volcanic rock called *basalt* that can cover hundreds of thousands of square miles. No flood basalt has erupted in recorded history, and geologists are still debating how they form. They once thought that the expanses of basalt resulted from rapid "floods" of lava because slow-moving lava would solidify before flowing so far. However, more recent research has shown that slow-moving lava can flow long distances if

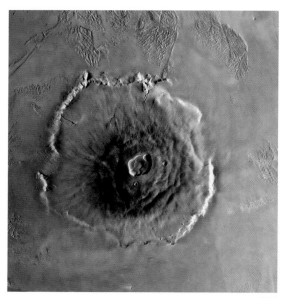

Olympus Mons on Mars is the largest volcano in the solar system. It measures more than 370 miles (600 kilometers) wide and rises about 16 miles (25 kilometers) above the surrounding plain.

it develops an insulating skin or crust of rock. Flood basalts include the Columbia River basalts in Washington, Oregon, and Idaho, the Deccan Traps in India, and the Paraná basalts in Brazil.

Studying volcanoes

The study of volcanoes offers many benefits. Erupting magma can carry material from deep underground to the surface, providing scientists with samples from Earth's interior. Hardened lava and ash deposits preserve evidence of major changes in Earth's history. Studying ancient volcanoes also helps scientists find new deposits of ores. The fluids moving through volcanoes can concentrate valuable metals in deposits called *veins.*

Perhaps most importantly, scientists study volcanoes to learn how to predict eruptions. This information can help reduce the damage and loss of life eruptions cause.

Predicting eruptions. To determine whether or when a particular volcano will erupt, volcanologists monitor *seismic activity* (earthquakes) near the volcano. They also watch for changes in the volcano's shape caused by pressure from magma below. They analyze the types and amounts of gases coming from vents. Changes may signal that an eruption is coming.

Modern volcanologists can state fairly confidently the probability that a particular volcano will erupt sometime in the next tens, hundreds, or thousands of years. With enough information, they can occasionally provide warnings a few days to a few hours before an eruption. Volcanologists are working to develop the ability to forecast an eruption a few years to a few months before it occurs. Warnings on this time scale would prove most useful to disaster planners.

Most volcanologists consider any volcano that has erupted in the last 10,000 years or so to be *active.* Some of them use the term *dormant* to describe an active volcano that is not currently erupting or showing signs of a coming eruption. Volcanologists label a volcano *extinct* if there is strong evidence it will never erupt again.

Describing eruptions. Volcanologists sometimes rate the explosive power of an eruption using a scale called the Volcanic Explosivity Index (VEI). The index ranks eruptions according to the volume of magma erupted, the amount of energy released, and the height of the eruption cloud. Most eruptions have VEI ratings from 0, for nonexplosive, to 8, for extremely explosive. Each number on the index represents a tenfold increase in explosive power or the volume of material erupted. For example, a VEI rating of 5 represents 10 times more power or eruption volume than a rating of 4.

Another common method for describing eruptions is based on how they produce pyroclasts. In this system, both *Plinian* and *Hawaiian* eruptions produce pyroclasts almost continuously. Plinian eruptions produce convective rise that can carry huge plumes of fine pyroclasts into the atmosphere. Hawaiian eruptions do not cause convective rise. Their pyroclasts are larger and stay closer to the ground, often landing while still molten. *Vulcanian* and *Strombolian* eruptions produce pyroclasts in bursts separated by periods of relative calm. Vulcanian eruptions produce convective rise, but Strombolian eruptions do not. Some volcanologists use classification schemes that have additional categories of eruptions.

Volcanoes in the solar system

Beyond Earth, planetary scientists have identified evidence of volcanic activity on the moon and on Mars, Venus, and Io, a large satellite of Jupiter. Much of the activity on these planets resembles volcanic activity on Earth. Scientists have identified basalt, a rock commonly erupted on Earth, in samples retrieved from the moon and in space probe observations of Mars. Many volcanoes on Mars and Venus resemble giant versions of shield volcanoes found on Earth. The shield volcano Olympus Mons on Mars ranks as the largest volcano in the solar system. It measures more than 370 miles (600 kilometers) in diameter and rises about 16 miles (25 kilometers) above the surrounding plain.

Io ranks as by far the most volcanically active body in the solar system. Space probes and telescopes have identified hundreds of volcanoes on Io, many of them active. Some eruptions on Io measure at least 350 Fahrenheit degrees (175 Celsius degrees) higher than the hottest eruptions on Earth. Other eruptions on Io involve sulfur rather than molten rock. Probes have recorded eruptions shooting sulfur as high as 310 miles (500 kilometers) above Io's surface.

Scientists studying Neptune's moon Triton have found evidence of a process called *cryovolcanism.* Cryovolcanism resembles volcanic activity but is by melted ice rather than magma. Volcanolike vents on Triton erupt liquid nitrogen far above the moon's frigid surface. Scientists think cryovolcanism may occur on other cold bodies in the outer solar system.   Scott K. Rowland

INDEX

How to use the index
This index covers the contents of the 2005, 2006, and 2007 editions.

Each entry gives the edition year and the page number or page numbers—for example, **Fibroblasts.** This means that information on this topic may be found on the page indicated in the 2007 edition.

When there are many references to a topic, they are grouped alphabetically by clue words under the main topic. For example, the clue words under **Fish** group the references to that topic under several subtopics.

The "see" and "see also" cross-references refer the reader to other entries in the index. For example, information on **Flu** will be found under the heading **Influenza.**

The indications (il.) and (ils.) mean that the reference on this page is to an illustration only, as in **Foam cells** in the 2007 edition.

A page number in italics means that there is an article on this topic on the page or pages indicated. For example, there is an Update article on **Fossil studies** on pages 215-219 of the 2007 edition. The page numbers in roman type indicate additional references to this topic in other articles in the volumes covered.

An entry followed by *WBE* refers to a new or revised *World Book Encyclopedia* article in the supplement section, as in **Fruit.** This means that there is a *World Book Encyclopedia* article on pages 275-279 of the 2007 edition.

ACKNOWLEDGMENTS

The publishers gratefully acknowledge the courtesy of the following artists, photographers, publishers, institutions, agencies, and corporations for the illustrations in this volume. Credits are listed from top to bottom, and left to right, on their respective pages. All entries marked with an asterisk (*) denote illustrations created exclusively for this yearbook. All maps, charts, and diagrams were staff-prepared unless otherwise noted.

6	Jacques Descloitres, MODIS Land Rapid Response Team, NASA/GSFC
7	Dale Debolt*
8	Zina Deretsky, National Science Foundation; NASA/JPL-Caltech
9	University of Wisconsin, Madison
10	Chandler Wilkerson, Institute for Molecular Design, University of Houston
12	© George Steinmetz, Corbis
16	AP/Wide World; © Daniel H. Sandweiss
17	© Walter Hupiu, EPA/Landov
18	© Daniel H. Sandweiss
19	AP/Wide World; © EPA/Peruvian National Institute of Culture/Landov
20	AP/Wide World; © Daniel H. Sandweiss
22	WORLD BOOK illustration by Bob Hersey
24	© Pilar Olivares, Reuters/Landov; © George Steinmetz
26	NASA/JPL-Caltech
28	WORLD BOOK illustration by Steven Karp
30	© Ann Ronan Picture Library/HIP/The Image Works
31	NASA/JHUAPL/SwRI
33	NASA/JPL-Caltech
34	NASA/ESA/A. Feild (STScI)
37	© Detlev Van Ravenswaay, SPL/Photo Researchers
38	© Don Dixon
40	© Steve Haddock, MBARI
41	© Deeble & Stone, OSF/Animals Animals; © Steve Haddock, MBARI; © Steve Haddock, MBARI
43	© Mark Harmel, Photo Researchers
44	AP/Wide World
45	© Steve Haddock, MBARI
46	© Steve Haddock, MBARI; © MBARI
47	© Steve Haddock, MBARI; © MBARI
48	Janet R. Voight; Claudia E. Mills; Claudia E. Mills; Ronald J. Larson
49	Dan E. Nilsson; © Paul A. Sutherland, SeaPics
50	© Steve Haddock, MBARI
51	© Steve Haddock, MBARI; © MBARI
52	© Steve Haddock, MBARI
53	Keith M. Bayha; © Ken Lucas, Visuals Unlimited
54-61	Conservation International
62	Zina Deretsky, National Science Foundation
65	Kalliopi Monoyios, University of Chicago
67	Beth Romey, University of Chicago
68	© Singh/Custom Medical Stock Photo
71	Library of the College of Physicians of Philadelphia
72-73	Christine Gralapp*
77	© TH Foto-Werbung/Photo Researchers
80	© Clem Spalding, Southwest Foundation for Biomedical Research
84	© Mark Garlick, Photo Researchers
87	WORLD BOOK illustration by Barbara Cousins
88	NASA/D. Padgett, IPAC & Caltech/ W. Brandner, IPAC/K. Stapelfeldt, JPL
89	© Mark Garlick, Photo Researchers
90	AP/Wide World; © Marli Miller, Visuals Unlimited
92-93	WORLD BOOK illustration by George D. Fryer, Bernard Thornton Artists
94	NASA/JPL-Caltech/P.S. Teixeira, Center for Astrophysics
95	WORLD BOOK illustration by Barbara Cousins; NASA/JSC Astrobiology Institute
96	University of Washington; Verena C. Wimmer et al., Southern Illinois University
97	© Mitsuaki Iwago, Minden Pictures
98	NASA
100	© PhotoAlto/SuperStock
102	WORLD BOOK illustration by Barbara Cousins
104	© Meyer/Custom Medical Stock Photo
105	WORLD BOOK chart by Bill and Judie Anderson
108-109	© SuperStock
110	© Phanie/Photo Researchers; © Chris Priest, SPL/Photo Researchers; © Uniphoto
112	Oak Ridge National Laboratory
114	© Roger Powell, Foto Natura/Minden Pictures; © NanoSphere/Schoeller Switzerland AG
115	U.S. Department of Energy
116	Paul Perrault*
117	© E.H. Sykes, P. Han, & B. A. Mantooth, Weiss Group/ Penn State University

319

119 © H. R. Bramaz; © ANT Photo Library
120 © Quantum Dot/Invitrogen; © David Wright, Vanderbilt University
121 © Robert Johnson, Johnson Group/University of Pennsylvania; © Fred Wudl, University of California at Santa Barbara
122 © Nanocor; © Hyperion Chemicals; © Science Museum, London/Science and Society Picture Library
123 © Pulickel Ajayan, Rennselaer Polytechnic Institute; © Craighead Group/Cornell University
124 © Craighead Group/Cornell University
125 Semtech
126 © Mirkin Group/Institute of Nanotechnology/Northwestern University
127 National Research Council of Canada; NASA/Ames Center for Nanotechnology
129 Caltech Archives
130 IBM
132 Foresight Nanotech Insitute
133 © Robert Hunter, MD & Akinori Shimada
134 © Lighting Research Center, RPI
135 © Michale Newman, Photo Edit
136 Christine Gralapp*
139 © Gregory Caputy, M.D.
140 Disc Interchange Service Company
141 Disc Interchange Service Company; Steve's Digicam Online, Inc.
142 Micron Inc.
143 Steve's Digicam Online, Inc.
144 © Eleanor Bentall, Corbis
145 © Reuters/Corbis
146 © David Lees, Corbis
148 © Lighting Research Center, RPI
149 LEDtronics, Inc.
150 Outside In, Ltd.
151 Traxon Technologies Ltd.
153 © Marsha Miller, University of Texas at Austin
155 Peggy Greb, USDA
157 © Erik Trinkaus, Czech Academy of Sciences
158 Sileshi Semaw, Stone Age Institute Indiana University
160 © Kenneth Garrett
161 University of Calgary
163 © Ronen Zvulun, Reuters/Corbis
165-166 © Chip Clark, Smithsonian Institution
167 ESA/DLR/FU Berlin (G. Neukum)
168-169 NASA/JPL
171 Ronak Shah & Jill Rathborne, Boston University
172 A. Kashlinsky & Group, NASA
174-175 Scientific Visualization Studio, GSFC/NASA
178 University of Wisconsin, Madison

179 NASA
181 AP/Wide World
182 © Glenn Williams, NOAA
183 F. Ganida-Haerrero et al., University of Murcia, Spain
187 Permission by Candlewick Press, Inc.
189 © John B. Carnett
191 MIT Media Laboratory
193 © Ian Austen
195 © Bryan & Cherry Alexander, Photo Researchers
196 © Dante Fenolio, Photo Researchers
199 © Jon Chase, Harvard University; © Jack Robinson, Hulton Archive/Getty Images
200 © John Eggitt, AFP/Getty Images; Stanford University
204 © David Muench, Corbis
206 IAEA
209 Jacob's School of Engineering, UCSD; © Lawrence Migdale, Photo Researchers
211 © Peter Parks, AFP/Getty Images
213 © Alfred Eisenstaedt, Time Life Pictures/Getty Images
215 © Gerald Mayr, et al./Senckenberg Research Institute
216 Zhongda Zhang, IVPP
219 University of Portsmouth, England
221 © Marsha Miller, University of Texas at Austin
222 © Mark Moffett, Minden Pictures
225 © Michael Bevis, Ohio State University
227 American Society of Interventional & Therapeutic Neuroradiology
228 Rehabilitation Institute of Chicago
231 © Etta D. Pisano, M.D./UNC School of Medicine
232 © David McCarthy, SPL/Photo Researchers
233 AP/Wide World
234 Durand-Wayland, Inc.
235 © Phil Schermeister, Corbis
237 AP/Wide World
238 © MBARI
239 AP/Wide World
243 Fig 1, Phys. Lett. Vol 94, 035505 (2005) by J. R. Gladden et al. Reprinted with permission, American Physical Society
245 © Mike Falco, Time Life Pictures/Getty Images
248 © Corbis/Bettmann
249 Yoshihiro Kawaoka, University of Wisconsin, Madison
251 AP/Wide World
253 NASA/JPL/University of Arizona/USGS; NASA/JPL/Corby Waste
254-256 NASA
257 TransCanada PipeLines Ltd.